12th International Conference on Residual Stresses 2025 (ICRS-12)

Detroit, Michigan, USA
20-23 October 2025

ISBN: 979-8-3313-3371-3

Printed from e-media with permission by:

Curran Associates, Inc.
57 Morehouse Lane
Red Hook, NY 12571

Some format issues inherent in the e-media version may also appear in this print version.

ASM International
9639 Kinsman Road
Materials Park, Ohio 44073-0002
USA

Phone: +1 440.338.5151

memberservicecenter@asminternational.org

Additional copies of this publication are available from:

Curran Associates, Inc.
57 Morehouse Lane
Red Hook, NY 12571 USA
Phone: 845-758-0400
Fax: 845-758-2633
Email: curran@proceedings.com
Web: www.proceedings.com

International Conference on Residual Stresses ICRS-12 2025

Proceedings from the 12th International Conference on Residual Stresses

October 20–23, 2025
Huntington Place
Detroit, Michigan, USA

Organized by

ASM INTERNATIONAL

Published by
ASM International®
Materials Park, Ohio 44073-0002
www.asminternational.org

SAN: 204-7586

ASM International®
Materials Park, OH 44073-0002
www.asminternational.org

Printed in the United States of America

The International Conference on Residual Stresses 2025
Organizing Committee

Local Organizing Committee, Conference Co-Chairs

Michael Prime, Los Alamos National Laboratory
Andrew Payzant, Oak Ridge National Laboratory
Michael Hill, University of California, Davis

International ICRS Board

Lyndon Edwards, Lyndon Edwards Engineering
Sabine Denis, University of Lorraine
Christoph Genzel, Helmholtz-Zentrum Berlin für Materialien und Energie
Stefanus Harjo, Japan Atomic Energy Agency
Thomas Niendorf, Universität Kassel
Andrew Payzant, Oak Ridge National Laboratory
Ru Lin Peng, Linköping University
Michael Prime, Los Alamos National Laboratory
Paolo Scardi, University of Trento
Olivier Thomas, Université de Toulo
Yandong Wang, University of Science and Technology Beijing
Philip Withers, The Henry Royce Institute

Contents

ICRS 12: Proceedings from the 12th International Conference on Residual Stresses, 2025
October 20–23, 2025; Detroit, Michigan, USA
https://doi.org/10.31399/asm.cp.icrs2025p0001

3D numerical modelling of residual stresses induced in surface milling

J.Rech, F.Valiorgue, S.Han
University of Lyon, Ecole Centrale de Lyon - ENISE, LTDS UMR5513, Saint-Etienne, France
joel.rech@enise.fr

L.Polly, M.Raffestin
MISUTECH, Saint-Etienne, France

Abstract

The relationship between surface integrity, particularly residual stresses, and fatigue life is well-established. A model known as the "3D hybrid multi-pass model" has been previously developed to predict the 3D residual stress fields generated by longitudinal turning over multiple revolutions until a steady state is reached. This model simulates the generation of residual stresses by accounting for equivalent 3D thermo-mechanical loads, which are calibrated using cutting force measurements, as they move over the machined surface. The aim of this paper is to adapt this methodology for surface milling, where multiple cutting edges generate the surface at a much higher frequency. A case study on the surface milling of 15-5PH martensitic stainless steel is presented

Extended Abstract

The fatigue life of safety components is a major concern for many economic sectors, including the aerospace and energy industries. The manufacturing processes, particularly the machining processes, involved in the production of these components are key factors influencing fatigue life [1]. Machining processes affect "surface integrity" [2], which includes microstructure, roughness, and residual stresses at the surface. [2-3] have summarized various studies that demonstrate how cutting processes can lead to significant changes in surface integrity. This paper focuses specifically on the generation of residual stresses. During a machining operation, residual stresses are generated by plastic deformation and high thermal gradients that occur simultaneously on the machined surface, while the rest of the part remains at ambient temperature [3]. The generation of residual stresses is governed by complex, non-linear, and highly coupled thermal and mechanical phenomena. For this reason, many authors have developed models to account for these complex and coupled mechanisms.

This article focuses on finish surface milling operations (Fig. 1). Surface milling is commonly used to achieve the desired flatness, surface roughness, and dimensions of many critical components, particularly in the aerospace industry. In finish surface milling, the milling tool removes small amounts of material from the part (typically between 50–500μm). While several experimental studies have investigated the residual stress state induced by surface milling [4-5], few have attempted to model the generation of residual stresses during the process [6-8]. Unfortunately, these models are unable to accurately predict residual stresses because they employ an explicit Lagrangian formulation, which does not allow for proper thermal and mechanical equilibrium to be computed on the machined surface. Only implicit models, by converging to specified mathematical residuals, are capable of estimating accurate residual stresses in the final state of the model. Moreover, these models can simulate only up to three successive cuts, which is insufficient to reach thermal steady-state conditions in milling (more than 50 cuts are needed, as shown by [9]). As a result, they cannot achieve a steady-state residual stress distribution. This limitation stems from the high computational time required to simulate the milling process. Milling involves high cutting speeds and very short contact durations (just a few milliseconds) between the cutting edge and the workpiece, which generate high-frequency thermal and mechanical loads on the machined surface. Additionally, the uncut chip thickness changes continuously. To accurately capture the variation in thermal and mechanical loads during a single cut, the time-step discretization must be very small (on the order of a few microseconds). When combined with the need to simulate numerous cuts, this results in prohibitively long computation times when using a 3D explicit formulation for chip formation.

Regarding side milling, several authors [10-13] have attempted to overcome this problem by using a numerical model based on a 2D orthogonal cutting configuration with a variable uncut chip thickness combined with circular movement. These works consider only up to two sequential cuts and do not account for mechanisms in the third perpendicular direction, limiting their applicability to real industrial use cases.

Other authors, such as [14-16], have developed 2D oblique analytical models that mimic side milling. While analytical models allow for fast calculations over many revolutions, their assumptions limit their generalizability.

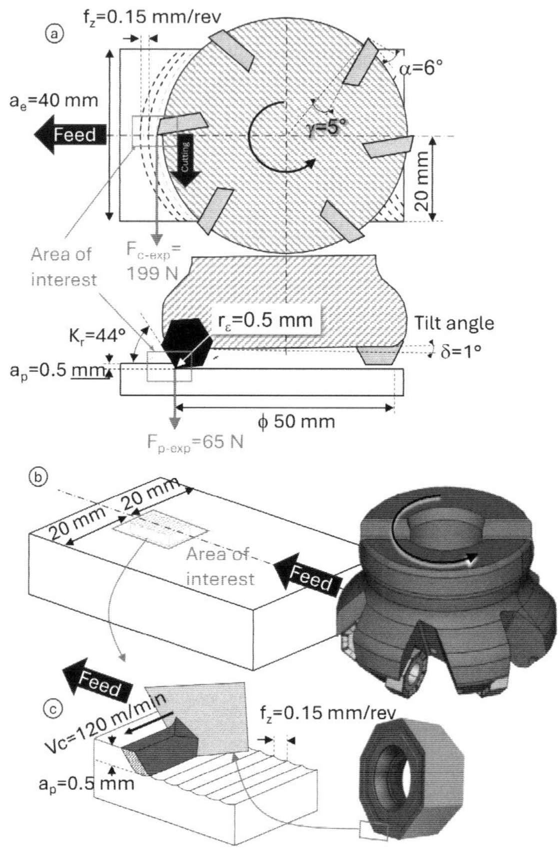

Figure 1: Technical description of the surface milling operation.

Thus, the high cutting speed and large number of teeth in the milling process, combined with the need to simulate a large number of cuts to reach a steady-state, require a different approach to obtain accurate residual stress predictions within a reasonable computation time.

A promising model [17], developed for turning with a single cutting edge, proposed a 3D model capable of simulating many revolutions to reach a steady-state in terms of residual stress. This model combines a two-step approach: the first step involves an explicit model of material removal to capture large strains and strain rates combined with high temperatures. This provides equivalent thermal and mechanical loadings applied by a single cut on the surface. This first step is called "hybrid," as the accuracy of the loadings can be calibrated when cutting force measurements are available. The second step is a 3D implicit Lagrangian model that moves the equivalent thermal and mechanical loadings over the machined surface during several cuts/revolutions to achieve a steady-state for residual stress generation. This modeling strategy has also been validated in reaming with solid tools having six teeth [18]. This methodology has the potential to be generalized to the modeling of any cutting process. Therefore, the objective of this paper is to investigate the possibility of predicting residual stress induced by surface milling using this modeling approach. Figure 2 presents the promising results obtained with this modeling approach for the case study presented in Figure 1.

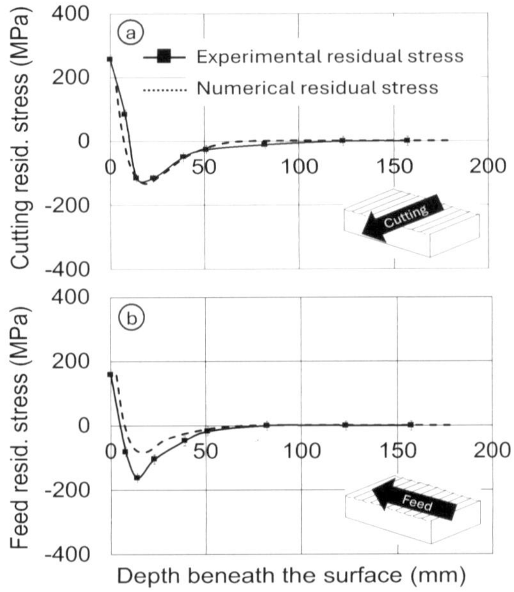

Figure 2: Residual stresses induced by surface milling. Experimental and numerical results.

References

[1] Smith, S. *et al.* "Effect of surface integrity of hard turned AISI 52100 steel on fatigue performance". *Materials Science and Engineering A,* Vol. 459, No1–2 (2007) p 337–346. https://doi.org/10.1016/j.msea.2007.01.011

[2] Jawahir, I.S. *et al.* "Surface integrity in material removal processes: Recent advances", *CIRP Annals,* Vol. 60, No. 2 (2011), p 603-626. https://doi.org/10.1016/j.cirp.2011.05.002

[3] Rech, J. *et al.*, *Workpiece Surface Integrity. In Machining: Fundamentals and Recent Advances*. Springer London, p 59–96.

[4] Perard, T. *et al.* "Experimental investigation on surface integrity in a face milling operation", *Procedia CIRP*, Vol. 108 (2022), p 400–405. https://doi.org/10.1016/j.procir.2022.03.063

[5] Maurotto, A. *et al.* "Surface integrity in dry milling of 304L steel: A parametric study", *Procedia CIRP*, Vol. 13 (2014), p 156–162, https://doi.org/10.1016/j.procir.2014.04.027

[6] Chen, Z. *et al.* "Machining-induced residual stress analysis and multi-objective optimization for milling process of Mg-Li alloy", *Measurement,* Vol. 204 (2022), p 112-127. https://doi.org/10.1016/j.measurement.2022.112127

[7] Wu, Q. *et al.* "Simulation analysis and experimental study of milling surface residuals stress of Ti-10V-2Fe-3Al", *Journal of Manufacturing Processe,s* Vol. 32 (2018), p 530-537. https://doi.org/10.1016/j.jmapro.2018.03.015

[8] Vovk, A. *et al.*, "Finite element simulations of the material loads and residual stresses in milling utilizing the CEL method", *Procedia CIRP,* Vol. 87 (2020), p 539-544. https://doi.org/10.1016/j.procir.2020.03.005

[9] Nemetz, A.W. *et al.* "FE temperature and residual stress prediction in milling inserts and correlation with experimentally observed damage mechanisms", *Journal of Materials Processing Technology* 256 (2018), p 98-107. https://doi.org/10.1016/j.jmatprotec.2018.01.039

[10] Rahul, Y. *et al.* "Methodology for prediction of sub-surface residual stress in micro end milling of Ti-6Al-4V alloy", *Journal of Manufacturing Processes,* Vol. 62 (2021), p 600-612. https://doi.org/10.1016/j.jmapro.2020.12.031

[11] Salashoor, M. *et al.* (2014) "Finite element simulation and experimental validation of residual stresses in high-speed dry milling of biodegradable magnesium-calcium alloys", *International Journal of Mechanical Sciences,* Vol. 80 (2014), p.153-159. https://doi.org/10.1016/j.ijmecsci.2014.01.010

[12] Li, J.L. *et al.* "An FEM study on residual stresses induced by high-speed end-milling of hardened SKD11", *Journal of Materials Processing Technology* 209 (2009), p 4515-4520. https://doi.org/10.1016/j.jmatprotec.2008.10.042

[13] Ullah, I. *et al.* "Numerical and experimental investigation on thermo-mechanically induced residual stress in high-speed milling of Ti-6AL-4V alloy", *Journal of Manufacturing Processes,* Vol. 76 (2022), p 575-587. https://doi.org/10.1016/j.jmapro.2022.02.039
[14] Lu, J. *et al.* "Analytical modeling of milling residual stress under different tool wear conditions", *The International Journal of Advanced Manufacturing Technology,* Vol. 127 (2023), p. 4253-4269. https://doi.org/10.1007/s00170-023-11715-4
[15] Wimmer, M. *et al.* "In-situ characterization and modelling of machining-induced residual stresses in peripheral milling of Ti-6Al-4V with rounded cutting edges", *Production Engineering*, Vol. 19 (2025), p 511-524. https://doi.org/10.1007/s11740-024-01323-w
[16] Wan, M., *et al.* "Theoretical prediction of machining-induced residual stresses in three-dimensional oblique milling processes", *International Journal of Mechanical Sciences,* Vol. 133 (2017), p 426-437. https://doi.org/10.1016/j.ijmecsci.2017.09.005
[17] Dumas, M. *et al.* (2021) 3D numerical modelling of turning-induced residual stresses – A two-scale approach based on equivalent thermo-mechanical loadings. *Journal of Materials Processing Technology* 297 (2021), 117274. https://doi.org/10.1016/j.jmatprotec.2021.117274
[18] Leveille, T. *et al.* (2024) 3D numerical modelling of residual stresses induced by reaming. *Journal of Manufacturing Processes*, 113:47-60. https://doi.org/10.1016/j.jmapro.2024.01.050

ICRS 12: Proceedings from the 12th International Conference on Residual Stresses, 2025
October 20–23, 2025; Detroit, Michigan, USA
https://doi.org/10.31399/asm.cp.icrs2025p0005

Residual Stress in Heterogeneous – Two-Phase Materials

Mohammed Belassel
Proto Manufacturing, 6150, Morton Industrial Parkway. Lasalle, Ontario, Canada
mbelassel@protoxrd.com

Abstract

Most materials can be considered heterogeneous, consisting of multiple phases. A simple example is steel, which contains ferrite and cementite (or carbides in general). Depending on the alloying elements, these phases can sometimes combine to form new compounds or structures within the steel. The amount of cementite can be significant, resulting in a multi-phase material.

When performing residual stress measurements using x-ray diffraction (XRD), all phases can be analyzed separately. These stresses can be macroscopic (bulk stresses) or microscopic (inter-phase stresses), depending on the loading conditions and processing history. Within the sampled (or diffracted) volume, the stress components can be triaxial, including all normal and shear stress components. Scientists have developed models based on Eshelby's equivalent inclusion method to calculate these microscopic stresses between phases. These considerations have, to some extent, been ignored by both industry and academia.

In this paper, we show that triaxial stress components can be measured via XRD and calculated using Eshelby-based models. The average stress across all phases corresponds to the macroscopic stress, and the boundary conditions at the surface remain valid.

Introduction

Stress measurement in two-phase or multi-phase materials is not new and has been applied to a variety of materials and composites, thanks to the X-ray diffraction (XRD) technique [1], which can separate the d-spacing measured in each phase, unlike other methods that are not based on diffraction [2, 3]. These phases are generally the result of heat treatment in alloy materials such as steel, aluminum, brass and titanium. Common phases in steel include ferrite, austenite, martensite, and carbides such as cementite. Other phases may exist as compounds composed of different elements bonded together to form distinct structures. In composite materials, these phases can be synthesized from various constituents and embedded within a matrix to produce a solid material with tailored mechanical properties.

XRD stress measurements depend on the quality of the collected diffraction peaks [4]. Small amounts of certain phases in a material may remain difficult to diffract and are sometimes undetectable [3, 5]. To address this challenge, scientists and engineers often turn to high-intensity sources such as rotating anodes or synchrotron [2, 3]. Additional challenges may arise in the presence of secondary phases in composites, which can lead to fluorescence from other phases. One solution is to use theoretical approaches such as finite element analysis (FEA) or micromechanical models based on Eshelby's theory [6]. Examples using the latter will be presented in this paper, along with measurements on a two-phase steel material. This subject should be revived to help new scientists in their endeavors.

Type of residual stresses in materials

1. <u>Single phase</u>

- **Type I, σ^{I}** (Macro Stresses): These stresses correspond to average values measured across multiple grains within the material, typically within a probed volume on the order of a few millimeters.
- **Type II, σ^{II}** (Micro Stresses): These stresses correspond to the average stress within individual grains of the material. The values vary across grains due to variations in crystallographic orientation and inherent anisotropy, both elastic and plastic, even in materials that appear isotropic on a macroscopic scale. The probed volume during measurement usually ranges from micrometers up to a few millimeters.
- **Type III, σ^{III}** (Sub-Micro Stresses): These stresses arise at the atomic scale and can vary across just a few interatomic distances within a single grain. They are primarily affected by crystallographic imperfections, such as dislocations, which locally disturb the internal stress field. Rather than causing a shift in diffraction

peak positions, this type of stress results in peak broadening due to the associated microstrain within the lattice.

2. Two-phase/multi-phase

In addition to the above definitions, the Type II stress in this case can be divided into two categories [7]: one contributing to the macro-stress from each phase, denoted as Type IIM (σ^{IIM}, pseudo-macrostress), which can be detected through peak shift; and another, denoted as Type IIm (σ^{IIm}, microstress), which refers to stress within the grains of each phase. Graph in (Figure 1) illustrates the different types of stresses in a two-phase material.

Pseudo-macrostress refers to stress generated by thermal deformation or plasticity, typically confined to interphase interactions. These stresses are usually self-equilibrated over the sample volume.

$$\sigma_{\alpha,\beta}^{IIM} = \langle \sigma_{\alpha,\beta}^{IIm} \rangle \quad (1)$$

$$\sigma_{\alpha,\beta}^{IIm} = \langle \sigma_{\alpha,\beta}^{III} \rangle \quad (2)$$

$$\sigma_{\alpha,\beta}^{I} = \sigma^{I} + \sigma_{\alpha,\beta}^{IIM} \quad (3)$$

$$f_{\alpha}.\sigma_{\alpha}^{I} + f_{\beta}.\sigma_{\beta}^{I} = \sigma^{I} \quad (4)$$

$$f_{\alpha}.\sigma_{\alpha}^{IIM} + f_{\beta}.\sigma_{\beta}^{IIM} = 0 \quad (5)$$

Where *f* is the volume fraction of the phase.
These equations are applicable to all stress results, normal and shear components.

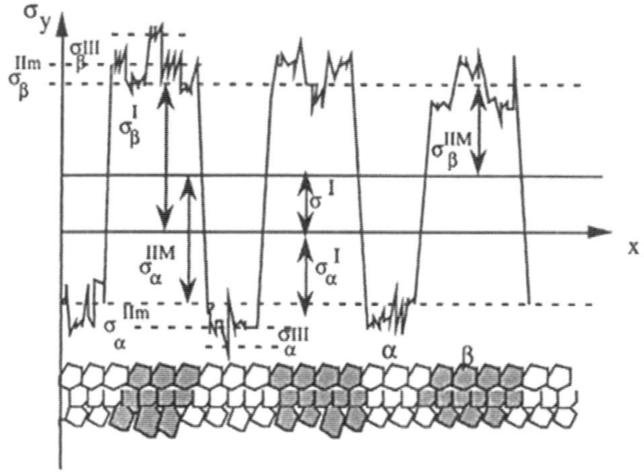

Figure 1: Illustration of types of stresses in a two-phase material [7].

Mechanism

Macro-stresses, known as Type I stresses (σ^{I}), are typically measured through relaxation-based methods such as hole drilling, slitting, contour analysis, or ultrasonic techniques. In contrast, Type II stresses (σ^{II}) occur at the intragranular level, within individual grains in individual phases. Single crystal techniques can be used to measure in a single grain material [8]. Type II stresses, or σ^{II}, arise from strain incompatibilities between phases and are self-equilibrated along the measurement direction, as shown in Equation (5) [4]. The strain incompatibilities can result from differences in thermal expansion or from distinct plastic responses due to varying mechanical properties of each phase. (Fig. 2) illustrates a stress-strain curve for a two-phase material exhibiting different behaviors, such as different yield stresses. After unloading, the resulting pseudo-macrostresses, also known as Type IIM stresses, are of opposite signs. In contrast, Type I residual stresses vanish upon unloading [9].

When a sample is loaded in a 4-point bending fixture, a through-thickness strain gradient develops due to pure bending between the inner loading points, see (Fig. 3). This bending induces tension on one surface and compression on the opposite, with a neutral axis in the middle. As a result, macroscopic residual stresses (Type I) are generated after unloading due to the gradient of plastic deformation. Differences in mechanical behavior between phases can lead to internal strain incompatibility generating Type IIM residual stresses in addition to Type I [3].

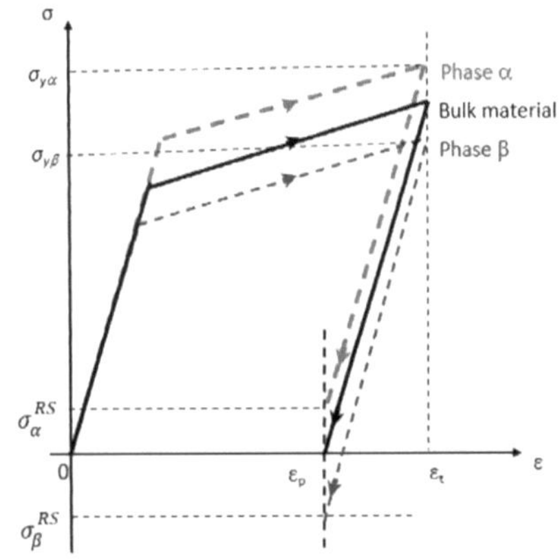

Figure 2: Stress-strain curve of a two-phase material [7, 9].

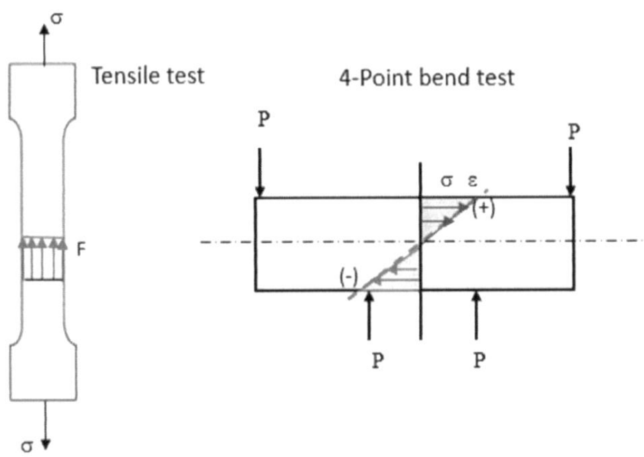

Figure 3: Tensile versus 4-point bend testing loading diagram showing uniform strain and strain gradient respectively.

The generation of the different types of residual stresses are summarized in *Table 1*. It illustrates the influence of the process on generation of stress types in a two-phase material.

Table 1: Types of residual stresses generated after deformation of a heterogeneous or two-phase material with different mechanical properties.

Type of loading /deformation	*Plastic deformation*	*Strain gradient*	*XRD peak shift*	*Type I (Macro)*	*Type II (Micro)*	*Measured via relaxation-based method*
Monotonic Uniaxial	✓	X	✓	X	✓	0
Bending Uniaxial	✓	✓	✓	✓	✓	✓
Heat treatment (Small sample)	X	X	✓	X	✓	0
Heat treatment (Large sample)	✓	✓	✓	✓	✓	✓
Machining	✓	✓	✓	✓	✓	✓
Shot peening / Rolling	✓	✓	✓	✓	✓	✓

✓: Present, X: Absent, 0: zero value

Boundary conditions

The stress field σ_{ij} in the material must satisfy the following equilibrium equations. [10]:

$$\frac{\partial \sigma_{ij}}{x_j} + F_i = 0 \tag{6}$$

Where i=1,2,3 corresponds to the three loading directions and F_i is the applied force. If no external loading is applied, then, F_i=0. For a free surface, the components of the stress tensor can be projected along the surface normal direction as follows:

$$T_i = \sigma_{ij} n_i = 0 \tag{7}$$

where n_i is the unit normal vector to the surface and Ti is the traction at the surface in the direction n. This condition ensures that no traction acts on the free surface. The resulting stresses in direction 3 (i.e. 0,0,1), all of the following stress components are zero:

$$\sigma_{i3} = 0, \sigma_{13} = 0, \sigma_{23} = 0, \sigma_{33} = 0 \tag{8}$$

In a multi-phase material, or more specifically in a two-phase material (for simplicity noted as α and β), the boundary condition at the interface can be expressed as:

$$T_i = \sigma_{ij}{}^{\alpha}\, n_i + \sigma_{ij}{}^{\beta}\, n_i = 0 \tag{9}$$

This condition ensures continuity of traction across the interface. However, each individual phase does not necessarily fulfill this condition on its own [11].

$$\sigma_{ij}{}^{\alpha}\, n_i \neq 0,\ \sigma_{ij}{}^{\beta}\, n_i \neq 0 \tag{10}$$

Although the sum of tractions vanishes at the surface, the tractions within each phase may be non-zero due to differences in their mechanical properties. This is important to understand when applying XRD techniques and is often not widely recognized by engineers. Furthermore, since the stress field can be triaxial, XRD measurements are capable of measuring all six components of the stress tensor [12].

Self-consistent model - Eshelby theory

Self-consistent micromechanical models are essential tools for understanding how multiphase heterogeneous materials respond to mechanical loads. These methods predict macroscopic material behavior by accounting for how individual phases interact within a surrounding matrix [13, 14].

The foundation was laid by Eshelby [6, 15], who analyzed the stress field caused by an ellipsoidal shaped inclusion subjected to plastic or thermal strain in an infinite elastic medium. The formulation assumes a uniform external stress and purely elastic material behavior. Voigt [16] and Reuss [17] introduced simpler averaging models assuming uniform strain and stress, respectively. Hill later showed that these approaches define the upper and lower bounds for effective elastic moduli, with their average offering a practical estimate [18].

Building on Eshelby's work, Kröner expanded the model to describe stress interactions in polycrystals [19]. The Mori-Tanaka method extends these ideas to materials containing inclusions like particles or fibers, allowing calculation of effective elastic properties [20].

While each model captures different micromechanical effects, the self-consistent model stands out for its ability to handle phase contrasts and anisotropy. By homogenizing local responses, it effectively predicts bulk material behavior in complex systems [21]. When applied to residual stress analysis, these models use the same elastic formulations but incorporate plasticity by introducing the secant or tangent modulus into the model [22].

Self-consistent models are versatile tools that can be applied to a wide range of materials subjected to various loading conditions. These models simulate the mechanical response at the grain or phase level by considering interactions among different phases or grain orientations. They are capable of predicting the internal stress distribution within each phase during loading and can also evaluate the residual stresses retained after unloading [13, 14]. This makes them particularly valuable for analyzing multiphase materials, textured polycrystals, and materials undergoing plastic deformation

The expression of stress inside the inclusion in a two-phase material with loaded matrix can be written as [7]:

$$\sigma_{ij}^{inc} = \sigma_{ij}^{I} + c_{ijkl}^{inc}(S_{klmn}^{-1} - I_{klmn})(\varepsilon_{mn}^{I} - \varepsilon_{mn}^{II}) \tag{11}$$

where

- σ_{ij}^{inc} is the stress in the inclusion;
- c_{ijkl}^{inc} is the stiffness tensor of the inclusion.
- S^{Esh} is the Eshelby tensor, a fourth-order tensor that depends on the elastic constants of the material and the geometry of the inclusion. This factor is often referred to as the "concentration factor".
- ε_{mn}^{I} is the macroscopic strain applied to the inclusion;
- ε_{mn}^{II} is the microscopic strain induced in the inclusion.

Using the homogenization approach, the expression for the effective elastic modulus can be written as follows:

$$C^{eff} = \sum_{i} f_i c^i A^i \tag{12}$$

where C^{eff} is the effective stiffness tensor of the medium made of all phases; c^i is the stiffness tensor of phase i; and A^i is the strain concentration tensor for each phase i, which accounts for the mechanical interaction between the phase and the effective medium [14]:

$$C^{eff} = \sum_{i} f_i c^i \, [c^i + C(S^{-1} - I)^{-1}][C^{eff} + C(S^{-1} - I)] \tag{13}$$

- C^{eff} is the effective stiffness tensor of the medium;
- c^i is the stiffness tensor of phase i.
- C is the stiffness tensor of the matrix.

Equation (13) can be solved iteratively since the effective stiffness tensor C^{eff} appears in both terms of the equation.

3. Plastic strain induced residual stress

When the source of residual stress is plastic deformation, and the second phase deforms elastically only, the total plastic strain of the material can be measured from stress-strain curve of the material [23]:

$$\varepsilon^P = f\varepsilon_1^P + (1-f)\varepsilon_2^P = (1-f)\varepsilon_2^P \tag{14}$$

Where ε^P is the total plastic strain applied to the material; ε_1^P is the plastic strain applied to phase 1 in the; and ε_2^P is the plastic strain applied to phase 2. The expression for the residual stress equation can be written as [3]:

$$\sigma_{ij}^{II} = C_{ijkl}^{S\_eff}(S_{klmn}^{-1} - I_{klmn})\varepsilon_{kmn}^{P} \tag{15}$$

C^{S_eff} is the secant effective modulus calculated at each level of plastic strain [22].

Substituting equation (14) into (15), the expressions for residual stresses in ferrite and cementite are:

$$\sigma_{ij}^{Fe} = C_{ijkl}^{S\_eff}(I_{klmn} - S_{klmn}^{-1})\frac{f}{1-f}\varepsilon_{mn}^{P} \tag{16}$$

$$\sigma_{ij}^{Fe3C} = C_{ijkl}^{S\_eff}(S_{klmn}^{-1} - I_{klmn})\varepsilon_{mn}^{P} \tag{17}$$

4. Thermal strain induced residual stresses

Identically, the residuals stress in cementite can be expressed as:

$$\sigma_{ij}^{II} = C_{ijkl}^{eff}(S_{klmn}^{-1} - I_{klmn})\alpha_i^*(T)\delta_{mn} \quad (18)$$

$$\alpha_i^* = (\alpha_i^{inc} - \alpha^M)\delta(T) \quad (19)$$

Where α_i^{inc} and α^M are the expansion coefficients of cementite and the matrix.

The coefficient of expansion of cementite is not linear, therefore, a discrete function must be applied. The stress expression of residual stress is:

$$\sigma_{ij}^{Fe3C} = (1 - f)C_{ijkl}^{eff}(S_{klmn}^{-1} - I_{klmn})\sum_i \alpha_i^*(T)\, dT\delta_{mn} \quad (20)$$

Experiment

Cylindrical tensile specimens, 5 mm in diameter, were fabricated and heat treated to produce either lamellar or spherical cementite. These samples were then subjected to uniaxial monotonic loading at various levels of plastic deformation, followed by unloading to induce Type IIM residual stresses between the phases. The expansion coefficient of ferrite was assumed constant for simplicity in the calculation, with a value of 12.10^{-6}/°C, while that of cementite was considered variable between 0 and 250 °C due to changes in its magnetic properties, as shown in (Fig.4).

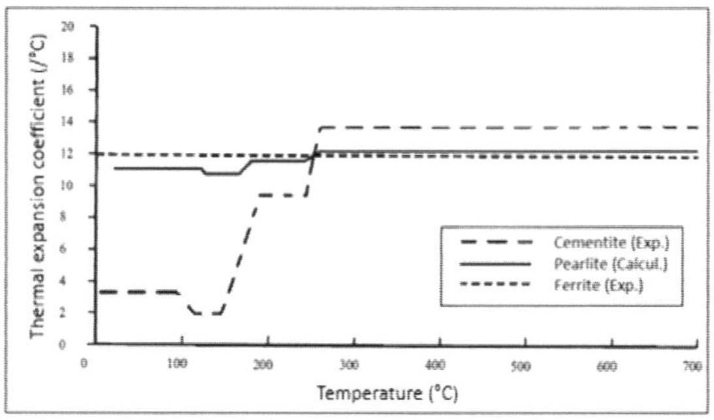

Figure 4: The thermal expansion coefficients of cementite and ferrite. In the calculation, the thermal expansion coefficient of ferrite was assumed constant, since its variation is not significant.

Residual stress measurements were conducted after heat treatment and plastic deformation in the axial, loading direction. Due to the low volume fraction of cementite in the pearlitic matrix, synchrotron radiation was employed for its high x-ray beam intensity. The synchrotron was used for samples subjected to plastic deformation. Residual stress was measured using the XRD-$\sin^2\psi$ technique. The measurement conditions were as follows: for the ferritic phase, Cr-Kα radiation was employed for the (211) lattice reflection at a 2θ angle of ~156°. For the cementite phase, x-rays with a wavelength of λ = 1.79 Å were used, and diffraction measurements were conducted at a 2θ angle of ~130°.

Results

The residual stress calculated using the self-consistent model after heat treatment is presented on (Fig. 5). At room temperature, the model predicts a low magnitude residual stress which aligns well with the X-ray diffraction (XRD) measurement results shown in (Fig. 6), except for the surface effects. It should be noted that no experimental stress measurements were performed on the cementite phase for these samples.

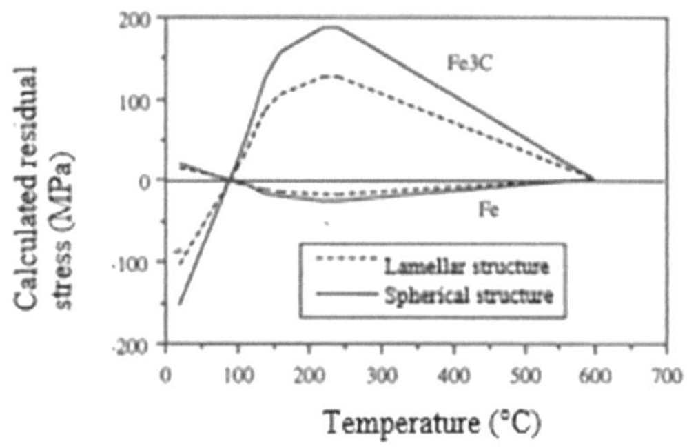

Figure 5: Stress evolution during cooling of ferrite and cementite in pearlitic steel, calculated using a self-consistent model. Data calculated from 600°C to room temperature.

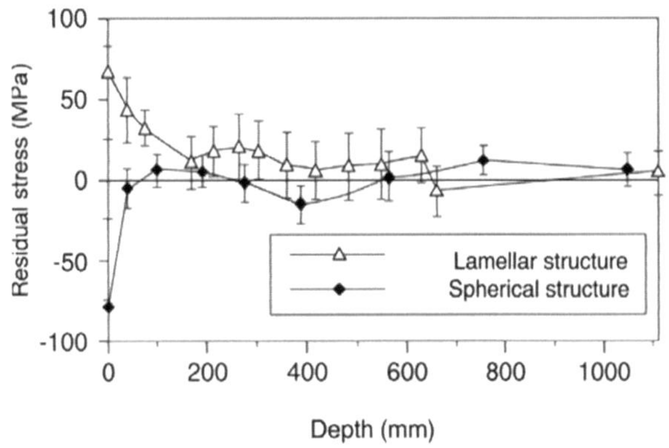

Figure 6: Residual stress evolution with depth after heat treatment in ferrite.

(Fig. 7) presents both XRD-measured and self-consistent model-predicted Type IIM residual stresses in pearlitic steel containing spheroidized cementite particles within a ferritic matrix. The curves with triangular data points shown represent experimental data obtained via XRD, whereas the "Model" curves without data points shown correspond to predictions from a self-consistent micromechanical model. The model estimates higher magnitude residual stresses than those measured by XRD. This difference is attributed to the brittle nature of cementite, which undergoes damage or fracture at plastic strain levels exceeding approximately 7%. Comparable findings have been reported in studies on tool steels, reinforcing the interpretation of cementite's limited ductility under such conditions [9].

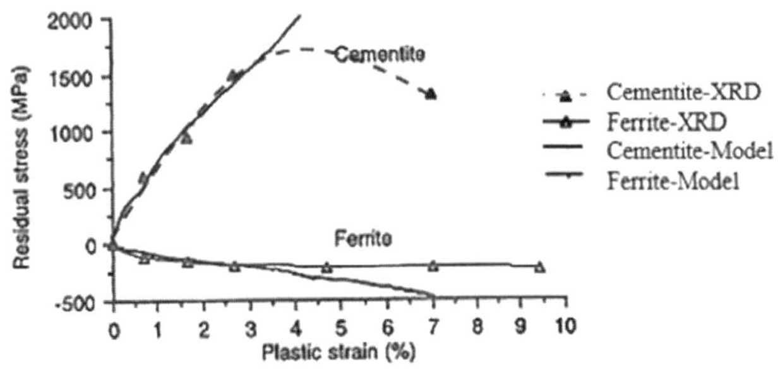

Figure 7: Type IIM residual stress evolution with plastic deformation of a pearlitic steel with spheroidal-shaped cementite. "XRD" indicates measurement using x-ray diffraction. "Model" indicates that a self-consistent model is applied to the steel [3].

In addition, triaxiality of residual stress can be observed in a two-phase material [12], including all the stress tensor components. The resultant average stress when considering all phases at the surface fulfills the boundary condition [11].

Conclusions

The origin of residual stresses in materials depends on both the microstructure and the processing history. In single-phase materials, XRD results can often be validated using relaxation methods. However, in multi-phase materials, significant residual stresses may arise due to mechanical property mismatches between phases, and XRD can be used to measure all stress tensor components in each phase present. If a multi-phase material is uniformly loaded, the internal stresses are typically Type II pseudo-macrostresses, self-equilibrated within the microstructure. On the other hand, when a strain gradient is introduced, both macroscopic (Type I) and pseudo-macroscopic (Type IIM) residual stresses can coexist. To analyze such internal stress states, micromechanical models based on Eshelby's inclusion theory offer a powerful framework for evaluating Type II stresses arising from phase interactions.

References

[1] I.C. Noyan, J.B. Cohen, "Residual Stress, Measurement by Diffraction and Interpretation", New York: springer-Verlag, 1987.

[2] R.A. Winholtz and J.B. Cohen, "Separation of the Macro- and Micro-Stresses in Plastically Deformed 1080 Steel," *Advances in X-Ray Analysis , Volume 32: Thirty-Seventh Annual Conference on Applications of X-ray Analysis, August 1-5, 1988, ,* vol. Vol 32, pp. pp. 341 - 353, 1988.

[3] M. Belassel et al., "Effect of Thermal and Mechanical Loading on the Generation of Macro and Micro Stresses in Eutectoid Steel," *Proceedings of ICRS4, Baltimore, Society for Experimental Mechanics Inc.,* pp. 392-401, 1994.

[4] V. Hauk, "Structural and Residual Stress Analysis by Non Destructive Method", Amesterdam: elsevier, 1997.

[5] R.A. Winholtz and J.B. Cohen, "Load Sharing of the Phases in 1080 Steel during Low-Cycle Fatigue," *Metallurgical Transactions A,* vol. 23A, pp. 341-354, 1990.

[6] J.D. Eshelby, "The determination of the elastic field of an ellipsoidal inclusion, and related problems," *Proc. Roy. Soc.,* vol. A241, pp. 376-396, 1957.

[7] M. Belassel, "Residual stress measurement and modelling of heterogeneous multiphase materials," *ASM Handbook,* vol. 25A, no. (To be published), 2025.

[8] M. Belassel, "Residual stress measurement in single crystal materials," *ASM handbook,* vol. 25A, no. (to be published), 2025.

[9] T. Hanabusa, J. Fukura and H. Fujiwara, "X-Ray Stress Measurement on the Cementite Phase in Steels," *Buleting of JSME,* vol. 12, no. 53, pp. 931-939, 1969.

[10] S. Nemat Nasser, M. Hori, Micromechanics: Overall Properties of Heterogeneous Materials, Amsterdam: Elsevier Science Publishers By., 1991.

[11] T. Hanabusa and H. Fujiwara, "Relation of shear between stresses ψ-splitting in X-ray unidirectionally Takao and microscopic measurement of stress deformed by layers.," *Zairyo (Journal of the Society of Materials Science, Japan),* vol. 31, no. 342, pp. 227-233, March, 1982.

[12] M. Belassel, J.L. Lebrun, and H. Ruppersberg, "Triaxial Elastoplastic Stresses in the Ferrite Phase of Pearlitic Steel and Their Influence on the Results Obtained from X-ray Stress Analysis," *4th European Conference on Residual Stresses (ECRS 4),* vol. 2, pp. 997-1002, 1996.

[13] T. Mura, "Fundamental Solutions in Elasticity," *Applied Mechanics, ASME.*

[14] D. François, A. Pineau, A. Zaoui, Comportement mécanique des matériaux, Paris: Hermes, 1991.

[15] J.D. Eshelby, "The elastic field outside an ellipsoidal inclusion," *Proc. Roy. Soc.,* vol. A252, pp. 561-569, 1959.

[16] W. Voigt, Lehrbuch der Kristallphysik, Leipzig: Lehrbuch der Kristallphysik, 1928.

[17] A. Reuss, Z. Angew, *Zeitschrift für Angewandte Mathematik und Mechanik,* vol. 9, pp. 49-58, 1929.

[18] R. Hill, "The Elastic Behaviour of a Crystalline Aggregate," *Proceedings of the Physical Society (Section A),* vol. A65, p. 349–354, 1952.

[19] E. Kröner, "Berechnung der elastischen Konstanten des Vielkristalls aus den Konstanten des Einkristalls," *Z. Physik,* vol. 151, pp. 504-518, 1958.

[20] T. Mori and K. Tanaka, "Average stress in matrix and average elastic energy of materials with misfitting inclusions," *Acta Metallurgica,* vol. Vol. 21, no. 5, p. 571–574, 1973.

[21] M. Bornert et al., "Self-consistent approaches and strain heterogeneities in two-phase elastoplastic materials," *Applied Mechanics Reviews,* vol. 47, no. 1, pp. 12-18, 1994.

[22] G.P. Tandon and G.J. Weng, "A Theory of Particle-Reinforced," *Transaction of ASME,* vol. 55, no. March 1988, pp. 126-134, 1988.

[23] M. Berveiller et al., "Theoretical Investigation of Ψ-Splitting After Plastic Deformation of Two-Phase Materials," *Textures and Microstructures,* vol. 14, pp. 151-156, 1991.

ICRS 12: Proceedings from the 12th International Conference on Residual Stresses, 2025
October 20–23, 2025; Detroit, Michigan, USA
https://doi.org/10.31399/asm.cp.icrs2025p0014

RESIDUAL STRESS USING X-RAY DIFFRACTION TECHNIQUES – UNIFIED APPROACH

Mohammed Belassel
Proto Manufacturing, 6150, Morton Industrial Parkway. Lasalle, Ontario, Canada
mbelassel@protoxrd.com

James Pineault, Michael Brauss
Proto Manufacturing, 12350, Universal Drive, Taylor. Michigan, USA
xrdlab@protoxrd.com, mbrauss@protoxrd.com

Abstract

Many residual stress measurements techniques using X-ray diffraction have been developed to optimize fabrication processes and enhance the fatigue resistance of components in various sectors, including automotive, aerospace, structural, and medical industries. The accuracy of residual stress measurement results is critical to the success of these components.

It has been scientifically established that stress measurement involves a series of steps: measuring d-spacings, determining strain in different directions, and then calculating stress using Hooke's law, applied to either isotropic or anisotropic materials. While these steps may appear different across techniques, they ultimately share the same assumptions and fundamental theory, leading to the same result.

In this paper, we demonstrated that all residual stress measurement techniques follow the same fundamental principles, with differences primarily arising from the geometry of the instrument used. Furthermore, any deviation from these steps will lead to erroneous results.

Introduction

Residual stress is a critical subject, with thousands of papers published to address diverse needs in both industry and science [1, 2]. Stress measurement using the x-ray diffraction (XRD) technique has been in use since the 1920s, and significant developments have been made to better understand it and extend its application across various industrial sectors [3]. A wide range of instruments have been developed, featuring different geometries and configurations, and incorporating various components, from X-ray sources to detectors and associated optics. The results obtained from these instruments are inherently linked to the assumptions made and the measurement parameters selected, including acquisition time, number of tilts, and the measurement direction on the component [4]. The triaxiality of residual stress in the material, as determined by XRD, is well-established and cannot be disregarded, except in rare cases where a thorough evaluation must be conducted [2, 5].

All stress measurements that employ diffraction are fundamentally based on determining the d-spacing, strain, and stress using Hooke's law. Regardless of how the theoretical framework is developed, these core steps, measuring lattice spacing changes, converting them to strain, and applying Hooke's law to derive stress, are always followed [6].

Stress measurement using XRD is influenced by several selected parameters. These include: anode type, depth of penetration, number of tilts, maximum tilt angle, spatial coverage, positive and negative tilting, sample positioning, sample orientation, and the theory applied to calculate the stress values [2]. Any differences in these selected parameters can lead to variations and, in some cases, inaccurate results. For example, the tilting angles employed during measurement have been shown to result in systematic variations in the residual stress results [7, 8]. Data collection parameters are dependent on the material being investigated and, for accurate and reliable results to be obtained, adjustments must be made according to the specific material and its microstructure.

Theory

All techniques using X-ray diffraction (XRD) for residual stress measurement rely on the distance between crystallographic planes, known as d-spacing, as a strain gauge. These methods are applicable only to crystalline, polycrystalline, or semi-crystalline materials. When the material is under tension, the d-spacing increases; under compression, it decreases [1]. The d-spacing is calculated from the measured Bragg angle (2θ) for a known X-ray wavelength (λ) using Bragg's law:

$$n\lambda = 2d sin\theta \quad (1)$$

The strain ε is calculated from d-spacing $d_{\varphi\psi}$ measured in each direction defined by φ and ψ and unstressed d-spacing (d_0) values, see (Fig 1). The expression of strain is calculated using the following relationship:

$$\varepsilon_{\varphi\psi} = \frac{d_{\varphi\psi} - d_0}{d_0} \quad (2)$$

The direction in spherical coordinate for general case can be expressed as:

$$n_i = \begin{pmatrix} sin\psi\ cos\varphi \\ sin\psi\ sin\varphi \\ cos\psi \end{pmatrix} \quad (3)$$

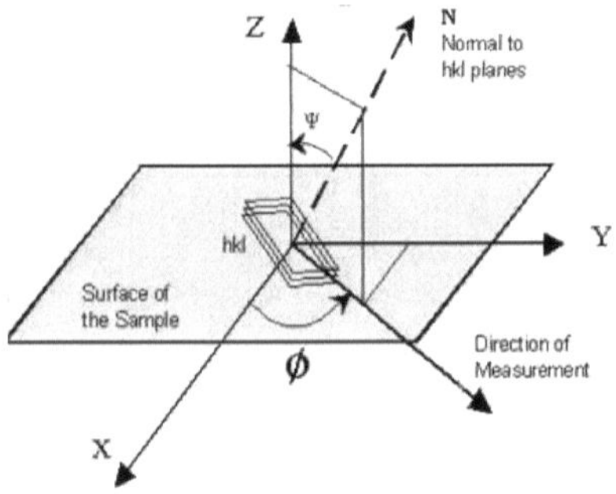

Figure 1: Definition of the axis and the direction of measurement.

For projection of directions located on a Debye ring in case of Cos-Alpha technique [9]:

$$n_i = \begin{pmatrix} cos\eta\ sin\psi_0 cos\varphi_0 + sin\eta\ cos\psi_0\ cos\varphi_0\ cos\alpha - sin\eta\ sin\ \varphi_0\ sin\alpha \\ cos\eta\ sin\psi_0\ sin\alpha\ sin\varphi_0 - \ sin\eta\ cos\psi_0\ sin\ \varphi_0\ cos\alpha + \ sin\eta\ cos\ \varphi_0\ sin\alpha \\ cos\eta\ cos\psi_0 + sin\eta\ sin\psi_0 cos\alpha \end{pmatrix} \quad (4)$$

Where

- $\eta = (\pi - 2\theta)/2$: is the half angle subtended by the incident and diffracted beams.
- ψ_0: specimen tilt angle (rotation around specimen axis)
- φ_0: azimuth angle in the specimen plane
- α: angle along the Debye ring (ring angle)

The tilting angles shown on (Fig. 2) are defined as: $\psi_1 = \psi_0 + \eta$, and $\psi_2 = \psi_0 - \eta$. In the same manner, these angles correspond to α=0° and α=180° on the Debye ring as shown on (Fig, 2).

The material is always assumed to have a triaxial stress and strain field. If assumptions of biaxial stress field are made, the stress or strain tensors can be reduced to 4 components. The strain in the direction (φ,ψ) or (φ,ψ,α) can be expressed in the following for independent directions:

$$\varepsilon_n = n_i n_j \varepsilon_{ij} \quad (5)$$

Where

- ε_n: normal strain in the direction of the unit vector n

- n_i, n_j: components of the direction unit vector n
- ε_{ij}: components of the strain tensor (a symmetric 2nd-order tensor)

The strain tensor can be calculated using Hooke's law:

$$\varepsilon_{ij} = S_{ijkl}\sigma_{kl} \qquad (6)$$

$$\varepsilon = \begin{pmatrix} \varepsilon_{11} & \varepsilon_{12} & \varepsilon_{13} \\ \varepsilon_{12} & \varepsilon_{22} & \varepsilon_{23} \\ \varepsilon_{13} & \varepsilon_{23} & \varepsilon_{33} \end{pmatrix}, \quad \sigma = \begin{pmatrix} \sigma_{11} & \sigma_{12} & \sigma_{13} \\ \sigma_{21} & \sigma_{22} & \sigma_{23} \\ \sigma_{31} & \sigma_{32} & \sigma_{33} \end{pmatrix} \qquad (7)$$

Where

- σ_{kl}: stress tensor (2nd rank), describes internal forces.
- S_{ijkl}: compliance tensor (4th rank), material property that relates stress to strain.

For isotropic elastic behavior, Equation (6) can be expressed as follows:

$$\varepsilon_{ij} = \frac{1+v}{E}\sigma_{ij} + -\frac{v}{E}(\sigma_{11} + \sigma_{22} + \sigma_{33})\delta_{ij} \qquad (8)$$

Where

- v: Poisson's ratio
- E: Young's modulus
- δ_{ij} is the Kronecker constant, $\delta_{ij} = 0$ if i≠j and $\delta_{ij} = 1$ if i=j.

Some of the different XRD techniques used for the measurement of residual stress are as follows:

- Multiple tilts using $Sin^2\psi$ technique as Multiple Exposure Technique, MET- $Sin^2\psi$.
- Two tilt method as Single Exposure Technique, SET- $Sin^2\psi$.
- Cosα with one Debye ring as SET - Cosα

Their constitutive equations are as follows [10]:

Cosα in single direction φ

Using the strain measured along the Debye rings at each α increment, the quantities a_1 and a_2 can be expressed as follows:

$$a_1 = \frac{1}{2}[(\varepsilon_\alpha - \varepsilon_{\pi+\alpha}) + (\varepsilon_{-\alpha} - \varepsilon_{\pi-\alpha}) \qquad (9)$$

$$a_2 = \frac{1}{2}[(\varepsilon_\alpha - \varepsilon_{\pi+\alpha}) - (\varepsilon_{-\alpha} - \varepsilon_{\pi-\alpha})] \qquad (10)$$

Where ε_α, $\varepsilon_{\pi+\alpha}$, $\varepsilon_{-\alpha}$, and $\varepsilon_{\pi-\alpha}$ are the measured strains from the four sectors of the Debye ring as shown on (Fig. 2).

Using equation (5) and (8), the following equations can be deduced:

$$a_1(\varphi = 0) = -\frac{(1+\upsilon)}{E}(\sigma_{11}sin2\psi_0 - 2\sigma_{13}cos2\psi_0)sin2\eta.\,Cos\alpha \qquad (11)$$

$$a_2(\varphi = 0) = \frac{2(1+\upsilon)}{E}\left(\sigma_{12}sin\psi_0 + 2\sigma_{23}cos\psi_0\right)sin2\eta.\,Sin\alpha \qquad (12)$$

Using equations (11) and (12) all stress components can be calculated. To accurately apply these equations to stress measurement in any direction, a set of two Debye rings must be used. A single Debye ring is not sufficient to solve for two stress components in the same equation, unless it has been verified that shear stresses σ_{13} and σ_{23} in the material are nonexistent. With this approach, the half space measurement condition covering both ψ^+ and ψ^- is fulfilled as shown in (Fig. 2).

Each component can be calculated from equations (11) and (12) using the slopes written in the following equations:

$$(\sigma_{11}) = -\frac{E}{(1+\upsilon)sin2\psi_0 sin2\eta}\frac{\partial a_1}{\partial Cos\alpha} \tag{13}$$

$$(\sigma_{12}) = \frac{E}{2(1+\upsilon).sin\psi_0.sin2\eta}\frac{\partial a_2}{\partial Sin\alpha} \tag{14}$$

$$(\sigma_{13}) = -\frac{E}{2(1+\upsilon)cos2\psi_0 sin2\eta}\frac{\partial a_1}{\partial Cos\alpha} \tag{15}$$

$$(\sigma_{23}) = \frac{E}{2(1+\upsilon).2cos\psi_0.sin2\eta}\frac{\partial a_2}{\partial Sin\alpha} \tag{16}$$

For accurate stress measurement in one direction, at least two directions (φ=0 and φ=180) are required to solve for $\sigma_{11}, \sigma_{12}, \sigma_{13}$ and σ_{23}. For triaxial stress measurements, at least 5 Debye rings are necessary for accurate measurement [10].

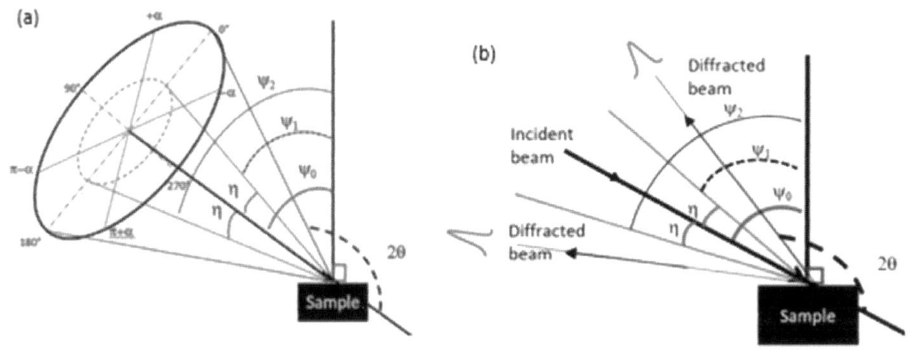

Figure 2: Illustration of, (a): Cosα technique geometry using one full Debye ring in a single exposure technique SET- Cosα and (b) two-tilt method or single exposure technique using 2 detectors (SET-Sin²ψ) [6].

Sin² ψ in any direction φ

The sin² ψ equation can be deduced from equations (5) and (8):

$$\varepsilon_{\varphi\psi} = \frac{1+\upsilon}{E}(\sigma_{11}cos^2\varphi + \sigma_{22}sin^2\varphi + \sigma_{12}sin2\varphi)Sin^2\psi - \frac{\upsilon}{E}(\sigma_{11}+\sigma_{22}) + \frac{1+\upsilon}{E}(\sigma_{13}cos\varphi + \sigma_{23}sin\varphi)Sin2\psi \tag{17}$$

Where the stresses in the direction of measurement φ for above techniques can be written as:

$$\sigma^{\varphi} = \sigma_{11}cos^2\varphi + \sigma_{22}sin^2\varphi + \sigma_{12}sin2\varphi \tag{18}$$

$$\tau^{\varphi} = \sigma_{13}cos\varphi + \sigma_{23}sin\varphi \tag{19}$$

Where σ^{φ} and τ^{φ}, are the normal and shear stress component in the direction φ. For these components, measurements must be performed using multiple tilts spread over ψ^+ and ψ^- tilt angles in the crystallographic. Equation (17) describes an ellipse, and its analysis requires nonlinear regression applied to the following expression:

$$\varepsilon_{\varphi\psi} = A \sin^2 \psi + B \sin 2\psi + C \tag{20}$$

The curve of d-spacing versus $\sin^2\psi$ is an ellipse. When the shear stresses $\tau^{\varphi} = 0$, the curve reduces to a straight line. The measurements direction should span the half crystallographic space required to solve for both normal and shear stress components σ^{φ}, τ^{φ} shown in equations (17), ((18) and (19). This half -space is illustrated on (Fig.3) where measurements should cover both sides (¼-space on each side) to ensure accurate fitting.

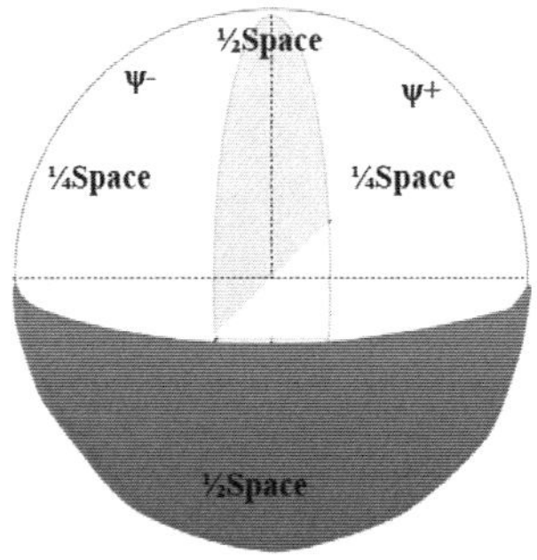

Figure 3: Crystallographic ½-space covered for stress measurement.

For two-tilts method (SET-$\sin^2\psi$), the slope from equation (17) can be written as:

$$\sigma^{\varphi} = \sigma_{11} = \frac{\partial \varepsilon_{\psi}}{\partial Sin^2\psi} \frac{E}{1+v} \tag{21}$$

In the same manner, triaxial stress tensor measurement requires three or more distinct directions to solve equation (18) and (19). However, additional measurement directions are often required to provide sufficient data for accurate results for both techniques. The more directions measured, the more robust and reliable the results become, especially in complex or anisotropic materials. This statement is applicable to most materials due to the triaxiality of stress field existing in the materials since most materials can be considered as heterogeneous [11].

Simulation

Simulations of stress using all aforementioned techniques, for the purpose of comparing and demonstrating the application of both theories, are presented in Table 1 for different type of stress fields and various stress components in a single direction using SET and MET. Stress component selection was based on the relationship between the different components in the same equation, $(\sigma_{11}, \sigma_{13})$ and $(\sigma_{12}, \sigma_{23})$.

Both MET-$\sin^2\psi$ and SET-$\sin^2\psi$ measurements were simulated using Cr-Kα (λ=2.291Å) x-radiation diffracting from the {hkl}{211} plane at 2θ=156.4°, $½S_2$=5.14.10^{-6} MPa^{-1}. The SET-$\sin^2\psi$ and SET-Cosα datasets were derived using a primary tilt angle ψ_0=+35° corresponding to effective detector angles to ψ_1=47° (α=180°) and ψ_2=23° (α=0°) in a dual-detector configuration.

Table 1: *Simulated residual stress using SET-Cosα, SET-Sin²ψ, and MET-Sin²ψ was used to illustrate the limitations of different components on the accuracy of results for the various techniques.*

Input data – Stress tensor (MPa)	SET- Cosα Stress (MPa)		SET-Sin²ψ Stress (MPa)		Difference Stress (MPa)		MET- Sin²ψ Stress (MPa)		MET- Sin²ψ Difference Stress (MPa)	
	σ_{11}	σ_{12}	σ_{11}	σ_{12}	σ_{11}	σ_{12}	σ_{11}	σ_{13}	σ_{11}	σ_{13}
$\begin{bmatrix} 500 & 0 & 0 \\ 0 & 500 & 0 \\ 0 & 0 & 0 \end{bmatrix}$	500	0	500	N/A	0	0	500	0	0	0
$\begin{bmatrix} 500 & 50 & 0 \\ 50 & 500 & 0 \\ 0 & 0 & 0 \end{bmatrix}$	500	50	500	N/A	0	0	0	0	0	0
$\begin{bmatrix} 500 & 50 & 50 \\ 50 & 500 & 0 \\ 50 & 0 & 0 \end{bmatrix}$	464	50	464	N/A	36	0	500	50	0	0
$\begin{bmatrix} 500 & 50 & 0 \\ 50 & 500 & 50 \\ 0 & 50 & 0 \end{bmatrix}$	500	21	500	N/A	0	29	500	0	0	0
$\begin{bmatrix} 500 & 50 & 50 \\ 50 & 500 & 50 \\ 50 & 50 & 0 \end{bmatrix}$	427	21	427	N/A	73	29	500	50	0	0
$\begin{bmatrix} 500 & 100 & 100 \\ 100 & 500 & 100 \\ 100 & 100 & 0 \end{bmatrix}$	427	43	427	N/A	73	57	500	50	0	0
$\begin{bmatrix} -579 & 0 & -149 \\ 0 & -873 & -2 \\ -149 & -2 & 0 \end{bmatrix}$*	-471	3	-471	N/A	108	3	-579	-149	0	0
$\begin{bmatrix} -873 & 0 & -2 \\ 0 & -579 & -149 \\ -2 & -149 & 0 \end{bmatrix}$**	-872	213	-872	N/A	1	213	-873	-2	0	0

*: These are real measured data on a steel sample, **: indicates the sample was rotated 90°.

When the tilting angle ψ_0 was varied from 5° to 95°, the normal stress value σ_{11} and the in-plane shear stress σ_{12} were also affected as shown in (Fig.4) [8].

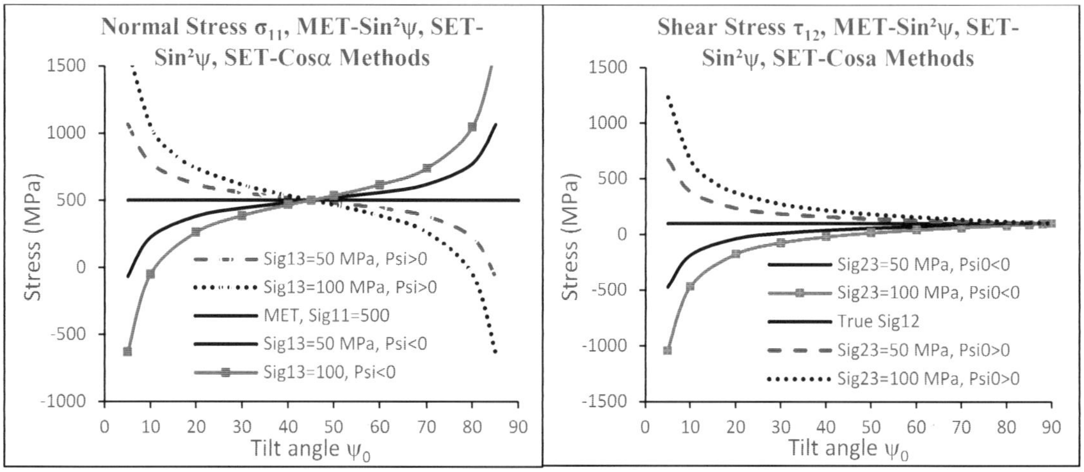

Figure 4: Simulated stress tensor terms σ_{11} and τ_{12} versus ψ_0, for both $\psi_0<0$ and $\psi_0>0$ values using: SET-Cosα,SET-Sin²ψ, and MET-Sin²ψ, for 2 different stress tensors inputs [8].

Discussion

The MET-Sin²ψ method yields accurate stress measurement results that are independent of the presence of shear stresses. In contrast, the SET-Sin²ψ and SET-Cosα methods provide matching and accurate results only when the out-of-plane shear stresses $\sigma_{13}$and, σ_{23} are zero. When these shear stresses are present, they influence the in-plane normal and shear components: σ_{13} affects σ_{11}, σ_{23} affects σ_{22}, and both σ_{13} and σ_{23} also impact σ_{12}. The resulting error in stress determination can be significant depending on the magnitude of shear stress in the material. The sign of the normal stress inferred from a d-spacing vs. Sin²ψ plot can be reversed if the analyzed branch of the ellipse is negative i.e. when ψ is negative.

To explain this effect when using SET-Sin²ψ and SET-Cosα, the black line tangent segments at various ψ angles (ψ_1, ψ_2) highlight the local slope of the curve, showing how the presence of shear stresses alters the slope across inclinations, (see Fig. 5). These slope variations reflect changes in measured strain values at different tilt angles. The discrepancy between the elliptical and linear responses demonstrates the influence of out-of-plane shear stresses on normal stress evaluation. The only mathematically robust way to eliminate the asymmetry or slope distortion caused by non-zero shear components is to set ψ_0=45°, corresponding to the center of the ellipse, see equations (11(12). This configuration equalizes the angular contribution and reduces slope differences. However, achieving ψ_0 = 45° is often not feasible due to mechanical constraints in the goniometer's tilting [12].

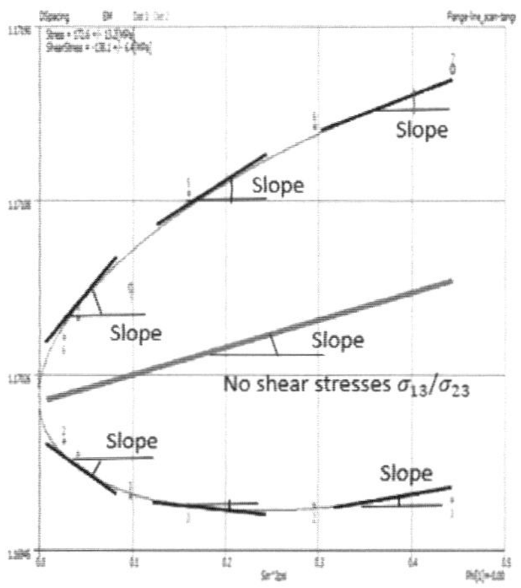

Figure 5: Variation of slope with goniometer inclination when out-of-plane shear stress is measured on steel.

Conclusion

Stress measurement using the XRD technique is widely applied in both academia and industry. With the emergence of new methods based on XRD, it is crucial to recognize that the underlying physics remains unchanged and that the original assumptions still apply. All developments made over the past several decades must be preserved and not overlooked. The triaxiality of stress in materials is a fact and must be considered and evaluated when performing stress measurement. A simplified calculation method can lead to erroneous results.

References

[1] I.C. Noyan, J.B. Cohen, "Residual Stress, Measurement by Diffraction and Interpretation", New York: springer-Verlag, 1987.

[2] V. Hauk, "Structural and Residual Stress Analysis by Non Destructive Method", Amesterdam: elsevier, 1997.

[3] Frank H. Chung, Deane K. Smith, Industrial Applications of X-Ray Diffraction, New York: Taylor and Francis, Sept. 22 1999.

[4] J. Lu, Handbook of Measurement of Residual Stresses, Lilburn: (Fairmont Press Inc. , 1996.

[5] T. Hanabusa, J. Fukura and H. Fujiwara, "X-Ray Stress Measurement on the Cementite Phase in Steels," *Buleting of JSME,* vol. 12, no. 53, pp. 931-939, 1969.

[6] M. Belassel, J. Pineault, M. Brauss, "Evaluation of x-ray diffraction stress instruments using cos-alpha and $\sin^2\psi$ techniques in relation to existing standards," *Advances in X-ray Analysis,* vol. 64, pp. 1-10, 2020.

[7] N. Peterson, et al., ", Assessment and Validation of Cosα Method for Residual Stress Measurement," *Proc. of the ICSP13,* pp. 80-86, 2017.

[8] M. Belassel, J. Pineault, M. Bolla, M. Brauss, "Introducing a triaxial stress field for residual stress measurements using cos-alpha techniques," *Proc. of ICRS11,* pp. 97-102, 2022.

[9] T. Sasaki et al., "Influence of image processing conditions of Debye Scherrer ring images in x-ray stress measurement using an imaging plate," *Adv. X-ray Anal,* vol. 40, p. 588–594, 1997.

[10] K. Tanaka, "[10] The cosα method for X-ray residual stress measurement using two-dimensional detector," *Bulletin of the JSME,* vol. 6, no. 1, pp. 1-15, 2019.

[11] T. Hanabusa and H. Fujiwara, "Relation of shear between stresses ψ-splitting in X-ray unidirectionally Takao and microscopic measurement of stress deformed by layers.," *Zairyo (Journal of the Society of Materials Science, Japan),* vol. 31, no. 342, pp. 227-233, March, 1982.

[12] M. Belassel, J. Pineault, & M. Brauss, "Updated Theory of the X-Ray Diffraction Residual Stress Measurement Comparing Cos-Alpha and $\sin^2\psi$ Techniques – Biaxial versus Triaxial," *Proceedings of the 14th International Conference on Shot Peening (ICSP-14),* 2022.

ICRS 12: Proceedings from the 12th International Conference on Residual Stresses, 2025
October 20–23, 2025; Detroit, Michigan, USA
https://doi.org/10.31399/asm.cp.icrs2025p0022

SI-traceability of XRD residual stress data by using Fe Kα radiation

Sebastian Send
Stresstech GmbH, Rennerod, Germany
sebastian.send@stresstech.com

Jonathan Mohan
Stresstech Inc., Pittsburgh, PA, USA

Mikko Palosaari
Stresstech Oy, Vaajakoski, Finland

Abstract

XRD residual stress analyses by means of the conventional $\sin^2\psi$-method in accordance with standards EN 15305 or ASTM E2860 require well-aligned and calibrated X-ray diffractometers. The qualification of an X-ray diffractometer is performed by measurements of a stress-free powder sample and a certified ILQ reference sample from round robin experiments. In order to obtain reliable qualification results, the traceability of residual stress data to SI units is optimal, which can be achieved via Standard Reference Material 660c (LaB_6) from NIST. However, in many cases, such as the traditional residual stress analysis of ferritic steel samples using Cr Kα radiation, the peak positions and intensities of LaB_6 are unfavorable for SI-traceability measurements. The technical difficulties arising from this condition can be overcome by the application of Fe Kα radiation providing a diffraction peak of LaB_6 close to the ones of common materials. This work quantitatively investigates the usability of Fe Kα radiation for XRD residual stress analyses including SI-traceability of stress-free powder samples based on LaB_6. Moreover, its potential for multi-phase residual stress determination in duplex steels is demonstrated as a laboratory application example.

Introduction

Residual stresses in polycrystalline materials can be precisely determined by means of X-ray diffraction (XRD) [1, 2]. Among various experimental approaches, the traditional $\sin^2\psi$-method, which resolves the direction-dependent lattice strains of the sample, was standardized [3, 4]. In this way, using the exact definition of strain and neglecting the normal stress component σ_{33} in the direction perpendicular to the surface, residual stresses can be measured without a-priori information about the unstrained material. The accuracy and reliability of the obtained results are mainly limited by mechanical imprecisions of the X-ray diffractometer. In this sense, adequate alignment and calibration of the device in accordance with standards [5, 6] are essential.

The European standard EN 15305 [3] defines a qualification procedure to monitor the performance of the used X-ray diffractometer. This method comprises a two-step process based on normal stress and shear stress measurements of a stress-free powder sample and a stressed, interlaboratory qualified (ILQ) reference material. The results are evaluated in terms of their deviations from zero stress (powder sample) as well as from a certified reference value (ILQ sample), respectively, which must not exceed the defined limits. However, in practical applications, the residual stress freeness of powder materials is just a simple, commonly accepted assumption lacking any experimental proof. If this condition is not fulfilled, subsequent residual stress analyses may obviously lead to erroneous results.

The process of device qualification becomes more reliable by involving a certified, stress-free standard, which enables traceability to the International System of Units (SI). Apart from this, in the course of their accreditation according to ISO/IEC 17025 and DIN EN ISO/IEC 17025 [7], the laboratories of Stresstech are obligated to ensure SI-traceability of their reported measurement results. A suitable material for the calibration of powder diffractometers is lanthanum hexaboride (LaB_6) certified by the National Institute of Standards and Technology (NIST), USA and distributed under the designation "Standard Reference Material 660c" [8, 9]. Unfortunately, the applicability of LaB_6 for SI-traceability of XRD residual stress data under usual laboratory conditions is limited.

This disadvantage manifests in the fact that the diffraction peaks of LaB_6 either occur at unfavorable positions or are not intense enough, e.g. when measuring residual stresses in ferritic steels with Cr Kα radiation based on the α-Fe 211 reflection at about 156° (2θ). In this case, Cr Kα radiation generates the LaB_6 311 and 320 reflections at 132.13° (2θ) and 167.04° (2θ), respectively, near the edges of the detectable angular range relevant for XRD residual stress analysis. Furthermore, due to its low multiplicity, the LaB_6 222 reflection at 145.35° (2θ) is too weak, thereby preventing routine usability of LaB_6 for device calibration with Cr Kα radiation. Similar experimental problems are obtained for Mn Kα radiation applied to measure residual stresses in austenitic steels based on the γ-Fe 311 reflection.
With regard to diffractometer calibration and SI-traceability of residual stress data by means of LaB_6, the limitations arising from standard laboratory conditions for residual stress measurement can be overcome by using Fe Kα radiation. Fe Kα radiation generates the LaB_6 410 reflection at 147.82° (2θ) being less than 3.5° away from the diffraction peaks of common materials like α-Fe, Al, Ni and WC, whereas γ-Fe, Cu and α-Ti diffract in angular distances between 8.9° and 11.5° from the LaB_6 410 reflection. In this work, it is shown that with the application of LaB_6 and Fe Kα radiation, SI-traceability of residual stress data can be achieved for a variety of materials. The results discussed in this paper are helpful to evaluate the quality of powder samples to be used for calibration and device qualification as per EN 15305. Finally, the simultaneous measurement of the phase contents and residual stresses in the ferritic and austenitic phases of duplex-type powders by means of Fe Kα radiation is considered.

Experimental

The performance of LaB_6 was experimentally proven by measuring the residual stresses in ten equivalent LaB_6 powder samples, based on which the capability of Fe Kα radiation to ensure SI-traceability under laboratory conditions could be exemplarily demonstrated for seven powders of pure α-Fe, Al, Ni, WC, γ-Fe, Cu and α-Ti. These powders were mixed with epoxy resin and cast inside a thin hollow cylinder with an inner diameter of 14 mm and an outer diameter of 20 mm [10]. An Xstress DR45 as manufactured by Stresstech Oy, Finland served for data acquisition and analysis using the $\sin^2\psi$-method and the modified χ-mode in accordance with EN 15305 for residual stress measurement. Two highly sensitive X-ray area detectors with an angular coverage of 17.7° (Δ2θ) each were applied to record strain data in standard mode [11] with eleven symmetrical tilt angles between −45° and +45° and equidistant $\sin^2\chi$-spacing.
Table 1 gives an overview of the investigated powders and the corresponding experimental conditions in terms of the diffraction angles 2θ of the measured reflections {hkl}, the maximum information depth z_0 as calculated from sample absorption of Fe Kα radiation in modified χ-mode and the maximum normal stress magnitude σ_{max}. This value is the acceptable limit for a qualified X-ray diffractometer as per EN 15305 (eq. 8a) and was accordingly extracted based on the X-ray elastic constant $\frac{1}{2}S_2$ of the material for the relevant reflection. The elastic constants of LaB_6 reported in [12] served to simulate $\frac{1}{2}S_2$ for the LaB_6 410 reflection using the Eshelby-Kröner model [13, 14], whereas for the other materials and reflections, the associated values of $\frac{1}{2}S_2$ were taken from [15].

Table 1: Experimental conditions for XRD residual stress analyses of various powders studied with Fe Kα radiation.

Material	{hkl}	2θ [°]	z_0 [μm]	σ_{max} [MPa]
LaB_6	{410}	147.82	2.3	31.7
α-Fe	{220}	145.82	8.1	17.2
Al	{400}	146.19	17.5	4.9
Ni	{222}	144.44	5.5	21.0
WC	{112}	145.24	1.0	53.8
γ-Fe	{222}	138.29	7.7	19.9
Cu	{222}	136.34	4.8	12.7
α-Ti	{203}	156.76	3.0	8.7

The diffraction signals of LaB_6, α-Fe, Al, Ni and WC were recorded at a distance of 45 mm from the sample surface after centering the X-ray detectors at a diffraction angle of about 145° (2θ) on a circular arc. In order to measure the diffraction signals of γ-Fe, Cu and α-Ti with sufficient background information at both sides of the peaks, the detectors were centered at diffraction angles of about 141° (2θ) for γ-Fe and Cu and about 156° (2θ) for α-Ti, respectively. For simplicity, the 2θ-scales of the detectors were calculated by assuming stress freeness of the powder samples and using the diffraction peak position of α-Fe for calibration in the 145° (2θ) and 141° (2θ) configurations and the diffraction peak position of α-Ti for calibration in the 156° (2θ) configuration. Figure 1 shows the normalized I(2θ) spectra of the different materials as measured with Fe Kα radiation from a 210 W (30 kV, 7 mA) Fe tube and a circular collimator providing an incident X-ray beam with a diameter of 1 mm. For optimal data quality, the Kβ peaks of the investigated materials were suppressed by mounting a 15 μm Mn foil in front of the detectors acting as a filter for Fe Kβ radiation.

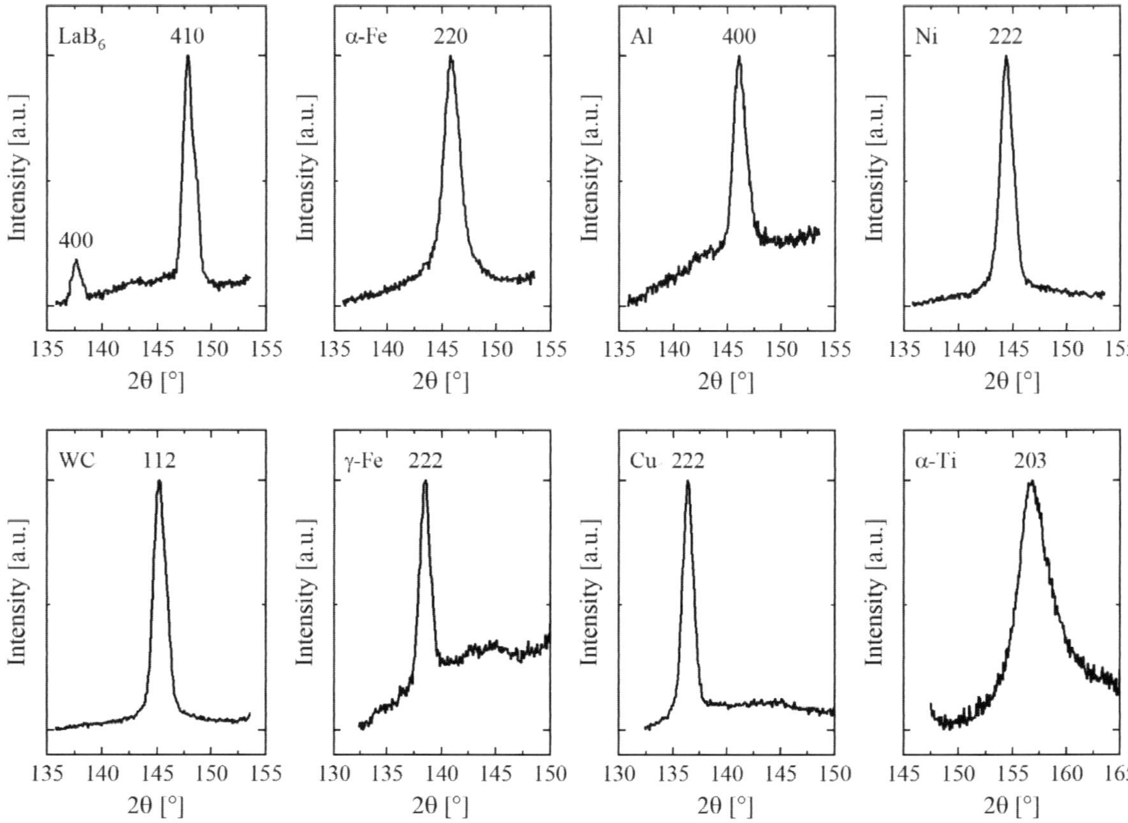

Figure 1: XRD signals and indexed reflections of the materials listed in Table 1 after exposure to Fe Kα radiation.

From a statistical point of view, quality indicators like peak height, signal-to-background ratio and integrated intensity of the diffraction peaks generated by Fe Kα radiation are adequate for reliable residual stress analyses of the individual powders. Favored by its high scattering power, WC provides the best measurement conditions among the materials in terms of quality indicators. The diffraction peaks of LaB_6, α-Fe, Al, Ni, WC, γ-Fe and Cu are detected with full widths at half maximum (FWHM) between 0.9° (LaB_6) and 1.4° (α-Fe), whereas the α-Ti 203 reflection occurs with a FWHM of 3.0°. Moreover, compared to the other materials, the I(2θ) spectrum of α-Ti shows a significantly higher noise level. This observed degraded data quality in the case of α-Ti is caused by the energy threshold selected for the measurement. On the one hand, the energy threshold must be high enough to suppress fluorescence radiation from α-Ti. On the other hand, it needs to be low enough to avoid cutting of diffracted intensity from the α-Ti 203 peak. In that respect, Fe Kα radiation provides sufficient data quality to enable precise residual stress measurements of α-Ti, which, however, does not approach the data quality obtained with Cu Kα radiation.

SI-traceability of the experimental results was ensured by recording statistically reliable data sets for each investigated material. For that purpose, two kinds of analysis methods were applied: In the first run, the repeatability of the residual stress data was determined by carrying out five successive measurements at the center of the circular sample surface under the same conditions without repositioning the sample. In the second run, the homogeneity of the material was analyzed by performing single residual stress measurements at four additional positions located in a distance of 2 mm from the center in the positive and negative x- and y-directions of the X-ray diffractometer. Following the definitions in EN 15305 for modified χ-mode, the positive x- and y-directions correspond to the measurement directions $\varphi = 0°$ and $\varphi = 90°$, respectively. In order to increase the accuracy of the acquired strain data by minimizing possible effects of coarse grain size or texture, the rotation capability of the Xstress DR45 was used to apply a rotation sweep between $\varphi = 0°$ and $\varphi = 180°$ for each tilt angle. Under this experimental condition, assuming isotropic stress states, complete Debye-Scherrer rings of the powder samples contributed to the final results.

Besides the normal stress σ evaluated with regard to its maximum acceptable magnitude σ_{max} for device qualification, the shear stress τ of the material was monitored. The corresponding maximum acceptable magnitude τ_{max} of the shear stress is given by EN 15305 (eq. 8b) and amounts to half the value of σ_{max}. For each measurement, the average residual stresses including their standard deviations were simultaneously calculated from the data via an elliptical least-squares fit of the strain curve $\varepsilon(\sin^2\chi)$ recorded in modified χ-mode. Taking the experimental results of the repeatability tests into account, the combined uncertainties of the individual data points were obtained as per [16].

Results

SI-traceability of the residual stress data gathered for the 145° (2θ) configuration was ensured by directly calibrating the X-ray area detectors with the LaB_6 410 reflection. To this end, the sample LaB_6-1 out of the set of ten LaB_6 samples (denoted by LaB_6-1, ..., LaB_6-10) was exposed to Fe Kα radiation at normal incidence in the center of its surface. The results of the subsequent homogeneity measurements for the normal stresses and the shear stresses of the investigated sample set are summarized in Fig. 2.

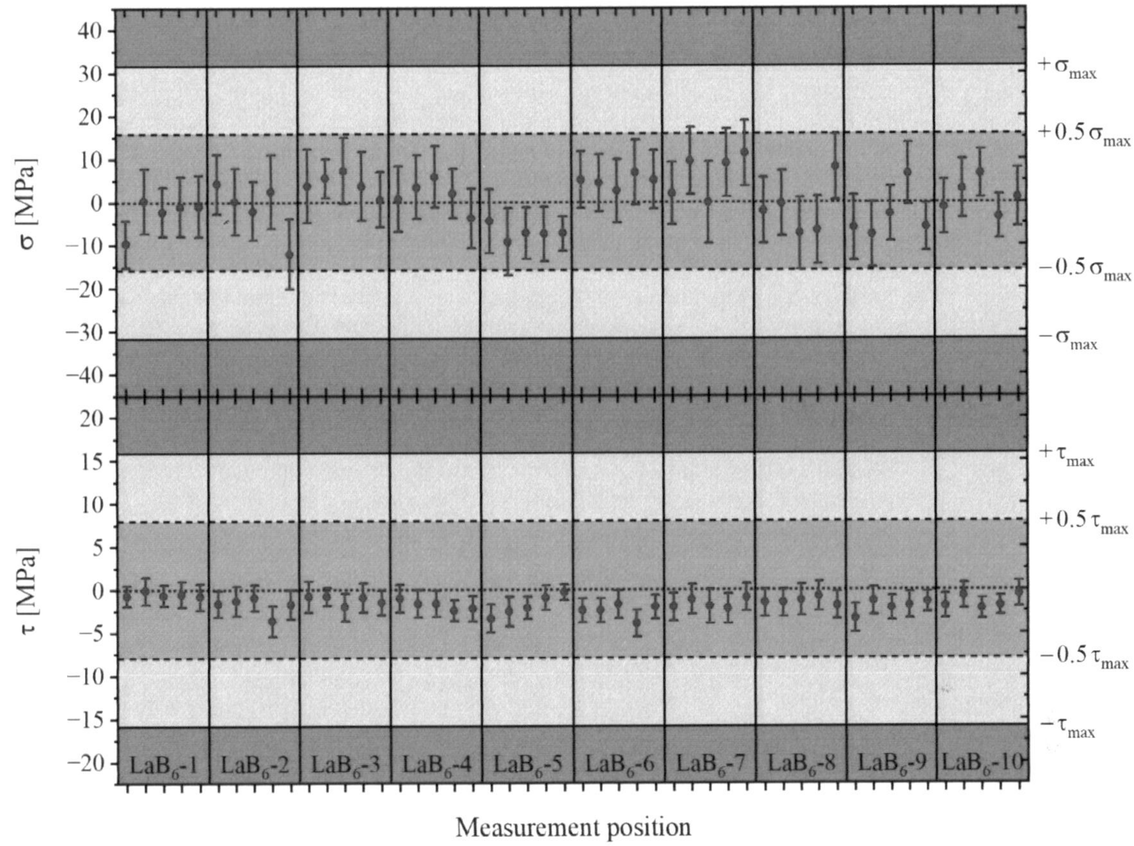

Figure 2: Normal stresses and shear stresses of ten LaB_6 samples measured at different positions along the surface.

For each specimen, the first data point corresponds to the mean normal stress and the mean shear stress of five repeat exposures at the center, respectively. The other four data points are the results of single exposures at the previously defined positions. With regard to the repeatability tests, the standard deviations of the mean normal stress values range from 0.5 MPa to 1.4 MPa with an average of 0.8 MPa. Significantly larger contributions to the combined measurement uncertainty of the normal stress are obtained in terms of the standard deviations of the elliptical strain data fits ranging from 4.0 MPa to 9.5 MPa with an average of 7.0 MPa. In this sense, neglecting the error of the associated X-ray elastic constant $\frac{1}{2}S_2$ involved in the calculation, the combined uncertainties of the measured normal stress values can be well approximated by the standard deviations of the fit. Equivalent observations are made in the case of the shear stresses being detected with a maximum repeatability standard deviation of 0.3 MPa and a mean deviation of 0.1 MPa, whereas the standard deviations of the least-squares fits range from 0.9 MPa to 2.0 MPa with an average of 1.5 MPa.

In order to classify the experimental results, appropriate color coding was defined based on the maximum acceptable stress values σ_{max} and τ_{max} for device qualification as per EN 15305 with a stress-free powder sample: Residual stresses with magnitudes less than 50% of the limit value (marked green) are considered to be of high quality. Residual stresses with magnitudes above that value but not exceeding the limit (marked yellow) are considered to be of moderate quality. Residual stresses with magnitudes above the limit (marked red), making the device qualification fail, are considered to be of low quality. Data points lying outside the acceptable ranges may be caused by either high local stresses present within the sample or an imprecisely aligned X-ray diffractometer.

In the case of LaB_6 certified as a stress-free standard for X-ray powder diffractometry, the measured residual stresses are of high quality for each individual sample and measurement position. The average normal stress values obtained from the homogeneity analyses vary between −7.1 MPa (sample LaB_6-5) and 6.4 MPa (sample LaB_6-7) with a mean sample standard deviation of 0.8 MPa. Taking the complete data set into account, the experimental mean normal stress of LaB_6 amounts to 0.2 MPa with a standard deviation of 0.8 MPa quantifying the overall scattering of the data points. The associated average shear stress values range from −2.5 MPa (sample LaB_6-6) to −0.5 MPa (sample LaB_6-1) with a mean sample standard deviation of 0.3 MPa. The overall average shear stress amounts to −1.6 MPa, corresponding to 10% of the acceptable minimum shear stress value $-\tau_{max}$, with a standard deviation of 0.2 MPa.
In summary, using the Xstress DR45 and Fe Kα radiation for XRD residual stress analyses after routine diffractometer alignment, the experimental results confirm the assumption that the certified LaB_6 standard is free of normal stresses. The discrepancy between the measured mean shear stress and zero stress within the calculated uncertainties indicates a small misalignment of the Xstress DR45, which, however, has no effect on the accuracy of the normal stress values. In any case, the application of LaB_6 enables reliable device qualification and provides a well-calibrated setup ensuring SI-traceability of absolute strain values measured in the 145° (2θ) configuration. Consequently, the identical setup as used to verify the performance of LaB_6 could serve to test the quality of the investigated α-Fe, Al, Ni and WC powders. The γ-Fe powder and the Cu powder were tested in the 141° (2θ) configuration, whereas the α-Ti powder was analyzed in the 156° (2θ) configuration after adequate calibration of the Xstress DR45 ensuring SI-traceability of the recorded strain data in both situations.
Figure 3 shows the experimental results of the homogeneity measurements for the various powders. In order to enable comparability between the different materials, the residual stress values were normalized by the maximum acceptable normal and shear stress magnitudes σ_{max} and τ_{max} of the corresponding material, respectively, and expressed in relative units. By analogy with the results depicted in Fig. 2, the first data point of each powder is the mean value of five repeat measurements carried out at the center of the circular sample surface prior to the homogeneity analysis. The error bars of the data points represent the combined measurement uncertainties calculated as per [16] including the repeatability standard deviation s_{rep} and the standard deviation s_{fit} of the strain data fit after conversion into relative units. Table 2 summarizes the results obtained after complete statistical analysis of the data sets in terms of the overall mean residual stress, repeatability, average fit accuracy and homogeneity of the materials considered.

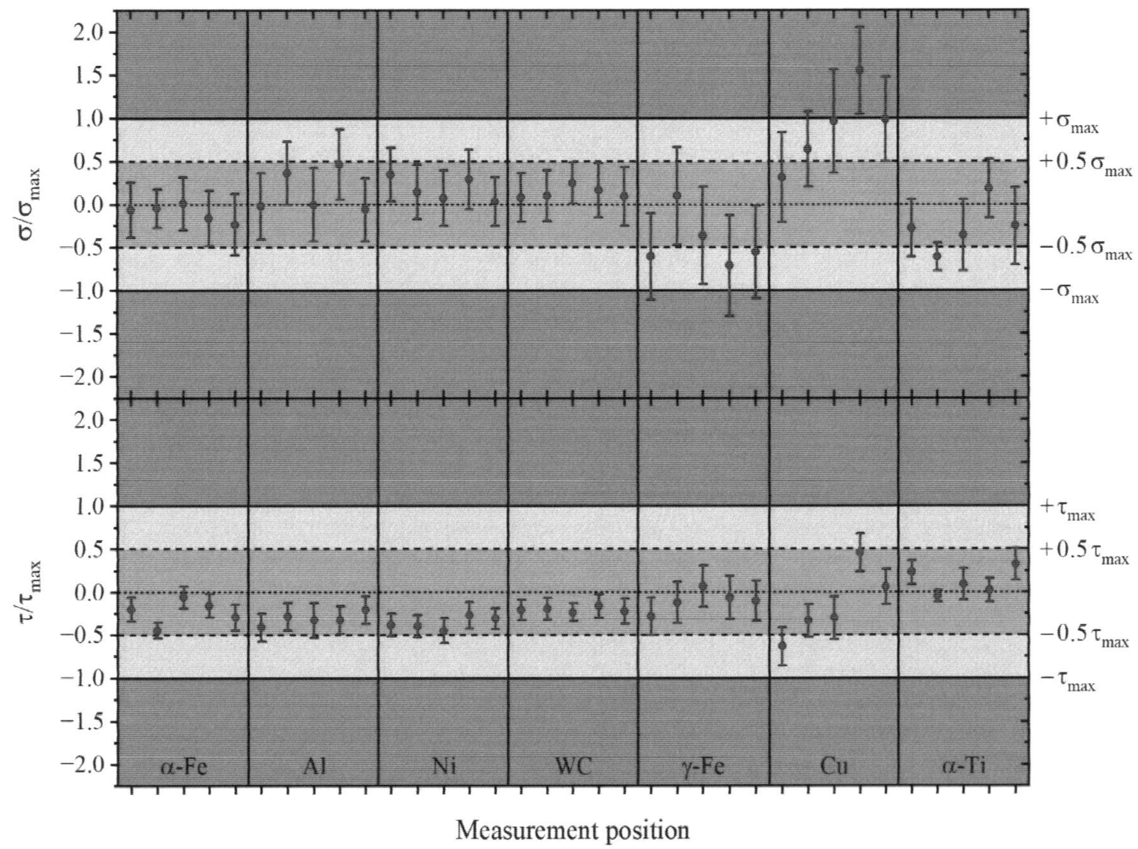

Figure 3: Normal stresses and shear stresses of various powders measured at different positions along the surface.

Table 2: Statistical results of the repeatability and homogeneity measurements (values normalized to σ_{max} and τ_{max}).

Material	Normal stresses				Shear stresses			
	<σ>	s_{rep}	s_{fit}	s_{hom}	<τ>	s_{rep}	s_{fit}	s_{hom}
LaB_6	+0.01	0.05	0.22	0.06	−0.05	0.01	0.05	0.01
α-Fe	−0.10	0.03	0.31	0.04	−0.23	0.01	0.13	0.07
Al	+0.15	0.10	0.39	0.11	−0.31	0.02	0.17	0.03
Ni	+0.18	0.03	0.32	0.06	−0.36	0.01	0.14	0.03
WC	+0.14	0.11	0.30	0.03	−0.20	0.01	0.13	0.01
γ-Fe	−0.42	0.13	0.55	0.14	−0.10	0.02	0.24	0.06
Cu	+0.89	0.06	0.51	0.21	−0.15	0.03	0.22	0.19
α-Ti	−0.26	0.14	0.34	0.13	+0.12	0.03	0.14	0.07

After directly calibrating the Xstress DR45 with LaB_6 in the 145° (2θ) configuration, the residual stress measurements along the surfaces of the α-Fe, Al, Ni and WC powders provided data points of highest quality without exception. In those cases, the magnitudes of the mean stresses do not exceed 20% for the normal stress and 40% for the shear stress, respectively. In terms of fit accuracy, which makes the highest contribution to the combined measurement uncertainty, the results for α-Fe, Ni and WC are similarly precise. However, none of those powders reaches the accuracy of a single residual stress analysis of LaB_6. On the other hand, with regard to uniformity quantified by the standard deviation s_{hom} of the average stress measured at different positions, the normal stresses of α-Fe and WC are more homogeneous than the normal stresses of LaB_6. For the Al powder sample, the residual stresses were determined with degraded accuracy as indicated by the larger standard deviations. This outcome originates from the lower data quality in the case of Al, which is caused by unfavorable structural properties (coarse grain size, texture) and statistical issues associated with the low-multiplicity Al 400 reflection (counting statistics, peak-to-background ratio).
Nevertheless, in relation to the combined uncertainty, the measured mean normal stress is compatible with zero stress for each of the powder samples investigated in the 145° (2θ) configuration. The experimental shear stress values are also of highest quality, with larger deviations from zero stress due to a small mechanical misalignment of the device. The metrological traceability of the results for the 145° (2θ) configuration was used to calibrate the Xstress DR45 in the other two configurations based on an unbroken chain of calibrations in accordance with DIN EN ISO/IEC 17025. In the 141° (2θ) setting, the α-Fe 220 reflection generated by Fe Kα radiation could directly be applied for that purpose, whereas in the 156° (2θ) setting, the WC 211 reflection generated by Fe Kβ radiation at an angle of 153.71° (2θ) was the only feasible option to achieve reliable device calibration with adequate accuracy. Consequently, in contrast to the situations of the 145° (2θ) and 141° (2θ) settings, the residual stress measurements of the α-Ti powder in the 156° (2θ) setting were carried out after removing the Kβ filter.
The statistical analyses of the data sets for the γ-Fe and the Cu powder recorded in the 141° (2θ) configuration reveal that in both cases, the precision of residual stress determination is significantly lower compared to the other powders. This loss of accuracy manifests in the enhanced standard deviations of the elliptical data fits and of the homogeneity measurements, which may arise from coarser grain sizes or less structural uniformity. The degraded fit quality and the larger scattering of the data points are observed for both the normal stresses and the shear stresses.
In the case of the γ-Fe powder, though most of the measured normal stresses are of moderate quality, the magnitude of the average normal stress lies in the high-quality range. Taking its low shear stresses and the level of the combined measurement uncertainties into account, the performance of the γ-Fe powder is sufficient to enable device qualification within the limits defined in EN 15305. For the normal stresses of the Cu powder, on the other hand, one data point of low quality with a magnitude of 55% above the limit σ_{max} and two data points of moderate quality close to this limit were obtained. Therefore, although the average normal stress lies in the moderate-quality range, it could not be proven by means of SI-traceability that the Cu powder is stress-free at all probed positions within the combined uncertainties. Moreover, with regard to the scattering of the residual stress data, the inhomogeneity of the Cu powder is unacceptably high, which may lead to erroneous calibration results. Under those conditions, the Cu powder does not fulfil the quality requirements for a reference material to be used for device qualification as per EN 15305.
Referring to the previous statements in the experimental part of this work, the accuracy of the measurement results for the α-Ti powder investigated in the 156° (2θ) configuration is limited by the conditions imposed by Fe Kα radiation. However, this constraint affects the repeatability of the residual stress measurement rather than the standard deviation of the least-squares fit, which is close to the ones of α-Fe, Ni and WC and significantly smaller than the ones of γ-Fe and Cu. The experimental normal and shear stresses of the α-Ti powder lie in the high-quality range with one exception and are adequately homogeneous. This result confirms that after metrological traceability using an unbroken chain of calibrations based on LaB_6, the average residual stress values of the α-Ti sample are compatible with zero stress within the combined measurement uncertainty. Accordingly, besides the α-Fe, Al, Ni, WC and γ-Fe powders, the α-Ti powder can be applied for qualification purposes as per EN 15305 and serve for direct calibration of the Xstress DR45 operated in the 156° (2θ) configuration with Fe Kα radiation.

Application

As demonstrated in the experimental and in the results part of this work, Fe Kα radiation enables XRD residual stress analyses of various materials with measurement accuracies depending on the microstructural properties of the samples. In that respect, Fe Kα radiation may be a suitable tool to investigate multi-phase materials, as long as the diffraction peaks of the constituents are spectrally separated. A common example from practice is duplex steel, which contains a two-phase microstructure of α-Fe and γ-Fe with well-defined volume fractions.

Conventional X-ray diffractometers for residual stress analyses equipped with a Cr X-ray tube simultaneously provide the α-Fe 211 reflection at 156.4° (2θ) generated by Cr Kα radiation and the γ-Fe 311 reflection at 148.9° (2θ) generated by Cr Kβ radiation. Due to the small tube output of Cr Kβ radiation, the α-Fe 211 peak is detected with an integrated intensity, which can be more than an order of magnitude higher than the one of the γ-Fe 311 peak, leading to statistical imbalance. Additionally, under the measurement conditions using the Xstress DR45 for residual stress determination, the diffraction peaks of ferritic and austenitic powders are relatively broad with FWHM ranging from 1.4° (γ-Fe 311) to 1.8° (α-Fe 211). In the case of bulk materials, especially for hardened steels, the FWHM typically increase to several degrees, which may result in a spectral overlap of those peaks and impede the residual stress calculation. On the other hand, the α-Fe 220 peak at 145.8° (2θ) and the γ-Fe 222 peak at 138.3° (2θ) generated by Fe Kα radiation occur with the same angular difference of 7.5° (Δ2θ), but with lower FWHM and better-balanced intensities. Consequently, it is expected that improved experimental conditions for XRD residual stress analyses of duplex steels can be obtained by using an Fe X-ray tube instead of a Cr X-ray tube.

In order to evaluate the capability of Fe Kα radiation to investigate duplex-type materials, two-phase microstructures were manufactured by mixing the ferritic powder 430L and the austenitic powder 316L with different phase fractions. For test purposes, seven samples with nominal volume fractions V_γ of γ-Fe between 80% and 2% and with the same geometry as used in the experiments on SI-traceability were created. In another application, those specimens serve as a set of internal reference samples for XRD measurements of retained austenite as per ASTM E975 [17]. Accordingly, the specimens were pre-characterized by means of the conventional four-peak method with the Xstress DR45 operated in an ω-mode setup with Cr Kα radiation and an incident angle of 45°. Following ASTM E975, the R-factors for the α-Fe 211, α-Fe 200, γ-Fe 220 and γ-Fe 200 peaks of the 430L and 316L powders were calculated from basic principles. The results for Cr Kα radiation, which are listed in the third column of Table 3 in terms of the mean retained austenite contents and their standard deviations, show a good agreement with the expected values.

Table 3: Nominal and experimental retained austenite values for the investigated sample set of 430L/316L mixtures.

Sample	V_γ nominal	V_γ measured (Cr Kα)	V_γ measured (Fe Kα)
1	80%	(81.9 ± 1.1)%	(82.3 ± 0.8)%
2	66.7%	(68.3 ± 1.2)%	(68.7 ± 1.6)%
3	50%	(51.5 ± 1.8)%	(53.3 ± 1.8)%
4	30%	(30.0 ± 1.5)%	(30.6 ± 1.5)%
5	15%	(14.8 ± 1.5)%	(15.9 ± 1.4)%
6	8%	(7.9 ± 0.6)%	(8.2 ± 0.5)%
7	2%	(2.2 ± 0.5)%	-----

Subsequently, this sample set was analyzed in modified χ-mode using Fe Kα radiation in the same experimental setup as for the SI-traceability measurements discussed in the previous section. Figure 4 shows the I(2θ) diffraction spectra of the various specimens acquired by one of the two detectors at normal incidence. For a residual stress determination carried out at the sample center in modified χ-mode, the complete data volume comprises two equivalent I(2θ) spectra per tilt angle. In the measured intensity distributions, the α-Fe 220 peak and the γ-Fe 222 peak were recorded with a FWHM of 1.1° each providing better conditions than in the case of the α-Fe 211 peak and the γ-Fe 311 peak generated by Cr Kα and Cr Kβ radiation, respectively. No Kβ filter was used in connection with the Fe tube for statistical reasons, so that the low-angle tail of the α-Fe 310 peak, generated by Fe Kβ radiation at 151.5° (2θ), appears in the spectrum. The variation of the γ-Fe volume fraction between 81.9% and 2.2% gives rise to the expected decrease of the γ-Fe 222 signal in favor of the α-Fe 220 signal. For the highest γ-Fe phase content (sample 1) and Fe Kα radiation, the intensity ratio between the γ-Fe 222 peak and the α-Fe 220 peak improves to about 3.0, whereas for Cr Kα and Cr Kβ radiation, an intensity ratio of only 0.4 between the γ-Fe 311 peak and the α-Fe 211 peak is observed. On the other hand, for low γ-Fe contents in the range of 2% (sample 7), the γ-Fe 222 peak cannot be resolved above the background signal, which limits the applicability of Fe Kα radiation under the defined experimental conditions.

In the first step of data analysis, the integrated intensities of the α-Fe 220 peak and the γ-Fe 222 peak were determined for the complete data set (i.e. for both detectors and each tilt) and used to calculate the phase contents of the samples based on simulated R-factors for the case of Fe Kα radiation. Compared to the four-peak method with Cr Kα radiation, this approach represents a two-peak method, whereby the diffraction signals of a larger number of grains are collected. Adequate experimental accuracy is ensured by exploiting the higher grain statistics and averaging the individual phase contents over the data volume. The final results of the retained austenite measurements with Fe Kα radiation obtained in this way are listed in the fourth column of Table 3. Due to the insufficient integrated intensity of the γ-Fe 222 peak, no data point could be ascribed to the 2% sample. Apparently, within the statistical uncertainties, the mean γ-Fe phase contents of the remaining samples coincide with the reference values from the four-peak method with Cr Kα radiation as per ASTM E975. Furthermore, in both cases, the standard deviations of the average retained austenite values have similar magnitudes, so that the two-peak method with Fe Kα radiation provides a measurement precision equivalent to the standardized method.

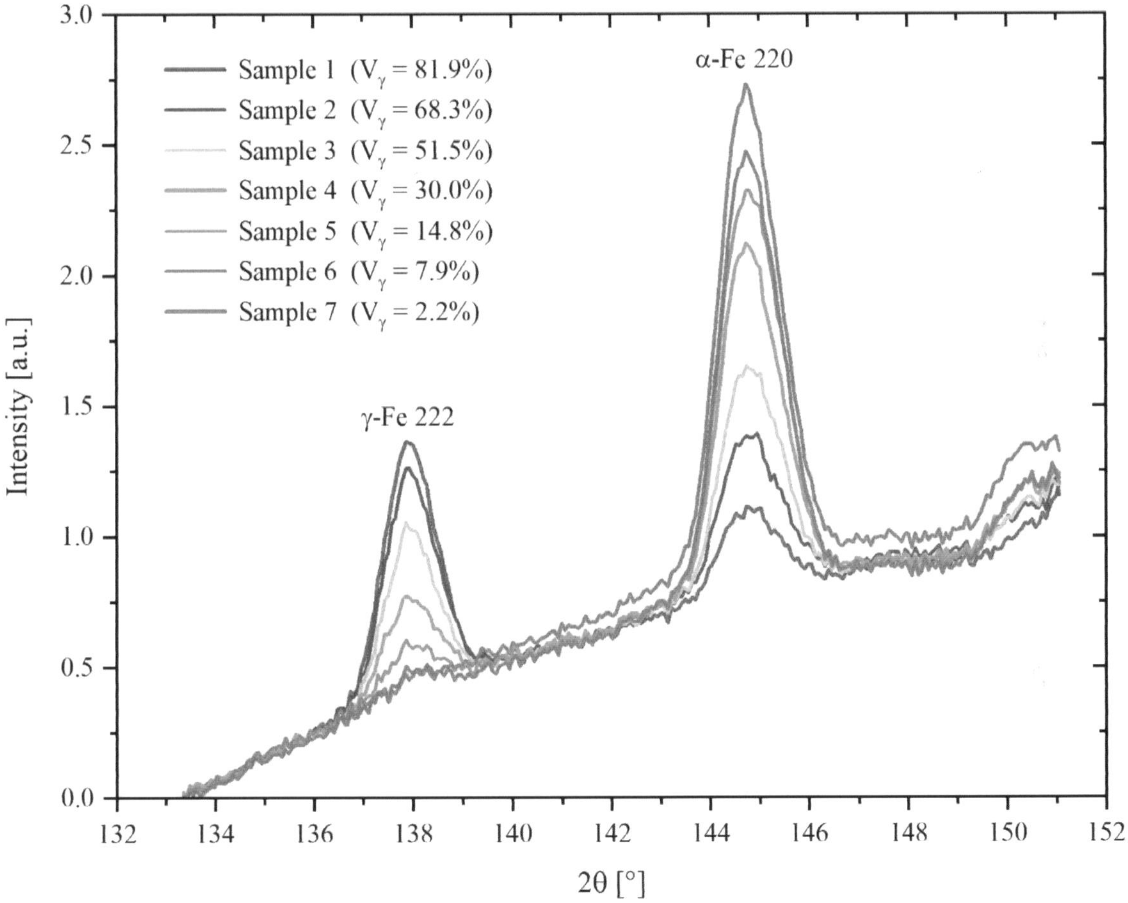

Figure 4: XRD signals of duplex-type powder samples with different phase contents generated by Fe Kα radiation.

In the second step of data analysis, the residual stresses of the α-Fe and the γ-Fe phase were simultaneously extracted from the same data set used for retained austenite determination. Figure 5 shows the normal stresses σ_α and σ_γ in both phases depending on the associated volume fractions V_α and V_γ of the specimens, respectively. As the individual data points originate from a single residual stress measurement, their error bars are solely given by the standard deviations of the elliptical strain data fits. The red curves in Fig. 5 are linear least-squares fits of the data to indicate trends in the experimental stress values of the two components.

For the α-Fe phase, the normal stresses could be calculated with a fit standard deviation of 4.5 MPa averaged over the complete data set. After normalization to the maximum acceptable normal stress, this value corresponds to a relative standard deviation of 0.26, which is even below the accuracy obtained in the previous homogeneity measurements of a pure α-Fe powder. However, only in the case of the sample with the highest α-Fe content of about 98% (sample 7), a normal stress value in accordance with zero stress is observed. In contrast, the sample with the lowest α-Fe content of about 18% (sample 1) gives a significant compressive stress of −35 MPa (close to twice the value of $-\sigma_{max}$) resulting from the high γ-Fe content. The trend of the data can be roughly described by a line with an intercept of −31 MPa and a slope of 0.22 MPa/%. Taking only specimens with usual phase contents of duplex steels between 30% and 70% into account (samples 2-4), the average normal stress of the α-Fe phase in this range is about −16 MPa.

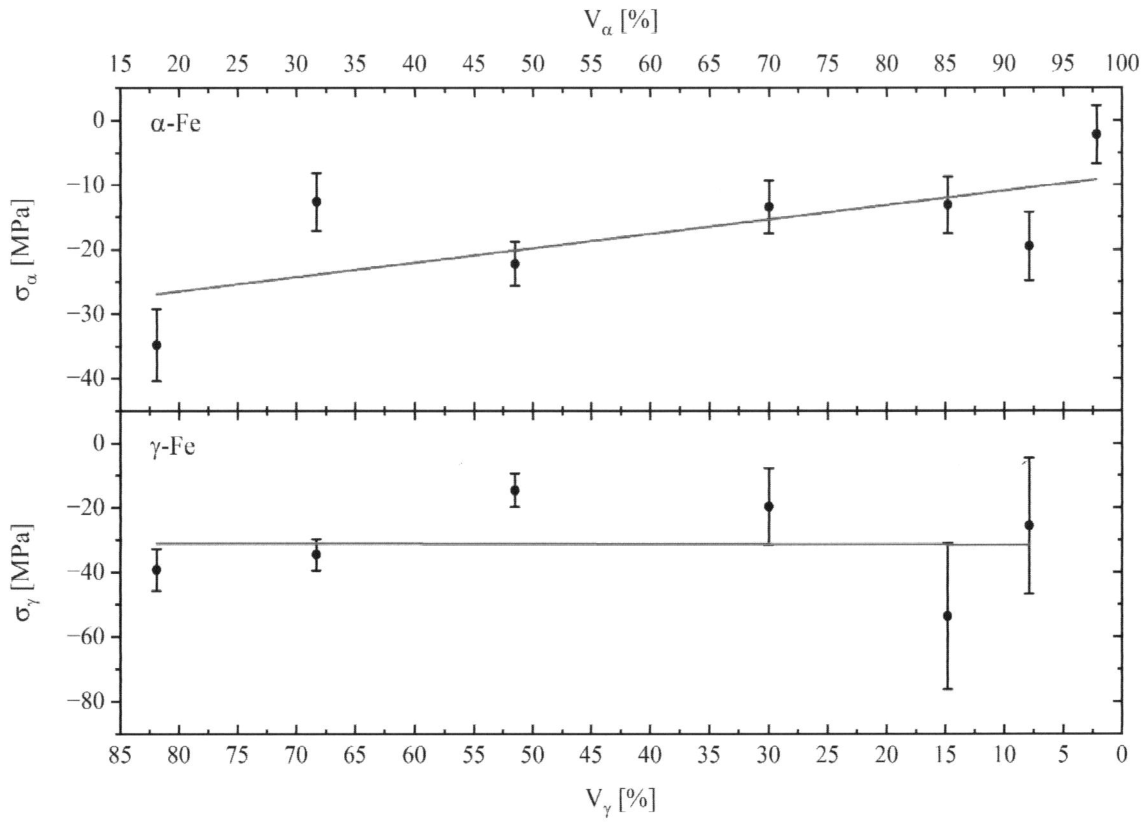

Figure 5: Normal stresses of the α-Fe phase (top) and of the γ-Fe phase (bottom) as functions of the phase contents.

In the case of the γ-Fe phase, on the other hand, the accuracy of residual stress measurement depends strongly on the signal-to-background ratio of the γ-Fe 222 peak. For high γ-Fe phase contents above 50%, the normal stresses could be determined with a mean fit standard deviation of 5.5 MPa (corresponding to a relative standard deviation of 0.28), which provides similar precision as for the case of the α-Fe phase. This value amounts to half the fit standard deviation calculated for the pure γ-Fe powder in the homogeneity test suggesting a fine-grained microstructure of the γ-Fe phase in the duplex-type powder. As a result of the decreasing signal-to-background ratio, the relative fit standard deviation increases to a value of 0.60 for the 30% sample being comparable with the accuracies of the previous measurements. The normal stresses in the γ-Fe phase of the 15% sample and the 8% sample were obtained with fit standard deviations of about 1.10 relative to the maximum acceptable normal stress. High compressive stress of about −40 MPa is observed for the 82% sample, which equals twice the value of $-\sigma_{max}$ and indicates a low-quality data point. With reducing γ-Fe content, the stress level decreases to the moderate-quality range for the 52% sample and the 30% sample. However, a clear trend of the normal stress values cannot be derived directly from the complete data set, which is reflected in the fact that the linear fit of the data gives a nearly horizontal line at a value of about −30 MPa.

It should be noted that in contrast to the investigations of various materials based on LaB_6, only single measurements were carried out in the center of the duplex-type powders to evaluate the capability of Fe Kα radiation for simultaneous phase content and residual stress analyses of multi-phase materials. More accurate results can be expected from more extensive studies of those specimens including repeatability and homogeneity tests.

Conclusions

In this work, the potential of Fe Kα radiation for XRD residual stress determination under laboratory conditions was investigated. Compared to conventionally applied X-ray sources, Fe tubes offer the advantage of flexible usability to measure a variety of common materials. Therefore, Fe Kα radiation enables the direct inclusion of LaB_6 as a certified, stress-free standard in the process of diffractometer qualification as per EN 15305. In particular, the basic assumption of zero stress in powder samples used for calibration purposes could be experimentally verified via repeatability and homogeneity measurements. With the exception of Cu, the analyzed powder samples provided high-quality residual stress data on the basis of LaB_6 allowing for metrological traceability to the SI. In this way, depending on the device configuration, the validity of the results was ensured either by direct calibration with LaB_6 or by indirect calibration with another powder having proven to be stress-free within the accepted limits. Regarding the reliability of reported residual stress measurement results, SI-traceability significantly contributes to an increase of credibility. Furthermore, the concept of using LaB_6 in connection with Fe Kα radiation facilitates the quality assessment of powder materials with respect to stress freeness and uniformity.

As an application example of Fe Kα radiation for multi-phase materials, the simultaneous phase content and residual stress analysis of duplex-type powder samples was considered based on a single standard residual stress measurement of each specimen in modified χ-mode. It could be shown that at least γ-Fe contents of 8% and higher can be determined correctly by means of a two-peak method after averaging the integrated intensities over the data volume. This approach gives the same accuracy as the standard four-peak method using Cr Kα radiation as per ASTM E975. For the lowest γ-Fe content of about 2%, the signal-to-background ratio of the γ-Fe 222 peak is not sufficient for strain and integrated intensity calculation. In the case of the α-Fe phase, precise residual stress values were obtained for the complete sample set, whereas for γ-Fe phase contents below 30%, the reduced signal height leads to a significantly degraded accuracy.

References

[1] Hauk, V., *Structural and Residual Stress Analysis by Nondestructive Methods*, Elsevier (Amsterdam, 1997). doi:10.1016/B978-0-444-82476-9.X5000-2

[2] Noyan, I.C., Cohen, J.B., *Residual Stress: Measurement by Diffraction and Interpretation*, Springer (New York, 1987). doi:10.1007/978-1-4613-9570-6

[3] EN 15305:2008, "Non-destructive testing – Test method for residual stress analysis by X-ray diffraction" (2008). doi:10.31030/1425472

[4] ASTM E2860:2020, "Standard Test Method for Residual Stress Measurement by X-Ray Diffraction for Bearing Steels" (2020). doi:10.1520/E2860-20

[5] EN 13925-3:2005, "Non-destructive testing – X-ray diffraction from polycrystalline and amorphous materials – Part 3: Instruments" (2005). doi:10.31030/9602423

[6] ASTM E915:2021, "Standard Practice for Verifying the Alignment of X-Ray Diffraction Instruments for Residual Stress Measurement" (2021). doi:10.1520/E0915-21

[7] ISO/IEC 17025:2017 & DIN EN ISO/IEC 17025:2018, "General requirements for the competence of testing and calibration laboratories" (2017/2018). doi:10.31030/2731745

[8] Black, D.R., *et al.*, "Certification of Standard Reference Material 660c for powder diffraction", *Powder Diffraction*, Vol. 35, No. 1 (2020), p 17-22. doi:10.1017/s0885715620000068

[9] National Institute of Standards & Technology, Gaithersburg, MD, USA, Certificate of Standard Reference Material 660c (29 October 2015), https://tsapps.nist.gov/srmext/certificates/660c.pdf, accessed June 2025.

[10] Send, S., *et al.*, "Non-destructive analysis of the tungsten carbide grain size by means of two-dimensional X-ray diffraction", *International Journal of Refractory Metals and Hard Materials*, Vol. 120 (2024), 106596. doi:10.1016/j.ijrmhm.2024.106596

[11] Send, S., *et al.*, "Fast residual stress determination by means of two-dimensional X-ray diffraction using the $\sin^2\psi$-method", *Proceedings of ICRS 11 – The 11th International Conference on Residual Stresses*, Nancy, France, March 2022, hal-04015041, https://hal.science/hal-04015041v1, accessed June 2025.

[12] Ivashchenko, V.I., *et al.*, "Electronic, thermodynamics and mechanical properties of LaB_6 from first-principles", *Physica B: Condensed Matter*, Vol. 531 (2018), p 216-222. doi:10.1016/j.physb.2017.12.044

[13] Eshelby, J.D., "The determination of the elastic field of an ellipsoidal inclusion, and related problems", *Proceedings of the Royal Society A*, Vol. 241 (1957), p 376-396. doi:10.1098/rspa.1957.0133

[14] Kröner, E., "Berechnung der elastischen Konstanten des Vielkristalls aus den Konstanten des Einkristalls", *Zeitschrift für Physik*, Vol. 151 (1958), p 504-518. doi:10.1007/BF01337948

[15] Eigenmann, B., Macherauch, E., "Röntgenographische Untersuchung von Spannungszuständen in Werkstoffen. Teil III", *Materialwissenschaft und Werkstofftechnik*, Vol. 27, No. 9 (1996), p 426-437. doi:10.1002/mawe.19960270907

[16] Joint Committee for Guides in Metrology, JCGM 100:2008, "Evaluation of measurement data – Guide to the expression of uncertainty in measurement", first edition (2008). doi:10.59161/JCGM100-2008E

[17] ASTM E975:2022, "Standard Test Method for X-Ray Determination of Retained Austenite in Steel with Near Random Crystallographic Orientation" (2022). doi:10.1520/E0975-22

ICRS 12: Proceedings from the 12th International Conference on Residual Stresses, 2025
October 20–23, 2025; Detroit, Michigan, USA
https://doi.org/10.31399/asm.cp.icrs2025p0032

Application of Contour Method for Residual Stress Analysis of Large and Complex Cross-Sections in Civil Engineering

Dénes Kollár, András Horváth, Balázs Kövesdi, Attila László Joó and István Völgyi
Department of Structural Engineering, Budapest University of Technology and Economics, Budapest, Hungary

Abstract

The study employs the contour method to determine two-dimensional longitudinal residual stress distributions in large and complex welded cross-sections widely used in the civil engineering industry. Measurements include T-joints, orthotropic steel decks with U-shaped stiffeners, and box sections with plate thicknesses varying between 4 mm and 16 mm, while maximum characteristic cross-sectional dimension is 600 mm. Results indicate stress peaks near welds exceeding the yield strength of the base material, while the effect of additional processes such as hot rolling, thermal cutting, and cold-forming are also evaluated. Statistical analysis confirms the reliability of contour method results, with standard deviations comparable to X-ray diffraction and hole-drilling techniques. Firstly, measurement results are applied for validating finite element models (FEM) developed for manufacturing simulation. In addition, novel residual stress models, satisfying the requirement for equilibrium of internal forces and bending moments, are developed for welded T-joints, OSDs, and box sections, incorporating manufacturing parameters such as heat input, filler material strength, and steel grade. These contour method-based residual stress models provide economic prospects for improving load-bearing capacity predictions of welded structures in advanced FEM-based design.

Introduction

Problem Statement

Welding is an essential fabrication process in civil engineering structures, yet it inevitably introduces initial imperfections, namely residual stresses and distortions. These imperfections arise from highly nonuniform thermal cycles caused by localized heating and rapid cooling, which induce microstructural changes, plastic strains, and transient thermal stresses [1]. As a result, welded joints often exhibit high longitudinal tensile residual stresses, frequently approaching or exceeding the yield strength of the base material [2]. Residual stresses and geometric imperfections significantly affect structural performance, including load-bearing capacity and stability resistance, while also increasing susceptibility to fatigue, brittle fracture, stress corrosion, and hydrogen-induced cracking [3,4]. Although their influence on structural performance is significant, current design standards and numerical models often depend on simplified assumptions and idealized residual stress distributions, which fail to account for the variability introduced by manufacturing processes. Consequently, the accurate characterization and modeling of these imperfections remain a major challenge in structural engineering. Addressing this issue requires the integration of advanced experimental techniques and numerical simulations to enhance the reliability of structural design and ensure safety. In this context, residual stress measurement methods, such as the contour method, play a crucial role in capturing the complexities of residual stress development arising from manufacturing processes. However, the large and often complex cross-sectional geometries commonly used in civil engineering structures present significant challenges for the practical application of such methods, introducing limitations in terms of accessibility, accuracy, and feasibility.

Theoretical Background

Residual stress measurement techniques reported in the international literature are generally classified into three categories: destructive, semi-destructive, and non-destructive. Mechanical methods, including both destructive and semi-destructive approaches, assess residual stresses by measuring deformations that occur upon material removal. In contrast, non-destructive techniques evaluate stress-related parameters without altering or damaging the specimen. Commonly used mechanical methods include sectioning, the contour method, and hole-drilling methods, while non-destructive techniques typically involve X-ray diffraction, neutron diffraction, and ultrasonic methods. This paper provides a concise overview of the theoretical background of the contour method applied in this study.

The destructive contour method, introduced by Prime [5] as a variation of Bueckner's superposition principle [6], provides two-dimensional maps of residual stress fields. Initially, the specimen contains equilibrium residual stresses from manufacturing processes (Step A, Fig. 1a). It is then cut along a predefined plane, commonly by wire electrical discharge machining (WEDM) to minimize additional stresses, releasing the residual stresses and causing elastic deformation (Step B, Fig. 1b). The resulting surface deformations are measured, often by coordinate measuring machines (CMM) [7] or alternative scanning methods [8]. Assuming linear elastic relaxation, the deformed surface is virtually restored to its original shape to obtain the residual stress field (Step C, Fig. 1c), typically limited in practice to the normal stress component σ_z on the cut plane.

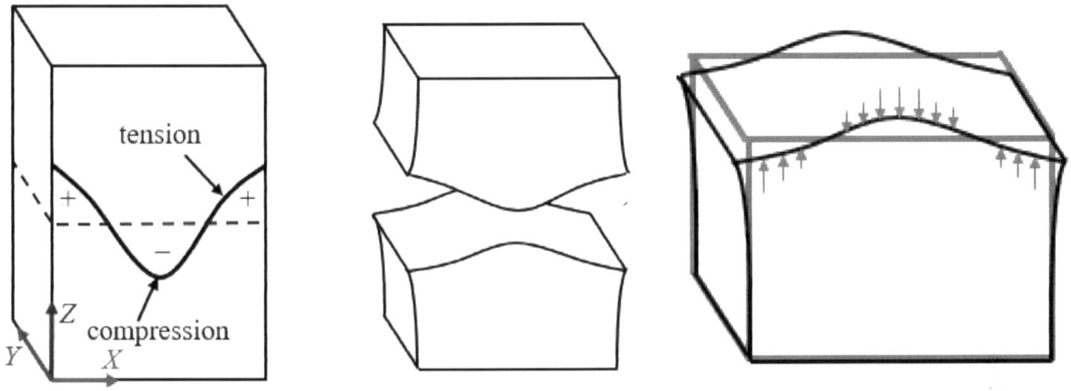

a) $\sigma^A(x, y, z) = \sigma^B(x, y, z) + \sigma^C(x, y, z)$ b) $\sigma^B(x, y, z)$ c) $\sigma^C(x, y, z)$

Figure 1: Background of contour method: a) initial residual stress state, b) deformed shape after cutting (elastic spring-back) and c) virtually forcing deformed surface back to its original shape.

Literature Review

The literature review highlights experimental findings related to residual stress measurements obtained through the contour method. Most previous studies focused on thick butt-welded plates or compact cross-sections with relatively small overall dimensions (typically less than ~500 mm).

For instance, Balakrishnan et al. [9] carried out residual stress measurements by both contour method and neutron diffraction technique in arc, laser and electron beam welded 30 mm thick SA508 steel specimens. The comparison of measured longitudinal residual stresses revealed a good agreement between the two approaches. On the other hand, DeWald et al. [10] applied contour method in the dissimilar metal weld region of axisymmetric pressurizer relief nozzles. Consistency of measured axial and hoop residual stresses confirmed the reliability of the method. In addition, Achouri et al. [11] introduced an incremental contour method to enhance residual stress measurements by progressively relieving stresses through multiple cuts, thereby reducing the likelihood of plastic redistribution. In the NeT TG6 benchmark specimen, this approach reduced normalized root mean square errors from 12.2% to 4.2% (relative to the total stress range) with four sequential cuts, demonstrating an improvement over the conventional single-cut method. Chu et al. [12] analyzed a stainless steel T-pipe joint after a stress relieving with a 10 mm overlay weld. The contour method results demonstrated that overlay repair welding introduced beneficial compressive residual stresses into the original weld. Murugan and Narayanan [13] used contour method and welding simulation to evaluate and compare longitudinal residual stresses at the mid-section of the flange of a welded carbon steel T-joint. Gadallah et al. [14] examined compressive residual stresses at the weld root for two cutting techniques, using the contour method with longitudinal cuts. They concluded that the applied cutting techniques influenced the distribution of mapped residual stresses, while examined boundary conditions showed a small influence on residual stresses.

According to Olson et al. [15], uncertainties in residual stress measurements using the contour method can reach approximately 10 MPa in aluminum specimens. The study also accounted for errors due to normally distributed noise in the measured displacement field, estimating them through a Monte Carlo approach. In addition, repeatability of contour method was also presented by Hill and Olson [16], showing standard deviation of about 20 MPa near the edges in quenched 7050 aluminum alloy bars. On the other hand, standard deviation of residual stresses was also approximated by using hole-drilling technique in [16] for aluminum friction stir weld (about 40 MPa) and shot peened spring steel (several hundred MPa in near surface measurements).

Research Aims

The current paper aims to demonstrate the applicability of contour method in the development of residual stress models for structures, with complex shapes and relatively large dimensions, used in the civil engineering practice. The study includes the analysis of T-joints, orthotropic steel deck parts with U-shaped stiffeners, and box sections, particularly due to the limitations of previous research. Earlier studies mainly relied on pointwise measurement techniques, lacking full-field two-dimensional stress distributions in critical cross-sections. To address this gap, the contour method is employed to obtain detailed residual stress maps after fabrication. A key contribution is the development of simplified residual stress models for T-joints and orthotropic decks that incorporate the effects of all relevant manufacturing processes, while also satisfying equilibrium conditions. Additionally, the novelty of the research lies in providing the first residual stress model for hybrid (combination of normal and high strength steel grades) welded box sections based on comprehensive residual stress measurement data. These research results could enhance fatigue design and evaluation of load-carrying capacity of some typical structural configurations.

Experimental Methods

Experimental program

A series of residual stress measurements [17-19] were conducted by the Department of Structural Engineering at Budapest University of Technology and Economics (BME) on representative structural elements manufactured under industrial conditions. On the one hand, the paper summarizes the methodology and main results of the previous contour method-based measurements for welded T-joints (lower flange of crossbeams) and orthotropic steel decks (OSDs) with U-shaped stiffeners, as important structural details of the Monostori Bridge (Fig. 2) on River Danube between Hungary and Slovakia. On the other hand, welded box sections, widely used in the civil engineering practice, composed of normal strength, high strength, and hybrid steel sections were analyzed. A total of three identical T-joints, four identical orthotropic steel deck segments and nine different box sections are investigated. Cross-sectional dimensions of the investigated specimens are summarized in Fig. 3. These specimens were subjected to a multi-step fabrication process including e.g., thermal cutting, cold-forming, tack welding, prestressing, welding, etc. all contributing to the final residual stress state. See [17–19] for further details on the T-joints, OSDs, and box sections.

Figure 2: Monostori bridge between Hungary and Slovakia during proof load test [17].

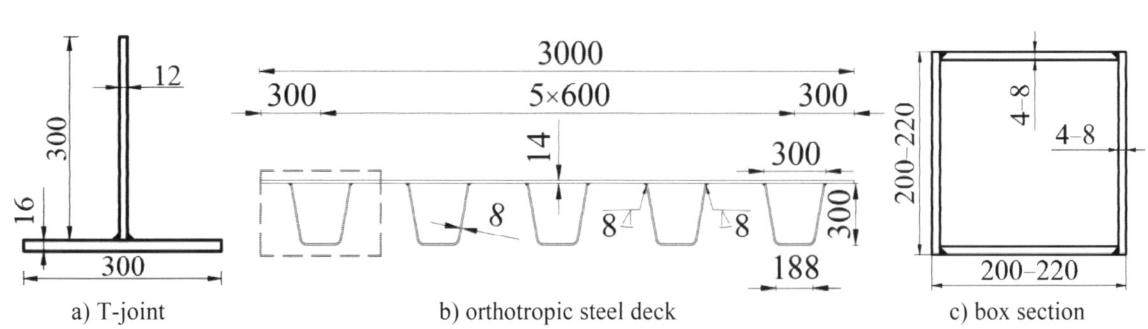

Figure 3: Representative cross-sectional dimensions of the specimens analyzed using the contour method.

All constituent plates and profiles used in the T-joints and orthotropic steel deck are made from structural steel grade S355J2+N, a standard material used in bridge engineering. For the box sections, the normal strength steel plates are fabricated from thermomechanically rolled fine-grained structural steel S355MC, characterized by a ferritic-pearlitic microstructure. In contrast, the high-strength steel plates utilize S700MC, a thermomechanically rolled and micro-alloyed steel grade with a ferritic-bainitic fine-grained microstructure in the as-received condition. The crossbeam web-to-flange joints (i.e., T-joints) are manually welded using single-pass, double-sided fillet welds with metal active gas (MAG) welding and solid wire (process 135). The effective heat input ranges from 1.40 to 1.62 kJ/mm, with a travel speed of 20–25 cm/min. Stiffener-to-deck plate joints are welded using automatic submerged arc welding (SAW, process 125) with tubular cored electrodes and a 45° single-bevel butt joint without root opening and a root face ≤2 mm. Each stiffener is welded simultaneously from both sides, with effective heat input between 2.68–3.06 kJ/mm and travel speed of 52–57 cm/min. For the box sections, the corner joints use a 45° single-bevel butt weld with a 1 mm root face, welded manually using MAG welding (process 135). One or two weld passes are applied depending on the plate thickness.

Details of Contour Method

Specimens are cut using brass wire with diameter of 150 μm and cutting speed of 0.08 mm/s. All the specimens are clamped during the operation. To ensure uninterrupted WEDM cutting following a wire breakage, the machine employs an automatic wire threading system. Upon detecting the break via dedicated sensors, the system re-threads the wire through the workpiece and resumes the cutting process. Three-dimensional distorted surfaces are measured by coordinate measuring machine (CMM), a moving bridge-type Zeiss UMC 850 is used (Fig. 4) with scanning probe head and a measurement range of 1200 mm × 850 mm × 600 mm. A ruby stylus with diameter of 6 mm is used. The manufacturer reported measurement accuracy of 3 μm + $L/200$ μm, resulting in approximately 6 μm over the L = ~600 mm measurement range (inserted in [mm] into the equation) with a resolution of 0.1 μm. Measurements are conducted in a temperature-controlled metrology laboratory with air temperature of 20-20.5°C; thus, minimizing temperature effects.

a) T-joint b) orthotropic steel deck c) box section

Figure 4: CMM measurement of representative welded structural elements with Zeiss UMC 850 [17-19].

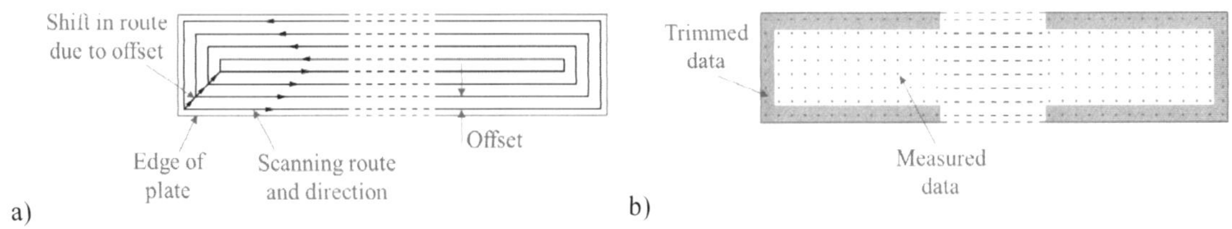

Figure 5: Scheme of a) CMM scanning route with offset, and b) data trimming at the edges of a plate [17].

The scanning route of the CMM is evaluated using the outline, i.e., the contour, of the cross-section with offsets of 0.5 mm (Fig. 5a). The measurement is conducted with different point densities for distinct regions. In the vicinity of the weld, average distance between adjacent points is 0.2 mm, while it is 0.5 mm in further regions. CMM measurements are carried out on both cutting surfaces (left and right side of WEDM cutting). Posterior evaluation of data is carried out in MATLAB [20]. Point clouds, as raw data after trimming at the edges (Fig. 5b), for two coherent cutting surfaces are shown in Fig. 6a and b, respectively with the indication of wire breakage at X = ~140 mm. Shear stress effects and anti-symmetric cutting errors are removed using averaging (Fig. 6c) of contour data on the two halves. Local cutting irregularities, such as wire breakage (potentially due to a non-conductive inclusion) and overburn, and measurement noise are removed by data smoothing after averaging. Thin-plate smoothing spline, a special case of a polyharmonic spline, is used in the software to interpolate surfaces over scattered data and smoothing (Fig. 6d). The name of the spline-based algorithm refers to the physical analogy of bending energy of a thin sheet on point constraints.

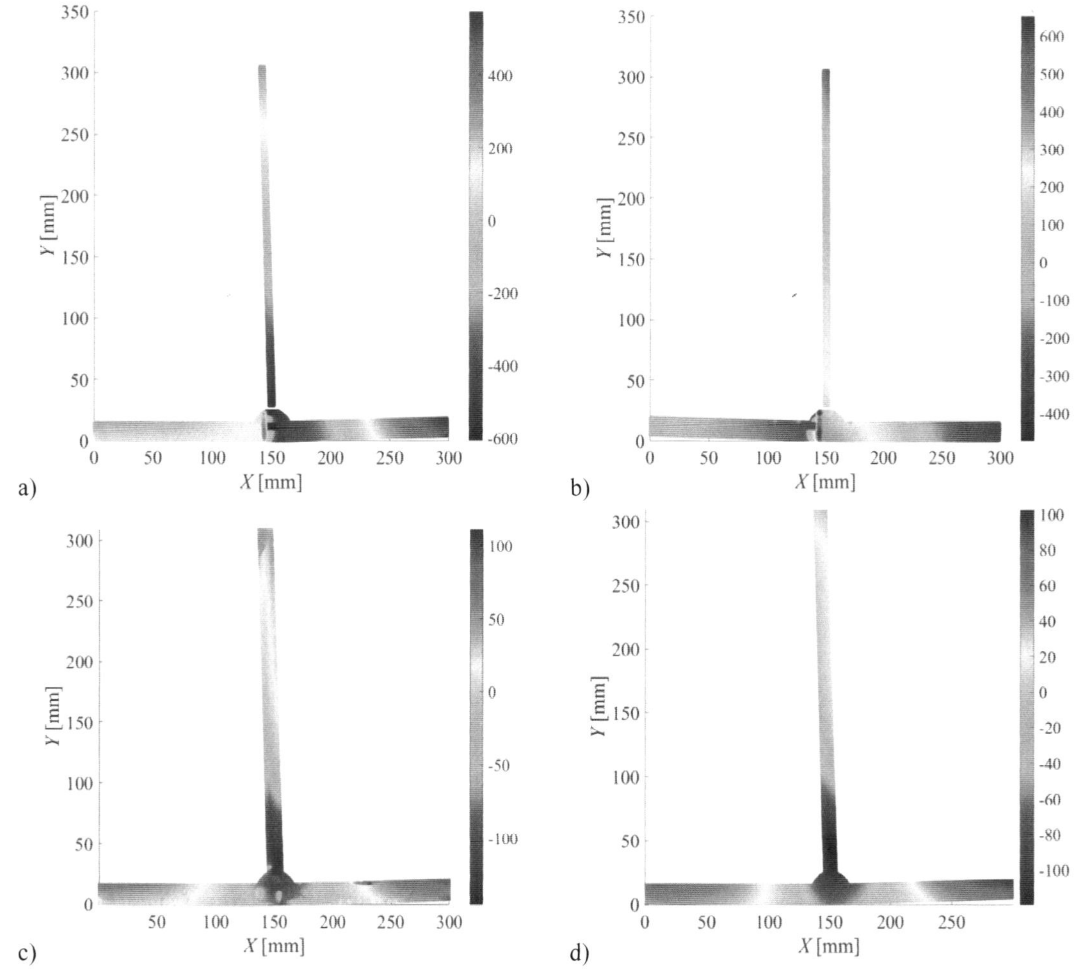

Figure 6: Point clouds a-b) as raw data for the coherent cutting surfaces (sides 1 and 2) of a welded T-joint after WEDM, c) after averaging, and d) after data smoothing. Z coordinates are in [µm] in accordance with [17].

Smoothed displacement data are mapped onto a three-dimensional finite element model as reversed displacements to compute the full two-dimensional residual stress distribution across the cutting surface. The model geometry is based on measured average thicknesses and includes initial welding-induced distortions (e.g., angular deformation) for accurate displacement mapping. Finite element simulations are performed in ANSYS [21] using 8-node solid elements (SOLID185) with enhanced strain technology, pure displacement element formulation and linear shape functions. A linear isotropic material model is applied with nominal elastic properties (Young's modulus E is 210000 MPa, Poisson's ratio ν is 0.3), assuming small deflection theory. Although slight variations in E may arise from local microstructural differences near welds, these effects are considered negligible on the macroscale and are therefore not taken into account in the inverse analysis of residual stresses. The finite element mesh of each structural configuration employs an average element size of 1–2 mm in the cross-section, with a nonuniform axial mesh consisting of 20 elements and a spacing ratio of 50, providing finer resolution near the WEDM cutting surface. The *MOPER,,,MAP command is used for mapping the measured displacements onto the mesh. Figure 7a shows the scaled deformed model of each configuration with boundary conditions applied to prevent rigid body motion. Figure 7b illustrates the distribution of longitudinal residual stresses, which decrease gradually with increasing distance from the cutting surface. A mesh sensitivity analysis was conducted to verify the chosen mesh scheme with respect to peak tensile and compressive residual stresses, as detailed in [17–19].

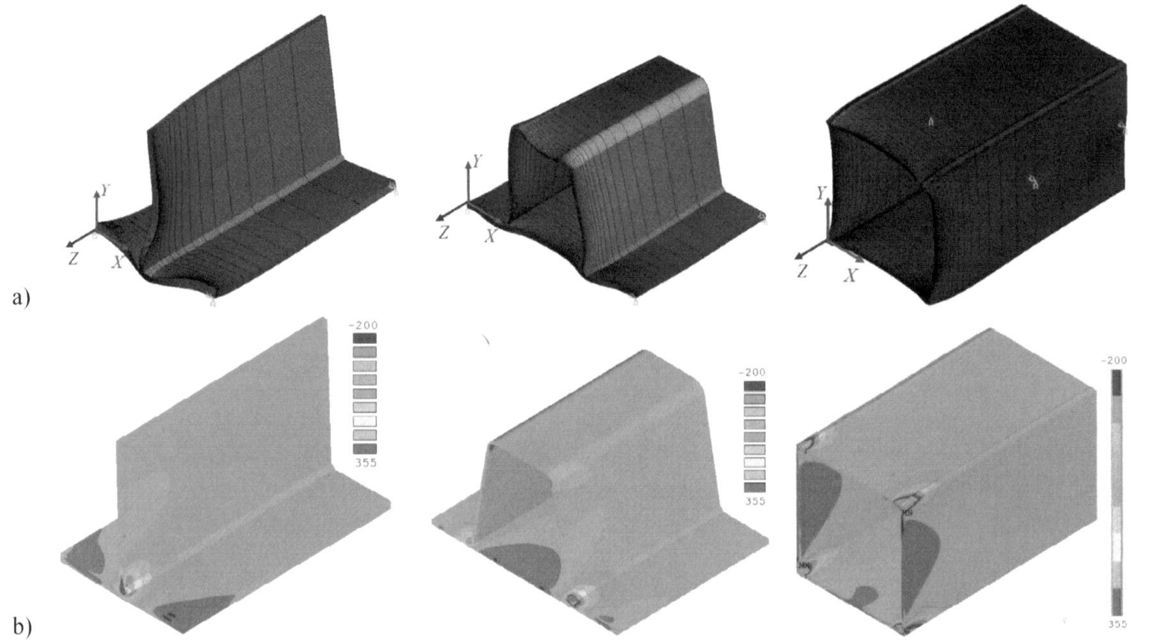

Figure 7: a) Finite element meshes with scaled displacements and additional boundary conditions to prevent rigid body motion, b) overview of longitudinal residual stresses σ_z [MPa] in the different models [17–19].

In line with the methodology introduced by Olson et al. [22], except for thin-plate smoothing spline instead of a bivariate surface fitting, the model error $U_{\text{model}}(x, y)$ coming from the smoothing model is assessed. Smoothing parameter p ($0 \leq p \leq 1$), basically provided by the software approximately based on the average spacing of the data sites, can be adjusted. The least-squares approximation by a linear polynomial is denoted by $p = 0$, while the thin-plate spline interpolant is characterized by $p = 1$. In order to determine the model error, residual stresses described by five different smoothing parameters are assessed: namely $p/3$, $p/2$, p, $2p$ and $3p$. The standard deviation of the estimated residual stresses is determined under the assumption of a Gaussian distribution for the model error. Figure 8a presents the spatial distribution of the model error for the T-joint depicted in Figure 6 (with wire breakage at $X = \sim 140$ mm). The mean model error across the analyzed cross-section is 7.8 MPa (corresponding to confidence interval of 68%). Figure 8b highlights the regions where the mean model error overlaps (blue) or does not overlap (red) the pointwise model error $U_{\text{model}}(x, y)$. Approximately 65.4% of points, uniformly distributed across the cross-section, satisfy this criterion, which is close to the expected 68% coverage, indicating that the smoothing model with parameter p provides reliable performance.

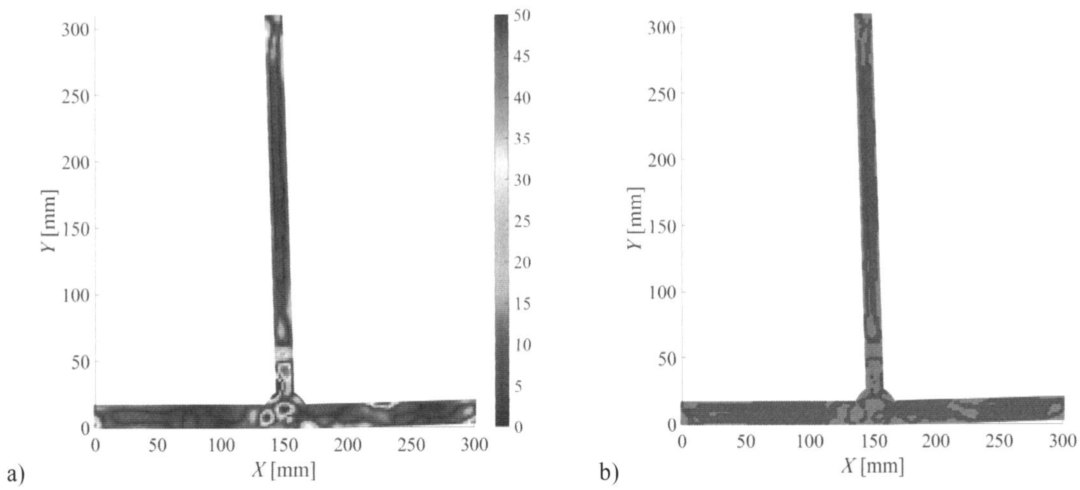

Figure 8: a) Pointwise model error [MPa] for a welded T-joint (corresponding to Figure 6), and b) points falling (blue) within the bound of the mean model error (confidence interval of 68%).

Results and Discussion

Longitudinal Membrane Residual Stresses

Longitudinal residual stresses are exported from ANSYS and analyzed in MATLAB for the assessment of membrane residual stresses ($\sigma_{rs,m,z}$). These could be easily implemented as a residual stress pattern for shell element models in the design. Membrane residual stresses, i.e., average residual stresses through the thickness t, were determined using Eq. 1. Nevertheless, internal longitudinal forces were summed ($\sum F_{i,z}$) in the vicinity of the welds, and average residual stresses are calculated by dividing it with the corresponding junction area (A_{joint}). This is a necessary technique when calculating membrane stresses because averaging through thickness is not straightforward in this area due to changes in geometry (Fig. 9). Numerical calculations are carried out in MATLAB; local coordinates x and y are applied in the constituent plates in order to determine membrane residual stresses in each cross-section.

$$\sigma_{rs,m,z}(x) = \frac{1}{t}\int_{-\frac{t}{2}}^{\frac{t}{2}} \sigma_{rs,z}(x,y)\, dy \tag{1}$$

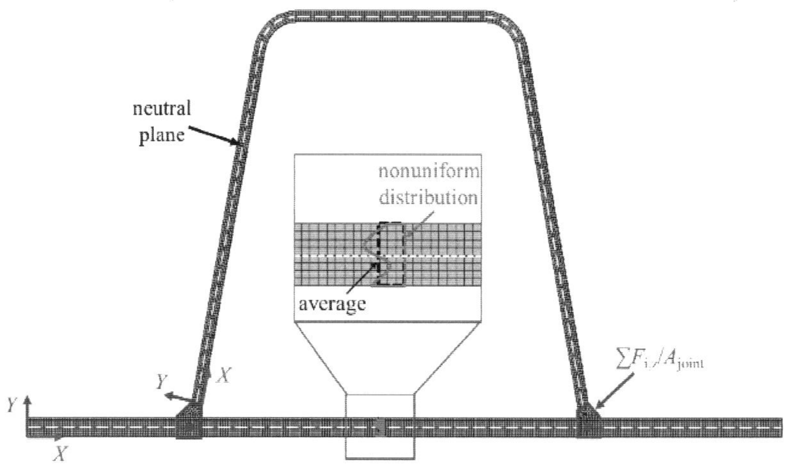

Figure 9: Demonstration of membrane residual stress calculation based on finite element model using solid elements for OSDs [18].

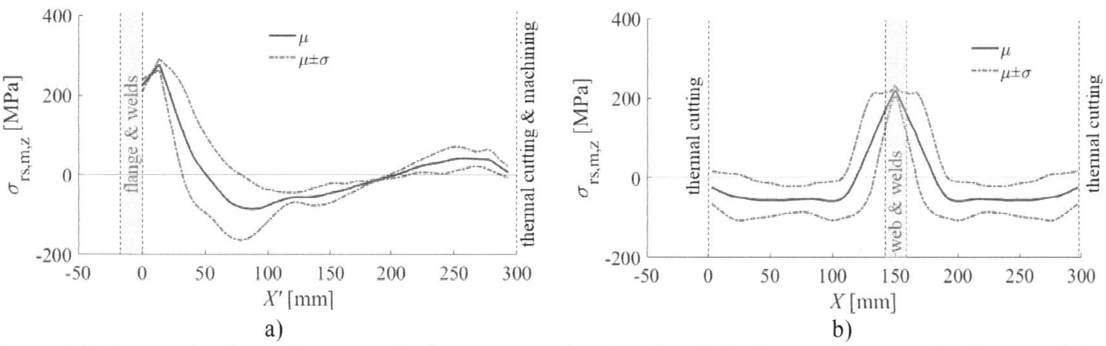

Figure 10: Longitudinal membrane residual stresses in the a) web and b) flange of T-joints [17]. Mean (μ) and standard deviation (σ) of residual stresses based on three specimens are plotted.

Longitudinal membrane residual stresses are evaluated for both the web and flange of welded T-joints (Fig. 10) using local coordinate systems (X and X'). Stress distributions revealed high tensile peaks near the welds with sharp gradients, while compressive stresses dominated regions further from the welds (except at the top of the web due to equilibrium of internal moments). To quantify variability, the mean (μ) and standard deviation (σ) of residual stresses are calculated at each location along the constituent plates. These values are visualized to assess measurement consistency and spatial variability. In the vicinity of the weld, defined by a polygonal region around the welds, lower scatter is observed due to averaging over a larger domain. A 2σ bandwidth ($\mu \pm \sigma$) was used to represent the spread of data. Results in the flange were mirrored before residual stress model development, ensuring a balanced representation of internal stresses.

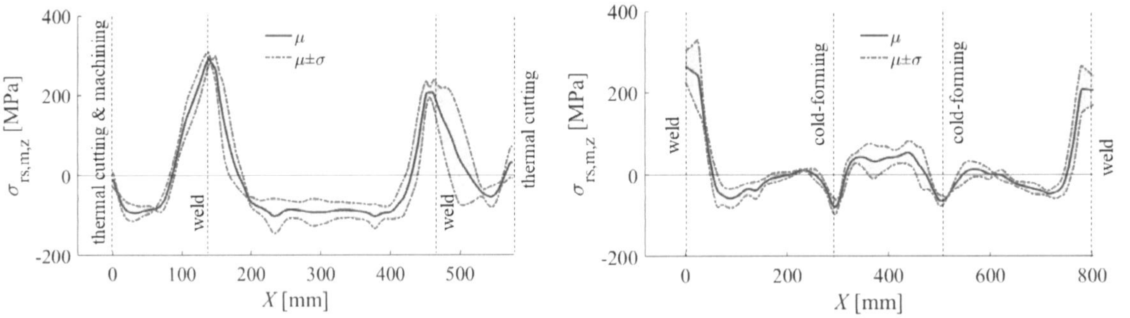

a) b)

Figure 11: Longitudinal membrane residual stresses in the a) deck plate and b) U-shaped stiffener of OSDs [18]. Mean (μ) and standard deviation (σ) of residual stresses based on four specimens are plotted.

The same methodology is used for the OSDs (i.e., in the deck plate and U-shaped stiffeners in Fig. 11). The 2σ bandwidth is applied to quantify uncertainty and measurement repeatability. The mean standard deviation is found to be 22 MPa in the stiffener and 34 MPa in the deck plate, which is consistent with the accuracy ranges reported [16,23,24] for other techniques such as X-ray diffraction (±30–37 MPa), neutron diffraction (±20–40 MPa) or hole-drilling technique (±12–40 MPa). These findings affirmed that the contour method used in this study offers a comparable or superior level of precision in repeatability and reliability. Considering its ability to resolve full-field residual stress fields is an advantage over most pointwise techniques.

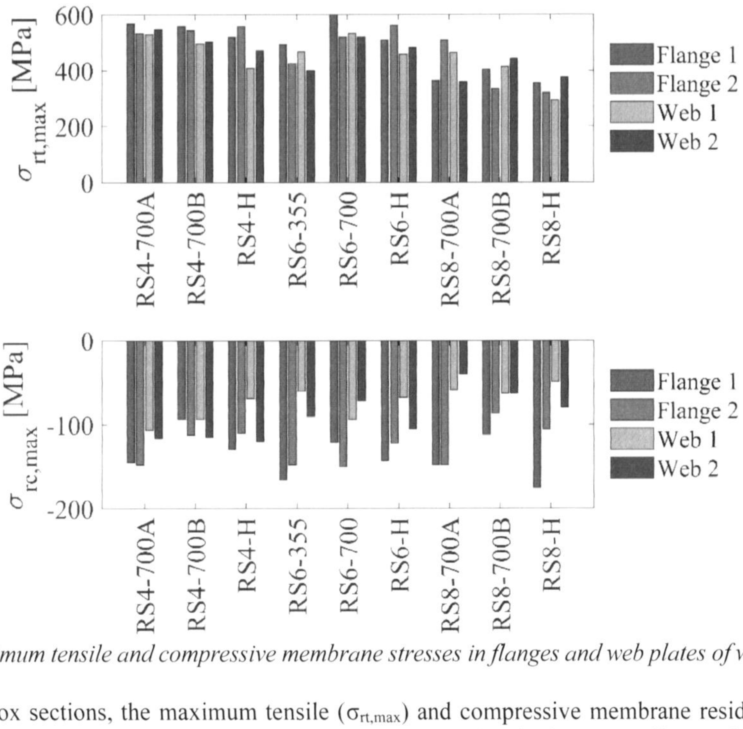

Figure 12: Maximum tensile and compressive membrane stresses in flanges and web plates of welded box sections [19].

In the welded box sections, the maximum tensile ($\sigma_{rt,max}$) and compressive membrane residual stresses ($\sigma_{rc,max}$) are determined (Fig. 12) for all nine specimens in both the flanges and web plates. Tensile membrane stress peaks ranged from 292 MPa (RS8-H with t = 8 mm and a hybrid cross-section) to 600 MPa (RS6-700 with t = 6 mm and high strength steel plates). While membrane residual stresses generally do not exceed the yield strength of the base material, localized regions do surpass the nominal yield strength (not shown in the paper; see [19] for details). Data scatter could be characterized by the range of maximum tensile stresses at the four corners. For instance, the range is ±19 MPa for RS4-700A with high strength steel plates, while it reached ±75 MPa for the hybrid cross-section RS4-H. The scatter observed in the figure reflects variations in the underlying sources of residual stresses. These include the pronounced effects of combining different steel grades within a single cross-section, as well as non-simultaneous welding of the four corners, where variations in heat input occur between weld passes.

Development of Residual Stress Models

Residual stress models are developed solely based on the experimental measurement results. For welded T-joints, the proposed model (Fig. 13a) follows a similar approach to the residual stress model for welded I-sections introduced by Schaper et al. [25], incorporating the steel grade through a modified yield strength $f_y \cdot \varepsilon$, where the material coefficient is defined as $\varepsilon = (235/f_y)^{1/2}$. This formulation accounts for reduced tensile membrane stresses near the welds, attributed to high stress gradients and decreased through-thickness stresses. The model also incorporates the plate thicknesses of the T-joint (t_{fl} and t_w). Expressions for the compressive and tensile residual stress magnitudes ($\sigma_{c,fl}$, $\sigma_{c,w}$, $\sigma_{t,w}$) and related geometrical parameters (c_{w1}, c_{w2}, c_{w3}) are provided in [17].

In addition, a simplified residual stress model is developed for an internal part of the orthotropic steel deck (Fig. 13b). The model takes thicknesses (t_{pl} and t_{st}), spacing between stiffeners (s_{st}), and yield strength (f_y) of the base material into account. Compressive residual stresses in the stiffener are $\sigma_{c,st1} = 0.15 f_y$, and $\sigma_{c,st2} = 0.21 f_y$, resulting in linear transition between tensile and compressive stresses in the vicinity of the weld. Tensile residual stress in the stiffener is calculated using $\sigma_{t,st} = 0.14 f_y$ (i.e., $2/3\sigma_{c,st2}$). The magnitude of compressive residual stresses in the deck plate is calculated using Eq. 2 based on equilibrium. More details can be found [18].

$$\sigma_{c,pl} = f_y \left(\frac{s_{st} - \sqrt{s_{st}^2 - 28 s_{st} t_{pl}}}{14 t_{pl}} - 1 \right) \quad (2)$$

The widths of the tensile stress blocks ($\alpha_1 t$ and $\alpha_2 t$) and the residual stress magnitudes (σ_{rt} and σ_{rc}) for box sections (Fig. 13c) are defined as functions of key welding parameters, including heat input, number of weld passes, and weld size. The model also incorporates the steel grade of both the filler metal and the base materials of the flange and web, enabling its application to hybrid box sections composed of a combination of normal and high strength steels. Further details are provided in [19].

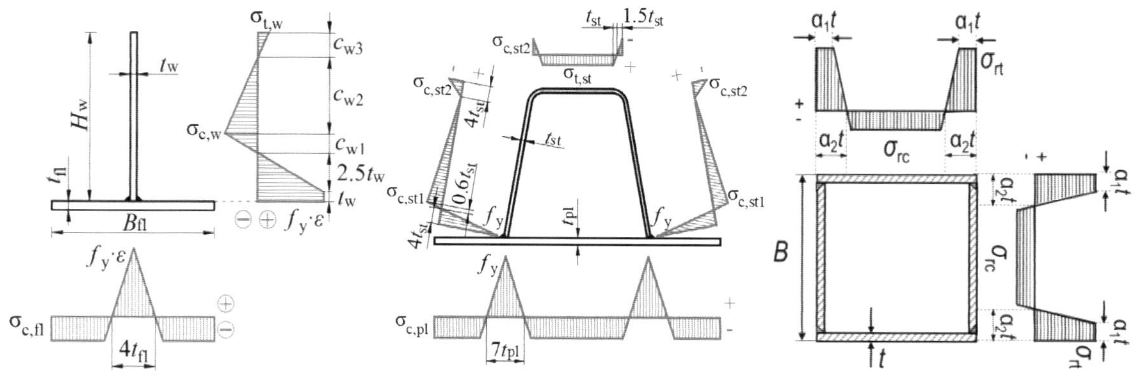

Figure 13: Proposed residual stress models of welded structural steel a) T-joints, b) orthotropic steel decks and c) box sections [17–19].

Conclusions

This paper summarizes the methodology and results of a comprehensive experimental investigation on longitudinal residual stresses in welded structural elements commonly used in civil engineering applications. Manufacturing-induced residual stresses are determined using the contour method with thin-plate smoothing spline and mapped via inverse linear elastic finite element analysis. The following key findings are revealed:

- The full-field residual stress measurements in this study demonstrated repeatability with a mean standard deviation of 22–34 MPa, which is comparable to the accuracy of previously reported diffraction techniques (20–40 MPa) and the hole-drilling method (12–40 MPa).
- The novel application of contour method for large and complex cross-sections enabled the development of residual stress models satisfying force and moment equilibrium, while also considering manufacturing specialties.
- In the case of normal strength steel, high strength steel, and hybrid box sections, a novel residual stress model was proposed incorporating welding technology parameters and material properties.

Future work will focus on validating and extending the proposed residual stress models through probabilistic manufacturing simulations and parametric studies, particularly in welded bridge components.

Acknowledgments

The presented paper has been supported by the Grant MTA-BME Lendület LP2021-06/2021 "Theory of new generation steel bridges" program of the Hungarian Academy of Sciences. The project supported by the Doctoral Excellence Fellowship Programme (DCEP) is funded by the National Research Development and Innovation Fund of the Ministry of Culture and Innovation and the Budapest University of Technology and Economics, under a grant agreement with the National Research, Development and Innovation Office.

References

[1] Radaj, D., *Heat Effects of Welding: Temperature Field, Residual Stress, Distortion*, Springer (Berlin, 1992)
[2] Gilles, P., *et al.*, "Robustness analyses of numerical simulation of fusion welding NeT-TG1 application: Single weld-bead-on-plate", *Int J Press Vessel Pip*, Vol. 8 (2009), p 3–12, doi:10.1016/j.ijpvp.2008.11.012
[3] European Convention for Constructional Steelworks, *Manual on Stability of Steel Structures: Part 2.2. Mechanical properties and residual stresses,* ECCS (Brussels, 1976)
[4] Masubuchi, K., *Analysis of Welded Structures*, Pergamon Press (Oxford, 1980)
[5] Prime, M.B., "Cross-sectional mapping of residual stresses by measuring the surface contour after a cut", *J Eng Mater Technol*, Vol. 123 (2001), p 162–168, doi:10.1115/1.1345526
[6] Bueckner, H., "The Propagation of Cracks and the Energy of Elastic Deformation", *Trans Am Soc Mech Eng*, Vol. 80 (1958), p 1225–1230
[7] Prime, M.B., DeWald, A.T., "The Contour Method". In: Schajer, G.S., editor, *Practical Residual Stress Measurement Methods*, Wiley (Chichester, 2013), p. 109–138. doi:10.1002/9781118402832.ch5
[8] Prime, M.B., *et al.*, "Laser surface-contouring and spline data-smoothing for residual stress measurement", *Exp Mech*, Vol. 44, No. 2 (2004), p 176–184, doi:10.1177/00144851039762
[9] Balakrishnan, J., *et al.*, "Residual stress distributions in arc, laser and electron-beam welds in 30 mm thick SA508 steel: A cross-process comparison", *Int J Press Vessel Pip*, Vol. 162 (2018), p 59–70, doi: 10.1016/j.ijpvp.2018.03.004
[10] DeWald, A.T., *et al.*, "Measurement of Welding Residual Stress in Dissimilar Metal Welds Using the Contour Method", *ASME 2011 Pressure Vessels and Piping Conference*, July 2011, Baltimore, MD, p 1599–1605, doi: 10.1115/PVP2011-57720
[11] Achouri, A., *et al.*, "The incremental contour method using asymmetric stiffness cuts", *Materials and Design*, Vol. 197 (2021), Paper 109268, doi:10.1016/j.matdes.2020.109268
[12] Chu, Q., *et al.*, "Introducing compressive residual stresses into a stainless-steel t-pipe joint by an overlay weld", *Metals*, Vol. 11, No. 7 (2021), p 1-11, doi:10.3390/met11071109
[13] Murugan, N., Narayanan, R., "Finite element simulation of residual stresses and their measurement by contour method", *Mater Des*, Vol. 30 (2009), p 2067–2071, doi:10.1016/j.matdes.2008.08.041
[14] Gadallah, R., *et al.*, "Residual stress measurement at the weld root of rib-to-deck welded joints in orthotropic steel bridge decks using the contour method", *Eng Struct*, Vol. 219 (2020), Paper 110946, doi:10.1016/j.engstruct.2020.110946
[15] Olson, M.D., *et al.*, "Assessment of weld residual stress measurement precision: Mock-up design and results for the contour method", *J Nucl Eng Radiat Sci*, Vol. 1, No. 3 (2015), doi:10.1115/1.4029413
[16] Hill, M.R., Olson, M.D., "Repeatability of the Contour Method for Residual Stress Measurement", *Exp Mech*, Vol. 54 (2014), p 1269–1277, doi:10.1007/s11340-014-9867-1
[17] Kollár, D., *et al.*, "Assessment of residual stresses in welded T-joints using contour method", *Thin-Walled Struct*, Vol. 190 (2023), Paper 110966, doi:10.1016/j.tws.2023.110966
[18] Kollár, D., *et al.*, "Development of residual stress model of orthotropic steel decks using measurements", *Structures*, Vol. 58 (2023), Paper 105601, doi:10.1016/j.istruc.2023.105601
[19] Horváth, A., *et al.*, "Distortions and residual stresses of NSS, HSS and hybrid box sections with single-bevel butt welds: An experimental study", *Thin-Walled Struct*, Vol. 208 (2025), Paper 112834, doi:10.1016/j.tws.2024.112834
[20] MATLAB R2020b. version 9.9.0.1453 (2020). Natick, Massachusetts, USA: The Mathworks Inc.
[21] ANSYS. version 19.5 (2020). Canonsburg, Pennsylvania, USA: ANSYS Inc.
[22] Olson, M.D., *et al.*, "Estimation of Uncertainty for Contour Method Residual Stress Measurements", *Exp. Mech.*, Vol. 55, No. 3 (2015), p 577-585, doi:10.1007/s11340-014-9971-2
[23] Lord. J.D., *et al.*, *A UK Residual Stress Intercomparison Exercise - An Examination of the XRD and Hole Drilling Techniques*, NPL Rep MATC(A)98, National Physical Laboratory (Teddington, 2002)
[24] Schajer, G.S., Whitehead P.S., "Hole Drilling and Ring Coring ". In: Schajer, G.S., editor, *Practical Residual Stress Measurement Methods*, Wiley (Chichester, 2013), p. 29–64. doi:10.1002/9781118402832.ch2
[25] Schaper, L, Tankova, T, Simões da Silva, L, Knobloch, M., "A novel residual stress model for welded I-sections", J Constr Steel Res, Vol. 188 (2022), Paper 107017, doi:10.1016/j.jcsr.2021.107

https://doi.org/10.31399/asm.cp.icrs2025p0042

The effect of low-strain pre-stretching at 77K on quenching residual stress and tensile properties of 6061 aluminum alloys

***Mingli Zhang*[a, b], *Kaixuan Gu*[a, b*], *Xiujie Zhao*[a, b], *Junjie Wang*[a, b]**

[a] *Key Laboratory of Cryogenic Science and Technology, Technical Institute of Physics and Chemistry, Chinese Academy of Sciences, Beijing 100190, China*

[b] *University of Chinese Academy of Sciences, Beijing 100049, China*

**Corresponding author. Tel./fax: +86 010 82543415.*

E-mail address: gukaixuan@mail.ipc.ac.cn (K. Gu)

ABSTRACT

The quenching residual stress in aluminum alloy thick sheets can be generated easily by the gradient temperature difference between the surface and interior caused by uneven cooling. It can cause distortion of components during the process of machining and service. Furthermore, residual stress concentration can reduce the mechanical properties, fatigue performance, corrosion resistance and so on. The pre-stretching is an effective method to release residual stress. The higher degree of plastic deformation during pre-stretching the greater the release of residual stress. However, great plastic deformation can be detrimental to the properties, due to the relative low plasticity of aluminum alloy. Considering the simultaneous increase in both strength and plasticity of aluminum alloy at cryogenic temperature, the effect of low-strain pre-stretching at 77K on quenching residual stress and tensile properties of 6061 aluminum alloy are investigated in the present. Residual stress of the sample after different treatments is tested by blind hole-drilling method. The effects between room temperature and cryogenic temperature pre-stretching is compared. It is revealed that low-strain pre-stretching at 77K can release the quenching residual stress more obvious, as well as improve the strength and plasticity of 6061 aluminum alloy.

Key words: Pre-stretching, Cryogenic temperature, Tensile property, Residual stress, 6061 aluminum alloys

Introduction

Quenching-induced residual stresses in aluminum alloy thick plate components represent a persistent and costly challenge in advanced manufacturing, particularly within the aerospace and transportation sectors [1-3]. These stresses originate from the severe thermal gradients established between the surface and core during rapid cooling, leading to non-uniform plastic deformation [4, 5]. The consequences are multifaceted and detrimental: components distort unpredictably during subsequent machining operations, compromising dimensional tolerances and increasing scrap rates, while in-service performance is undermined through accelerated fatigue crack initiation, reduced corrosion resistance, and premature failure under load [6-8].

Pre-stretching has long been established as a primary mechanical method for mitigating quenching residual stresses in aluminum components [9]. Its efficacy fundamentally relies on inducing controlled plastic deformation to redistribute and relax residual stress in the sample. In general, the greater the degree of plastic deformation, the better the residual stress release effect during pre-stretching at room temperature. However, the pursuit of significant stress relief through high-strain pre-stretching at RT inevitably damages the mechanical properties of aluminum alloy as well as results in surface roughening and microcrack initiation.

Recent insights into the cryogenic mechanical behavior of aluminum alloy offer a promising avenue to circumvent this limitation. It is well-documented that aluminum alloys exhibit a concurrent and often significant enhancement in both strength and ductility at cryogenic temperatures (e.g., 77K, liquid nitrogen temperature) compared to RT [10-12]. This unique cryogenic response, attributed to mechanisms such as suppressed dynamic recovery and altered dislocation mobility. It suggests that low strain pre-stretching at cryogenic temperatures may achieve excellent residual stress relief without the performance losses associated with high strain deformation at RT. However, the specific application of pre-stretching at cryogenic temperature as a method for releasing residual stress in aluminum alloys has largely remained unexplored. There is no clear answer regarding the quantitative comparison between cryogenic pre-stretching and traditional RT pre-stretching in terms of stress relief and their impact on the final tensile properties of alloys.

Therefore, we present a systematic investigation into the efficacy of low-strain pre-stretching at 77K for mitigating quenching residual stresses in 6061 aluminum alloy thick sheet, while concurrently evaluating its influence on the resulting tensile properties. The residual stress profiles within samples subjected to different post-quench treatments (untreated, pre-stretched at RT, pre-stretched at 77K) are tested using the standardized blind hole-drilling method.

Subsequently, the properties of samples pre-stretched at RT and cryogenic temperature (77 K) are investigated at equivalent low strain levels.

Materials and Methods

The rolled 6061 aluminum alloy plates with the thickness of 12mm were used in this work. Tensile samples were sectioned parallel to the rolling direction, with dimensions shown in Figure 1.The samples were first solution treated at 530°C for 2 h and then quenched into room-temperature water (S). Subsequently, pre-stretching was performed at room temperature (RT) and 77K (CT) using an electronic universal testing machine (UTM5305X) to strains of 1% and 2%. Finally, samples were subjected to artificial peak aging at 175°C for 9 hours (A). The tensile properties were evaluated for samples subjected to the following five distinct treatments: SA, SRT1%A, SCT1%A, SRT2%A, and SCT2%A. Tensile testing was conducted in accordance with GB/T 228.1-2010. Three samples were tested for each treatment condition, and the average value was reported as the final result. Fracture morphology of the sample was also detected by using Scanning Electron Microscope (SEM) (Hitachi SU1510).

Residual stress in the samples after different pre-stretching treatments was measured using the blind-hole drilling method according to ASTM E837-2013 standard. The measurement locations are shown in Fig. 2. A hole with a diameter of 1.803 mm was drilled using an RS-200 drilling device equipped with a high-speed (20,000−400,000 rpm) air turbine drill. The drilling process comprised 20 incremental steps, each with a depth of 0.05 mm. Strains after each drilling increment were recorded by the D4 Data Acquisition system. The stress distribution from the surface to a depth of 1 mm within the sample was obtained using the Hole-drilling Residual Stress Calculation Program (Version 3.30) software.

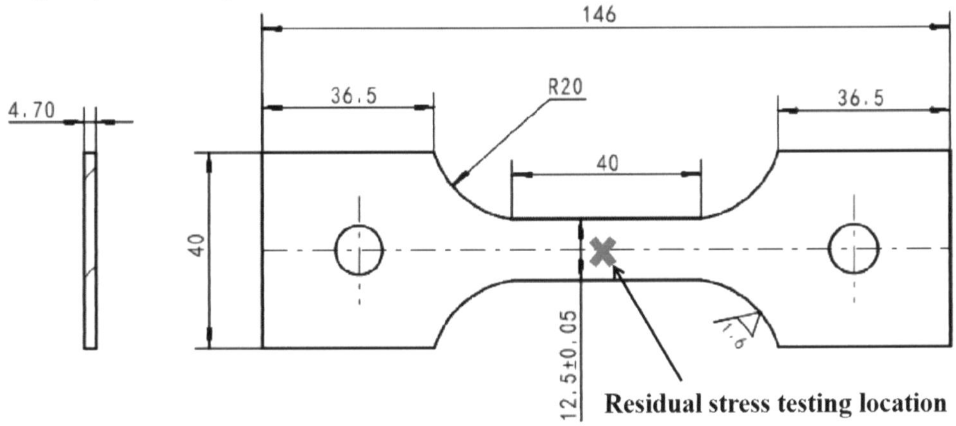

Figure 1: Dimensions of the uniaxial tensile-test sample (mm)

Table 1 Chemical compositions of 6061 aluminum alloy (wt. %).

Mg	Si	Cu	Cr	Zn	Mn	Al
1.1	0.5	0.3	0.23	0.15	0.03	Bal

Results and Discussion

Residual stress in samples subjected to different treatments was measured using the blind-hole drilling method, as shown in Fig. 2. The maximum surface stress values are listed in Table 1. It can be seen that the S sample exhibited the maximum residual stresses at the surface, with values of -49 MPa and -102 MPa in the X and Y directions, respectively. This gradient originates from inhomogeneous plastic deformation during the quenching process, which results in compressive thermal stresses exceeding the yield strength due to rapid surface cooling. The residual stress magnitude decreased gradually with depth into the sample before stabilizing, as shown in Fig. 2(a), consistent with the attenuated thermal gradient and reduced plastic constraint in the sample.

After 1% pre-stretching at RT (SRT1%), the surface residual stress significantly decreased , as shown in Fig. 2(b). The maximum stresses on the surface of the sample in the X and Y directions were 38 MPa and 13 MPa, respectively, which decreased by 22.4% and 87.3% compared to the S sample. This relaxation occurs because applied tensile strain promotes dislocation glide and rearrangement, partially annihilating residual stresses through plastic deformation. When the pre-stretching strain increased to 2% (SRT2%), the surface residual stress decreased further and exhibited a more uniform distribution compared to the sample with 1% strain , as shown in Fig. 2(c). The maximum stresses on the surface of the sample in the X and Y directions were 26 MPa and 21 MPa, respectively, which decreased by 22.4% and 87.3% compared to the S sample. Higher strain introduces greater plastic deformation, enabling more dislocation reorganization and stress homogenization via cross-slip and grain-boundary interactions.

Pre-stretching at 77K resulted in more significant residual stress reduction and a more uniform stress distribution compared to pre-stretching at RT at the same strain level, as shown in Fig. 2(d, e). After 1% pre-stretching at 77K (SCT1%), the maximum stresses on the surface of the sample in the X and Y directions were 23MPa and 14MPa, respectively, which decreased by 53.1% and 86.3% compared to the S sample. When the deformation increased to 2% (SCT2%), the maximum stresses in the X and Y directions on the surface of the SCT2% sample were 18 MPa and 12 MPa, respectively, which decreased by 63.3% and 88.2% relative to the S sample. Cryogenic conditions suppress dynamic recovery, increasing dislocation density and storage energy. This enhances stress relaxation through two mechanisms: (1) inhibited thermal activation prevents dislocation climb/annihilation, forcing more uniform strain distribution; (2) higher flow stress at 77K enables deeper penetration of plasticity into stress-concentrated zones.

In summary, pre-stretching at 77K was more effective than that at RT in reducing residual stresses and achieving a more uniform stress distribution. Furthermore, pre-stretching with a 2% strain can eliminate residual stress more completely compared to a 1% strain as it exceeds the critical threshold for pervasive plasticity, enabling near-complete reorganization of residual stress fields.

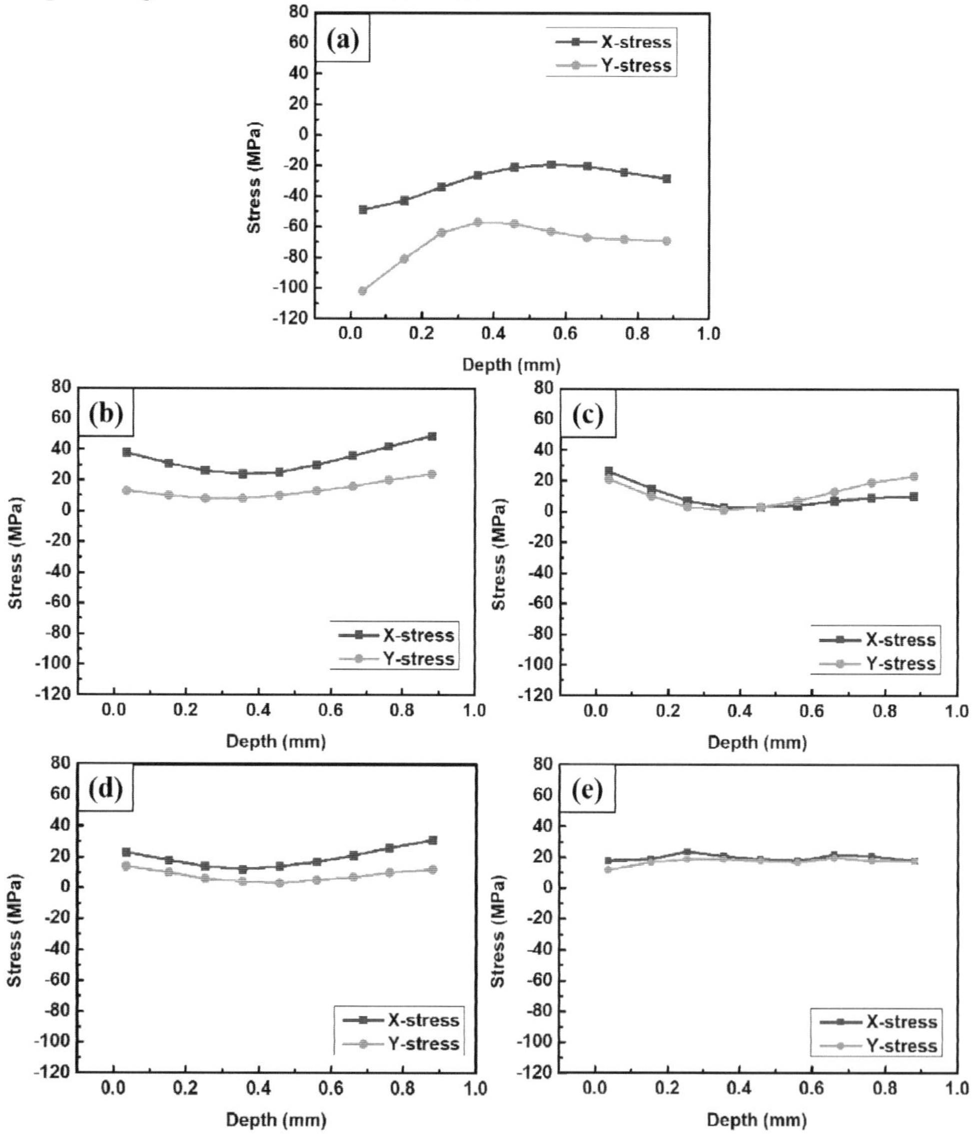

Figure 2: Residual stress distribution at the center of the surface of the sample after different treatments: (a) S; (b) SRT1%; (c) SRT2%; (d) SCT1%; (e) SCT2%.

Table 2 The experimental values of the maximum residual stress on the surface of the sample

sample	X-stress	Y-stress
S	-49	-102
SRT1%	38	13
SRT2%	26	21
SCT1%	23	14
SCT2%	18	12

Subsequently, we conducted artificial peak aging treatment on the samples after different pre-stretching treatments and evaluated their tensile properties, as shown in Fig. 3. The corresponding tensile properties are shown in Table 2. The tensile strength, yield strength and elongation of the SA sample were 294.68 MPa, 251.05 MPa and 24.80%, respectively. These properties reflect the equilibrium precipitate structure formed during peak aging after solution treatment.

The final mechanical properties were altered when pre-stretching at different temperatures and strain levels was introduced between solution quenching and peak aging. The tensile strength, yield strength and elongation of the SRT1%A sample were 300.51MPa, 266.91MPa and 24.22%, respectively, which were 5.83MPa and 15.89MPa higher than those of the SA sample. This strengthening arises because pre-deformation introduces dislocations that provide heterogeneous nucleation sites for precipitates during aging, refining their distribution and increasing number density, thereby enhancing Orowan strengthening. The tensile strength, yield strength and elongation of the SCT1%A sample were 308.49MPa, 274.09MPa and 24.78%, respectively, which were 13.81MPa and 23.04MPa higher than those of the SA sample. It can be seen that pre-stretching with 1% deformation can slightly increase the tensile strength and yield strength of the sample. In addition, pre-stretching at 77K has a more significant effect on improving the tensile properties than pre-stretching at RT. The superior strengthening at 77K is attributed to higher, more uniformly distributed dislocation density preserved due to suppressed dynamic recovery at cryogenic temperatures. This creates a denser network of nucleation sites, further refining the precipitate structure and enhancing barrier strength against dislocation motion.

When the pre-stretching deformation increased to 2%, the tensile strength and yield strength of the SRT2%A and SCT2%A samples decreased compared to those with 1% deformation. Their tensile strength is basically consistent with that of SA samples, while the yield strength still remains higher than that of SA samples. It can be seen that when the pre-stretching deformation degree is 1%, the tensile properties of the sample can be improved, while the deformation degree increasing to 2% has a weaker effect on improving tensile properties. This suggests that excessive pre-strain (2%) may: (1) Promotes the formation of dislocation cells/subgrains, reducing the number of nucleation sites within them and thereby decreases the density of the precipitation; (2) Partially consume the matrix supersaturation, reducing available solute for optimal precipitation; (3) introduces large stress concentrations that are detrimental to tensile properties. Notably, cryogenic pre-stretching consistently outperformed RT pre-stretching at equivalent strains. The strength of SCT1%A and SCT2%A samples remained higher than their counterparts at RT. This persistence highlights the critical role of the quality of the dislocation substructure introduced during pre-deformation. The finer, more stable dislocation structure generated at 77K maintains its effectiveness as a nucleation sites even after higher strains or during aging treatment. In summary, an optimal pre-strain exists (1%) for enhancing aging strength. During pre-stretching at 77K, the generation of higher-density dislocations provides more fine precipitates for second-phase strengthening, which further improves strength. However, excessive pre-stretching (2%) reduces this enhancement effect due to coarsening of the precipitated phase and a decrease in supersaturation.

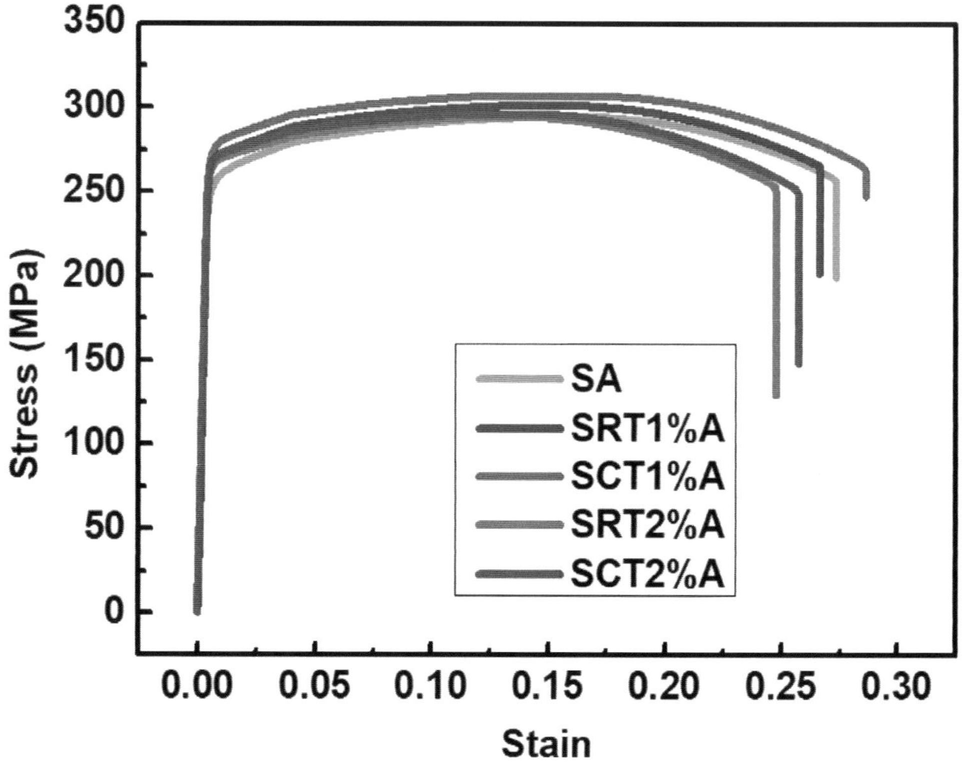

Figure 3: Engineering stress-strain curve of 6061 aluminum alloy after different treatments.

Table 3 Uniaxial tensile properties under different treatments of 6061 aluminum alloy.

Sample	Tensile strength (MPa)	Yield strength (MPa)	Elongation (%)
SA	294.68 ± 0.44	251.05 ± 0.82	24.80 ± 0.16
SRT1%A	300.51 ± 0.93	266.91 ± 0.04	24.22 ± 0.42
SCT1%A	308.49 ± 0.96	274.09 ± 0.85	24.78 ± 0.58
SRT2%A	294.72 ± 0.41	265.70 ± 0.70	24.48 ± 0.10
SCT2%A	296.26 ± 0.65	267.72 ± 0.51	24.40 ± 0.84

The tensile fracture surfaces of 6061 aluminum alloy exhibit a ductile dimple morphology under all conditions, consistent with its inherent plasticity, as shown in Fig. 4. However, the number and morphological characteristics of the dimples directly correlate with the mechanical properties. The SA sample features relatively large dimples, which are sparse and shallow, as shown in Fig. 4(a).

After 1% pre-stretching at RT treatment before peak aging, the dimple size decreases while the density increases, with a large number of small dimples, as shown in Figure 4 (b). Pre-stretching induced refinement of precipitates promotes uniform void nucleation, resulting in an increase in dimple density. The reduced dimple size indicates enhanced resistance to void coalescence, contributing to the increased strength of the sample. Compared to the SRT1%A sample, the SCT1%A sample fracture surface exhibits a further increase in the number of small dimples, as shown in Fig. 4(c). This is attributed to the cryogenic pre-stretching inducing a higher density of finer precipitates, resulting in more void nucleation sites. The increased number of small dimples requires more energy for crack propagation, explaining the further enhancement in strength for the SCT1%A sample.

When the pre-stretching increases to 2%, the dimple size on the SRT2%A fracture surface further decreases, and the number of small dimples increases significantly, as shown in Fig. 4(d). Excessive pre-stretching (2%) induces severe localized stress concentration, which accelerates void nucleation and ultimately causes a slight decrease in strength. Due to the more uniform overall deformation of aluminum alloy at cryogenic temperatures, the fracture morphology of the SCT2%A sample resembles that of SCT1%A, as shown in Fig. 4(e). Even at 2% strain, the suppressed recovery at 77 K maintains the uniformity of dislocation slip, preventing excessive stress concentration. Therefore, the samples subjected to cryogenic pre-stretching retain higher strength even under excessive deformation (2%).

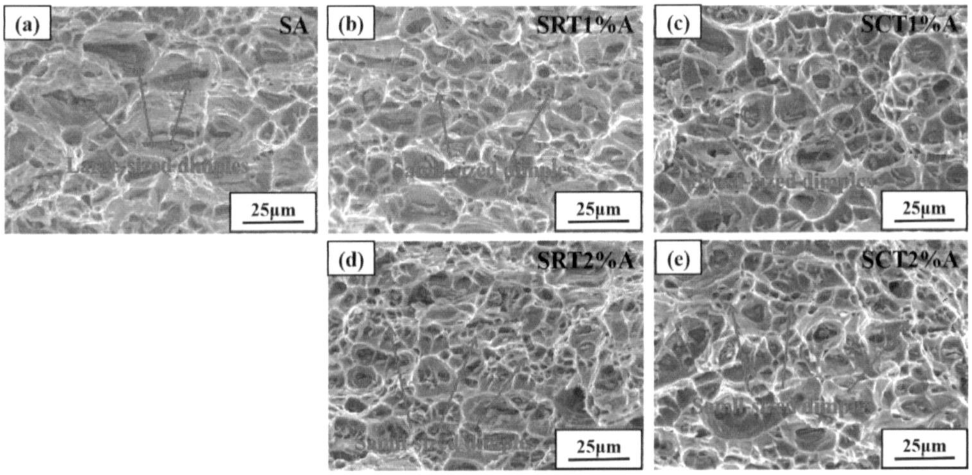

Figure 4: Tensile fracture morphology of 6061 aluminum alloy with different treatment: (a) SA; (b) SRT1%A; (c) SRT2%A; (d) SCT1%A; (e) SCT2%A.

Conclusions

This work investigated the residual stress and tensile properties of 6061 aluminum alloy following pre-stretching at different strain levels (1% and 2%) and temperatures (room temperature and 77K). The main conclusions are as follows:

(1)Pre-stretching proved to be an effective method for reducing residual stress. With the temperature decreasing from RT to 77K and the strain level increasing from 1% to 2%, the reduction in residual stress becomes more significant and the stress distribution becomes more uniform. It can be seen that increasing the pre-stretching strain or decreasing the pre-stretching temperature (to 77K) can effectively reduce the residual stress generated by quenching treatment in 6061 aluminum alloy.

(2)With the increase in pre-stretching deformation, the tensile strength and yield strength of the sample first increase and then decrease. The highest strength of the sample is obtained when the pre-stretching deformation is 1%. Moreover, when the strain is the same, the strength of samples pre-stretching at 77K is higher than that at RT.

References

[1] Lados, D.A., Apelian, D., Wang, L., "Minimization of residual stress in heat-treated Al–Si–Mg cast alloys using uphill quenching: Mechanisms and effects on static and dynamic properties," Mater. Sci. Eng. A, Vol. 527, No. 13-14 (2010), p 3159-3165. doi: 10.1016/j.msea.2010.01.064

[2] Wang, Q.C., Wang, L.T., Peng, W., "Thermal stress relief in 7050 aluminum forgings by uphill quenchings," Materials Science Forum. Trans Tech Publications Ltd, Vol. 490 (2005), p 97-101. doi: 10.4028/www.scientific.net/MSF.490-491.97

[3] Simencio, E.C.A., Canale, L.C.F., Totten, G.E., "Uphill quenching of aluminium: a process overview," International Heat Treatment and Surface Engineering, Vol. 5, No. 1 (2011), p 26-30.

[4] Sun, Y., Jiang, F., Zhang, H., Su, J., Yuan, W.H., "Residual stress relief in Al−Zn−Mg−Cu alloy by a new multistage interrupted artificial aging treatment," Mater. Design, Vol. 92 (2016), p 281-287. doi: 10.1016/j.matdes.2015.12.004

[5] Lim, H.J., Ko, D.H., Ko, D.C., "Reduction of residual stress and improvement of dimensional accuracy by uphill quenching for Al6061 tube," Metallurgical and materials transactions B, Vol. 45 (2014), p 472-481. doi: 10.1007/s11663-013-9997-3

[6] Zhang, M., Gu, K., Weng, Z. et al. "Residual Stress Evolution of 7050 Aluminum Alloy during Thermal Processing and Its Effects on Processing Deformation and Mechanical Properties," J. of Materi Eng and Perform, Vol. 33, p 11467 – 11483 (2024). doi: 10.1007/s11665-023-08816-2
[7] Lados, D.A., Apelian, D., Wang, L., "Minimization of residual stress in heat-treated Al–Si–Mg cast alloys using uphill quenching: Mechanisms and effects on static and dynamic properties," Mater. Sci. Eng. A, Vol. 527, No. 13-14 (2010), p 3159-3165. doi: 10.1016/j.msea.2010.01.064
[8] Qu, S.G., Lou, H.S., Li, X.Q., Kuang, T.R., Lou, J.Y., "Effect of heat-treatment on stress relief and dimensional stability behavior of SiCp/Al composite with high SiC content," Mater. Design, Vol. 86 (2015), p 508-515. doi: 10.1016/j.matdes.2015.07.044
[9] Zhang, Y.H., Li, S.Y., Cheng, X.L., Zhao, X.H., Su, Z.H., "Effect of heat-treatment on stress relief and dimensional stability behavior of SiCp/Al composite with high SiC content," Heavy Machinery, Vol. 86 (2015), p 508-515. doi: ," Mater. Design, Vol. 3 (2024), p 1-9. doi: 10.13551/j.cnki.zxjxqk.2024.03.001
[10] Zhang, M.L., Gu, K.X., Ran, P., Liu, B.S. Weng, Z.J., Wang, J.J., "Cryogenic deformation behavior of Al–Zn–Mg–Cu alloy and its sensitivity to temperature and strain rate," Mater. Sci. Eng. A, Vol. 893 (2024), p 146137. doi: 10.1016/j.msea.2024.146137
[11] Cheng, W.J., Liu, W., Fan, X.B., Yuan, S.J., "Cooperative enhancements in ductility and strain hardening of a solution-treated Al–Cu–Mn alloy at cryogenic temperatures," Mater. Sci. Eng. A, Vol. 790 (2020), p 139707. doi: 10.1016/j.msea.2020.139707
[12] Chen, X.S., Fan, X.B., Xu, A.J., Yi, Z.X., Yuan, S.J., "Formability enhancement and deformation mechanism of 2195 Al-Li alloy sheet in cryogenic forming, " Mater. Charact., Vol. 196 (2023), p 112577. doi: 10.1016/j.matchar.2022.112577

ICRS 12: Proceedings from the 12th International Conference on Residual Stresses, 2025
October 20–23, 2025; Detroit, Michigan, USA
https://doi.org/10.31399/asm.cp.icrs2025p0049

Prediction of quenching residual stress of Al-Zn-Mg-Cu alloy during quenching and cryogenic treatment

***Mingli Zhang*[a, b], *Kaixuan Gu*[a, b*], *Xiujie Zhao*[a, b], *Junjie Wang*[a, b]**

[a] *Key Laboratory of Cryogenic Science and Technology, Technical Institute of Physics and Chemistry, Chinese Academy of Sciences, Beijing 100190, China*

[b] *University of Chinese Academy of Sciences, Beijing 100049, China*

**Corresponding author. Tel./fax: +86 010 82543415.*

E-mail address: gukaixuan@mail.ipc.ac.cn (K. Gu)

ABSTRACT

Al-Zn-Mg-Cu alloys are widely used in the aerospace industry due to their excellent properties. The majority of heat-treatable aluminum alloys are treated sequentially by solution, quenching and aging, which gains a good combination of properties by precipitating secondary phases from supersaturated solid solution matrix. As a result of the different levels of thermal expansion and cold contraction of the sample during rapid cooling, a high level of residual stress is generated. Residual stress can induce distortion during subsequent machining. In recent years, cryogenic treatment is adopted to reduce residual stress. The stress-strain relationship of materials at cryogenic temperatures is one of the most critical parameters for determining the accuracy of simulation results. Therefore, this work investigates the stress-strain relationship of Al-Zn-Mg-Cu alloy at different cryogenic temperatures and strain rates, and establishes a cryogenic constitutive equation according to the Arrhenius model. The residual stress distribution of the sample after different treatments is predicted with and without the cryogenic temperature constitutive, respectively by the method of finite element simulation. The simulation results are also compared with the experimental measurements, which prove that the cryogenic temperature constitutive equations are essential for the accuracy of the finite element simulation results.
Key words: Residual stress, Cryogenic treatment, Quenching, Constitutive equation, Al-Zn-Mg-Cu alloy

Introduction

Al-Zn-Mg-Cu alloys are widely used in aerospace industry due to its low density, high strength, excellent corrosion resistance and high hot-working property [1-3]. The standard thermomechanical processing route for optimizing these properties involves sequential solution, rapid quenching, and artificial aging treatment. While solution treatment dissolves soluble phases, quenching retains solute in a supersaturated solid solution (SSSS), enabling subsequent precipitation of fine, strengthening metastable phases (e.g., η') during aging. However, this rapid quenching process introduces a significant technological challenge: non-uniform thermal contraction generates substantial residual stresses within the component [4]. These stresses are detrimental, often manifesting as unacceptable distortion during final machining operations, compromising dimensional accuracy and increasing scrap rates [5-7].

To mitigate quenching-induced residual stresses, cryogenic treatment (typically below -150 °C) has emerged as a promising post-quenching intervention [8-10]. While its effectiveness in reducing residual stress is recognized, the behavior of stress evolution and release at these cryogenic temperatures has not yet been fully characterized. Crucially, the stress-strain constitutive relationship at cryogenic temperatures is a paramount parameter dictating the fidelity of finite element (FE) simulations used to predict residual stress distributions and optimize treatment processes [4]. Inaccurate constitutive models inevitably lead to unreliable simulation outcomes, hindering the effective design and implementation of cryogenic treatment.

Therefore, we present a comprehensive experimental investigation of the stress-strain behavior of Al-Zn-Mg-Cu alloys across a range of cryogenic temperatures and strain rates. Building upon this data, we establish a physically-based cryogenic constitutive equation utilizing the widely applicable Arrhenius-type model to describe the material's flow stress response. Crucially, we then integrate this newly developed cryogenic constitutive model into a finite element framework to predict the residual stress distributions within samples subjected to different treatments. In order to rigorously validate this method, these finite element predictions are systematically compared with experimental residual stress measurements. This comparative analysis unequivocally demonstrates the indispensable role of accurate cryogenic constitutive relationships in achieving reliable FE simulation results for residual stress prediction in Al-Zn-Mg-Cu alloys undergoing cryogenic treatment.

Materials and Methods

The material used in this work was rolled Al-Zn-Mg-Cu alloy plates with the thickness of 40 mm. The chemical composition of the aluminum alloy was listed in Table 1.

Table 1 Chemical compositions of 7075 aluminum alloy (wt. %).

Zn	Mg	Cu	Fe	Zr	Si	Cr	Mn	Al
6.02	2.31	2.04	0.11	0.1	0.03	0.009	0.003	Bal

The dimensional size of the tensile sample is shown in Figure 1. Tensile tests were conducted in a Universal Tensile Testing Machine (UTM5305X) according to the standard of GB/T 228.1-2010. The samples were first solution treated at 465℃ for 1 h and then quenched into room-temperature water (S). Subsequently, uniaxial tensile experiments were conducted at five different temperatures (-80 °C, -120 °C, -160 °C, -190 °C) and three different strain rates (0.00025 s-1, 0.001 s-1, 0.01 s-1).

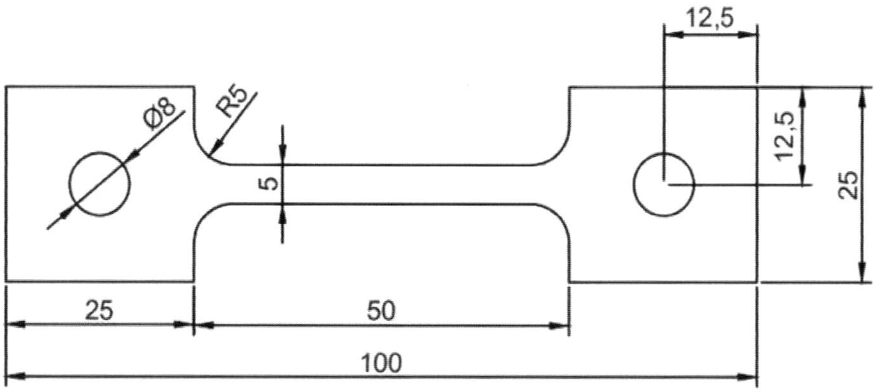

Figure 1: Dimensions of the uniaxial tensile-test sample (mm)

Samples used for residual stress tests were cut into cuboid with the size of 40×40×40 mm3, and the residual stress test position is shown in Figure 2(a). The C3D8T element in ABAQUS software was used to simulate the heat transfer and stress evolution of residual stresses in cubic samples after quenching (S) and cryogenic (C) treatment. The path of residual stress predicted by finite element simulation is shown in Figure 2 (b).

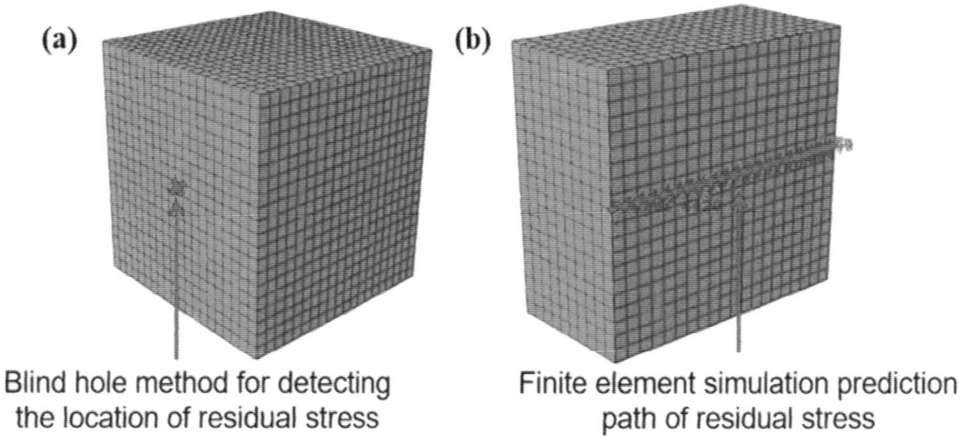

Figure 2: Residual stress testing location: (a) Blind hole method testing location; (b) Finite element simulation prediction path

Results and Discussion

This work aims to establish a constitutive equation to investigate the stress-strain relationship of 7075 aluminum alloy at different temperatures and strain rates, in order to accurately predict the stress changes inside the sample during hot and cryogenic cycling treatment. In the phenomenological model, the Arrhenius equation has been successfully applied to predict the flow stress of materials [11]. The true stress-strain curves of 7075 aluminum alloy at different temperatures (-80 °C, -120 °C, -160 °C, -190 °C) and strain rates (0.00025 s^{-1}, 0.001 s^{-1}, 0.01 s^{-1}) are shown in Figure 3. In this model, the relationship between strain rate, true stress and temperature is described by the following equations.

$$\dot{\varepsilon} = AF(\sigma)exp[-Q/(RT)] \quad (1)$$

$$where\ f(x) = \begin{cases} \sigma^{n'} & \alpha\sigma < 0.8 \\ exp(\beta\sigma) & \alpha\sigma > 1.2 \\ [sinh(\alpha\sigma)]^n & for\ all\ \sigma \end{cases}$$

(2)

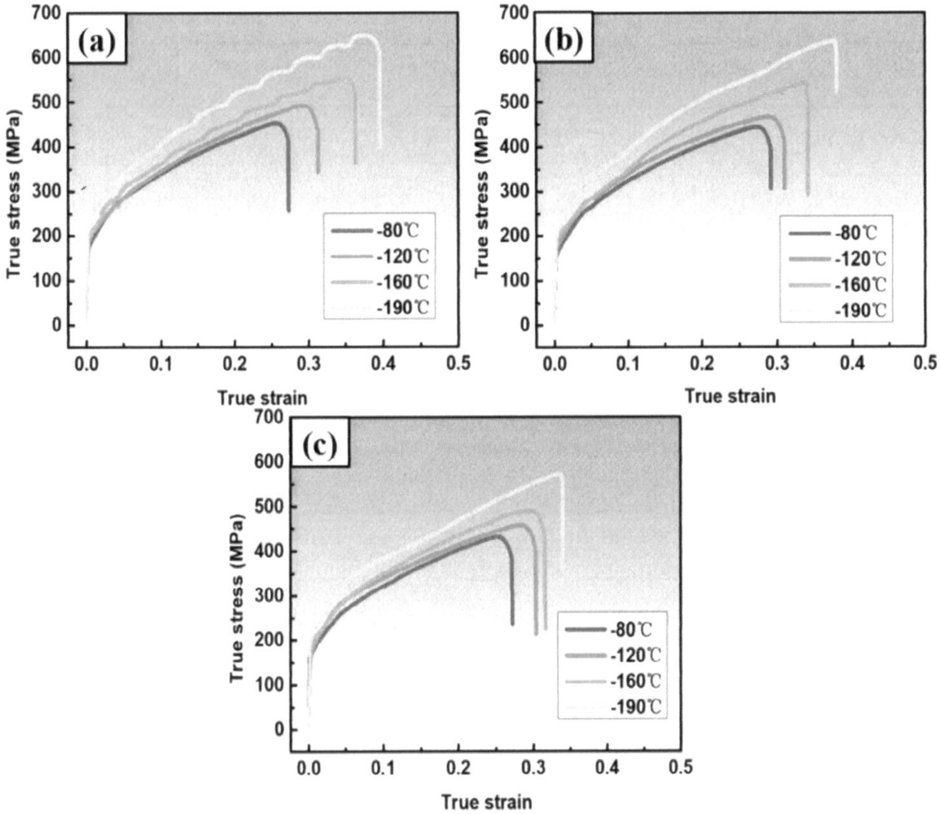

Figure 3: True stress-strain cave of 7075 aluminum alloy at different temperatures with same strain rate: (a) 0.00025 s^{-1}; (b) 0.001 s^{-1}; (c) 0.01 s^{-1}.

The Arrhenius equation cannot accurately describe the variation of material flow stress with strain after dynamic recrystallization, so strain compensation is usually introduced to improve the traditional Arrhenius equation. Zener and Hollomon proposed a strain rate factor (Zene Hollomon factor) for compensation, which can be represented by the Zene Hollomon parameter to express the coupling effect between temperature and strain rate, as shown in the following equation.

$$Z = \dot{\varepsilon}exp(\frac{Q}{RT}) = A[sinh(\alpha\sigma)]^n \quad (3)$$

Therefore, the flow stress can be written as a function of Zener-Hollomon parameter, and the constitutive equation can be summarized as,

$$\sigma = \frac{1}{\alpha}ln\{(Z/A)^{1/n} + [(Z/A)^{2/n} + 1]^{1/2}\} \quad (4)$$

In Eqs. (1-4), $\dot{\varepsilon}$ is the strain rate (s-1), T is the absolute temperature (K), R is the universal gas constant (8.31 J·mol-^{1}·K^{-1}), Q is the apparent activation energy of hot deformation (J·mol^{-1}), σ is flow stress (MPa), A, α and n are the material constants. In order to consider the effects of temperature and strain rate on the calculation of the above material parameters, a fourth-order polynomial function of strain is introduced to correct the material constants and accurately predict the flow stress behaviour of 7075 aluminium alloy, as shown in formula 5. The fitting curve is shown in Figure 4.

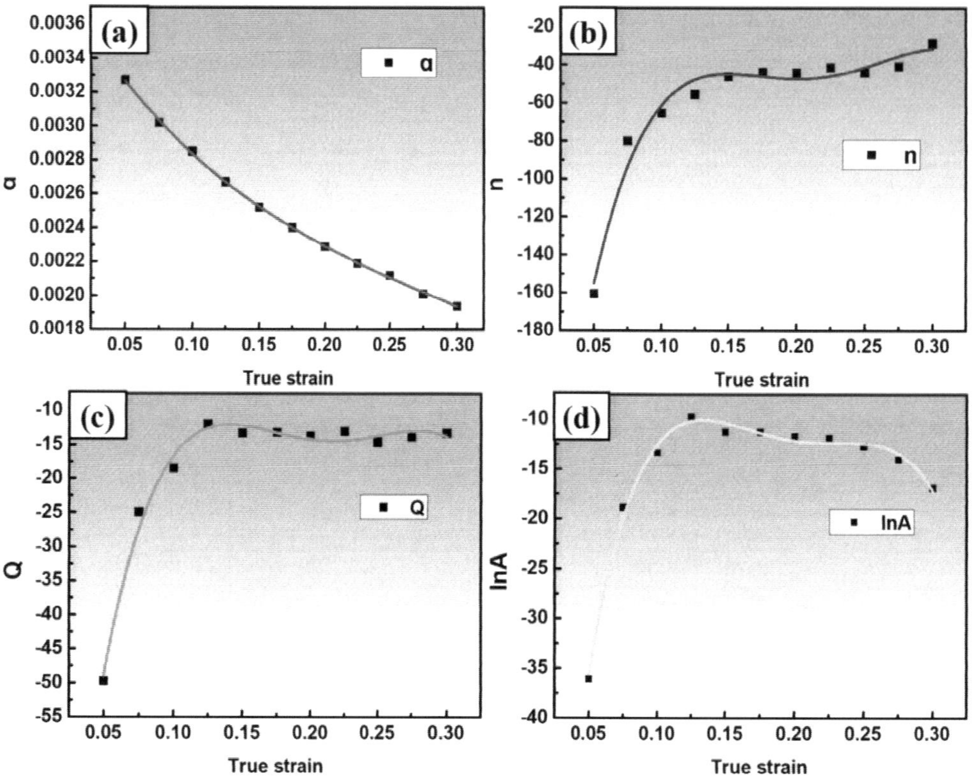

Figure 4: Relationship between α, n, Q, lnA and true strain.

$$\alpha(\varepsilon) = 0.00458 - 0.02231\varepsilon 0.109854\varepsilon^2 0.245413\varepsilon^3 0.186998\varepsilon^4$$

$$n(\varepsilon) 0.73668945.452019\varepsilon + 7606.8694\varepsilon^2 - 25493.183\varepsilon^3 + 30847.878\varepsilon^4$$

$$Q(\varepsilon) = -30.06284 + 432.9875\varepsilon - 3048.609\varepsilon^2 + 8568.223\varepsilon^3 - 8396.855\varepsilon^4$$

$$lnAL\varepsilon) = -22.2838 + 292.19933\varepsilon - 2194.5744\varepsilon^2 + 6806.85482\varepsilon^3 - 7846.8829\varepsilon^4 \quad (5)$$

To enhance the accuracy of the base model and provide a more realistic description of the flow stress behavior of 7075 aluminum alloy during cryogenic deformation, a novel constitutive model is proposed to improve upon the previous model. This new model comprehensively accounts for the coupled effects of deformation temperature, strain rate, and strain. A $K(\dot{\varepsilon}, T)$ function is established through numerical programming, incorporating both strain rate and deformation temperature. Based on the strain-compensated Arrhenius constitutive equation, a further refinement was performed, as represented by Equation (6):

$$\sigma = K(\dot{\varepsilon}, T) \cdot \frac{1}{a} ln\{(\frac{Z}{A})^{\frac{1}{n}} + [(\frac{Z}{A})^{\frac{2}{n}} + 1]^{1/2}\} \quad (6)$$

The above test and calculation results were imported into Origin software and fitted, with the results shown in Figure 5. Finally, a function $K(\dot{\varepsilon}, T)$ is established for strain rate and deformation temperature, where

$$K(\dot{\varepsilon}, T) = 0.97514 - 44.85687\dot{\varepsilon} + 0.00104T + 3399.1453\dot{\varepsilon}^2 + 0.06746\dot{\varepsilon}T - 4.64475 \times 10^{-6}T^2 \quad (7)$$

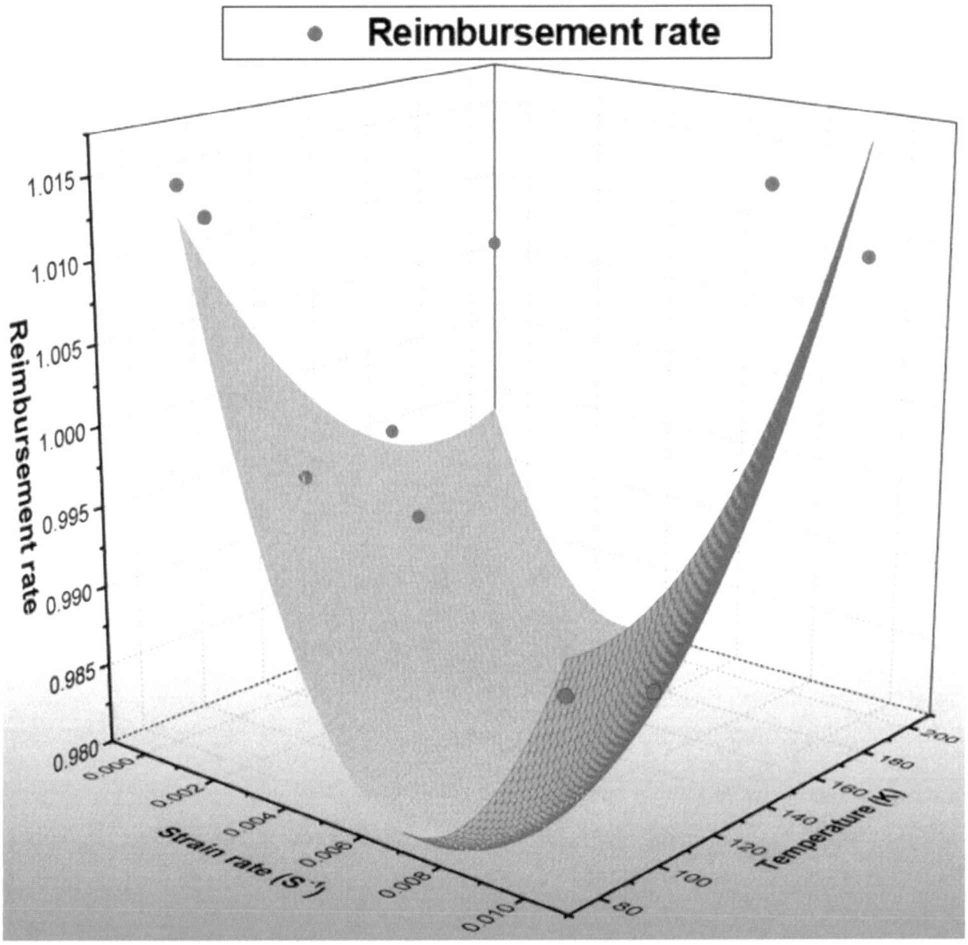

Figure 5: K(ε̇, T) function fitting.

The K-function is coupled with the previously established Arrhenius constitutive model. The modified constitutive model is then used to recalculate the flow stress values under various cryogenic deformation processes. These predicted values are represented by the solid black symbols in Figure 6. It can be observed that the results calculated by the modified model demonstrate excellent agreement with the experimentally measured values.

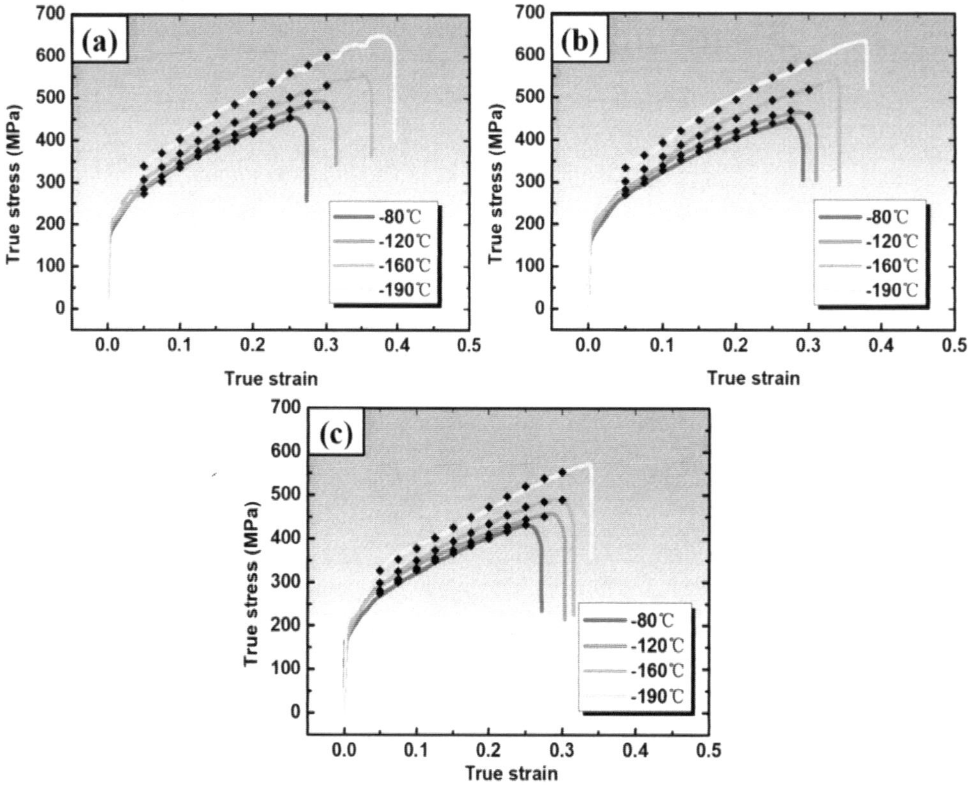

Figure 6: Comparison of the calculated flow stress with the experimental results: (a) 0.00025 s^{-1}; (b) 0.001 s^{-1}; (c) 0.01 s^{-1}.

Subsequently, the residual stresses in samples subjected to S (solution treated and quenched) and SC (solution treated, quenched, and cryogenic) treatment were predicted using ABAQUS software and measured using the blind-hole drilling method. The maximum surface residual stress values are shown in Table 2. The predicted residual stress distribution for the S sample is shown in Figure 7(a, d), while the corresponding experimental residual stress measurement is shown in Figure 7(g). It can be seen that the calculated and experimentally measured surface residual stresses for the S sample are in good agreement. It is demonstrated that finite element simulation is an efficient and accurate method for predicting the internal residual stress distribution in aluminum alloys.

The predicted residual stress distribution after SC1 treatment, incorporating the cryogenic constitutive equation for 7075 aluminum alloy, is shown in Figure 7(b, e). In contrast, the predicted distribution after SC2 treatment without cryogenic constitutive equation is shown in Figure 7(c, f). The experimentally measured surface residual stress for the SC sample is shown in Figure 7(h). The predicted value of SC1 exhibits an error of less than 2% compared to the measured value of SC. However, the predicted value of SC2 shows a significantly larger error of 24.8% relative to the measured value. Therefore, it is evident that the cryogenic constitutive equation is crucial for accurately predicting both the magnitude and distribution of residual stresses in samples after thermal-cryogenic cycling treatments.

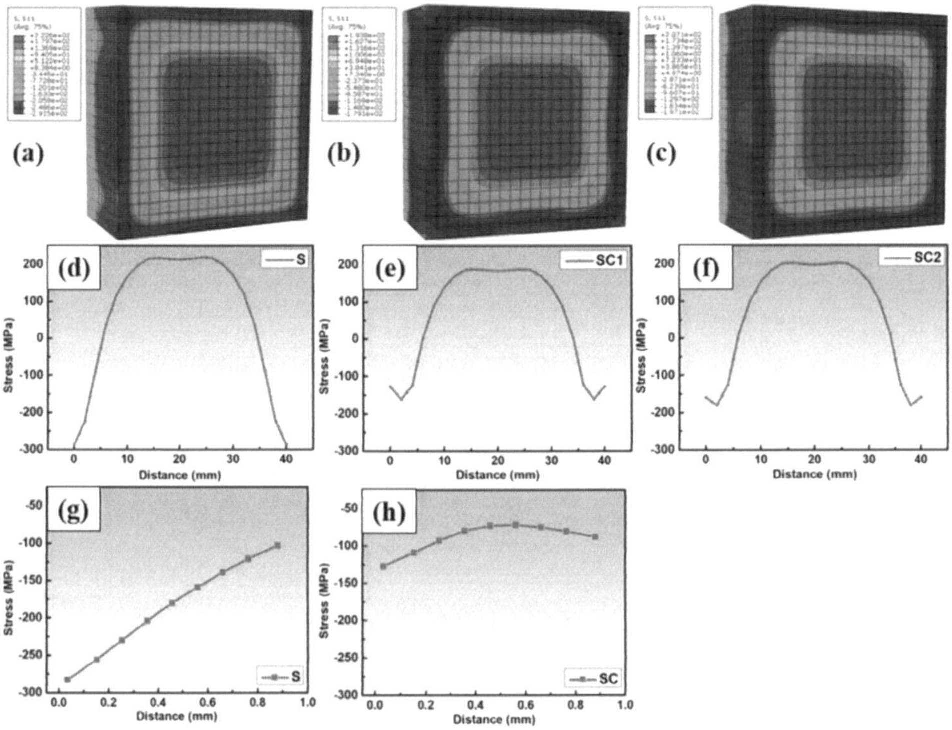

Figure 7: Finite element simulation and blind hole method for detecting residual stress distribution after different treatments: (a, d, g) S; (b, e) SC1; (c, f) SC2; (h) SC.

Table 2 Simulated and experimental values of the maximum residual stress on the surface of the cuboid

sample	Maximum surface compressive stress (simulation)/MPa	sample	Maximum surface compressive stress (experiment)/MPa
S	-286	S	-283
SC1	-131	SC	-129
SC2	-161		

Conclusions

This study systematically investigated the cryogenic deformation behavior of Al-Zn-Mg-Cu alloy (7075 aluminum alloy) and established an Arrhenius constitutive model to accurately predict the quenching residual stresses under cryogenic treatment. The key findings and contributions are summarized as follows:

(1) A strain-compensated Arrhenius constitutive model is developed, which combines a fourth-order polynomial function to describe the evolution of the material parameters (α, n, Q and A) with the true strain. Further refinement was achieved by introducing the coupling function $K(\dot{\varepsilon}, T)$, which improved the predictive accuracy of the constitutive model under cryogenic temperature conditions.

(2) This work provides a validated numerical framework for optimizing cryogenic treatments in aerospace aluminum alloys. By enabling precise prediction of residual stress distributions, the model supports the design of thermal-cryogenic cycling treatments to minimize distortion and improve machining accuracy.

References

[1]Fan, S.T., Deng, Y.L., Zhang, Y., Huang, X.Y., Zhang, X.M., "Homogenization of 7050 plates by a novel differential temperature rolling," Mater. Manuf. Process., Vol.33, No. 16 (2018), p 1822-1829. doi:10.1080/10426914.2018.1476758

[2]Song, Y.F., Ding, X.F., Zhao, X.J., Xiao, L.R., Guo, L., "The effect of stress-aging on dimensional stability behavior of Al-Cu-Mg alloy," J. Alloy Compd., Vol.718 (2017), p 298-303. doi: 10.1016/j.jallcom.2017.05.183

[3]Sarha, A.A.D., Zalnezhad, E., Hamdi, M., "The influence of higher surface hardness on fretting fatigue life of hard anodized aerospace AL7075-T6 alloy," Mater. Sci. Eng. A, Vol.560 (2013), p 377-387. doi: 10.1016/j.msea.2012.09.082.
[4] Zhang, M., Gu, K., Weng, Z. et al. "Residual Stress Evolution of 7050 Aluminum Alloy during Thermal Processing and Its Effects on Processing Deformation and Mechanical Properties," J. of Materi Eng and Perform, Vol. 33, p 11467 – 11483 (2024). doi: 10.1007/s11665-023-08816-2
[5] Lados, D.A., Apelian, D., Wang, L., "Minimization of residual stress in heat-treated Al–Si–Mg cast alloys using uphill quenching: Mechanisms and effects on static and dynamic properties," Mater. Sci. Eng. A, Vol. 527, No. 13-14 (2024), p 3159-3165. doi: 10.1016/j.msea.2010.01.064
[6] Qu, S.G., Lou, H.S., Li, X.Q., Kuang, T.R., Lou, J.Y., "Effect of heat-treatment on stress relief and dimensional stability behavior of SiCp/Al composite with high SiC content," Mater. Design, Vol.86 (2015), p 508-515. doi: 10.1016/j.matdes.2015.07.044
[7] Sun, Y., Jiang, F., Zhang, H., Su, J., Yuan, W.H., "Residual stress relief in Al–Zn–Mg–Cu alloy by a new multistage interrupted artificial aging treatment," Vol.92 (2016), p 281-287. doi: 10.1016/j.matdes.2015.12.004
[8] Liu, X.Z., Gu, K.X., Weng, Z.J., Wang, K.K., Cui, C., Guo, J., Wang, J.J. "A Review on Deep Cryogenic Treatment of Aluminium Alloy," Mater. Rep., Vol.34, No. 3 (2020), p 6. doi: 1005-023X(2020)34:2A<03172:LHJSLC>2.0.TX;2-Q
[9] Niu, X.M., Huang, Y., Yan, X.G., Chen, Z., Yuan, R.Z., Zhang, H.D., Tang, L., "Optimization of Cryogenic Treatment Parameters for the Minimum Residual Stress," Mater. Eng. Perform., Vol.30 (2021), p 9038-9047. doi: 10.1007/s11665-021-06136-x
[10] Wei, L.J., Wang, D.W., Li, H.S., Xie, D., Ye, F., Song, R.K., Zheng, G., Wu, S.J., "Effects of Cryogenic Treatment on the Microstructure and Residual Stress of 7075 Aluminum Alloy," Metals, Vol.8, No. 4 (2018), p 273. doi: 10.3390/met8040273
[11] Lin, Y.C., Chen, X.M., "A critical review of experimental results and constitutive descriptions for metals and alloys in hot working," Mater. Des., Vol.32 (2011), p 1733–1759. doi: 1016/j.matdes.2010.11.048

ICRS 12: Proceedings from the 12th International Conference on Residual Stresses, 2025
October 20–23, 2025; Detroit, Michigan, USA
https://doi.org/10.31399/asm.cp.icrs2025p0057

Effect of High-Pressure Cooling and Cutting Speed on Residual Stresses Generated During Turning of an Advanced Wrought Nickel-Based Superalloy (AD730®) using PCBN Tools

Henry Boyle[a*], Jinming Zhou[b], Zhe Chen[c], Jonas Holmberg[d], Rachid M'Saoubi[b,e], Alex Graves[f], Lina Rogström[g], Alexis Bohlin[h], Stefan Cedergren[h], Susanne Norgren[b,f] and Ru Peng[a]

[a]Division of Engineering Materials, Department of Management and Engineering, Linköping University, Linköping, Sweden,
[b]Division of Production and Materials Engineering, Lund University, Lund, Sweden,
[c]Siemens Energy, Finspång, Sweden,
[d]Research Institutes of Sweden (RISE), Mölndal, Sweden,
[e]Materials and Technology Development, Seco Tools AB, Fagersta, Sweden,
[f]Sandvik Coromant AB, Stockholm, Sweden,
[g]Nanostructured Materials, Physics Department, Linköping University, Linköping, Sweden,
[h]GKN Aerospace, Trollhättan, Sweden

Abstract

Ni-based superalloys components are utilised in some of the most hostile engineering environments. The exceptional high-temperature performance of these alloys presents significant challenges during manufacturing. Machining is a critical process, enabling the precise shaping of forgings into functional components. During material removal, plastic deformation of the workpiece material and friction generate substantial heat and high stresses, which can affect the integrity of both tool and workpiece. CBN (cubic boron nitride) tools are commonly employed for the finish turning of high-value Ni-based superalloy components. High-pressure cooling (HPC) is implemented to prevent premature tool failure by cooling the cutting tool and aiding chip breakage. In the study presented, the influence of HPC - and cutting speed on the resulting workpiece residual stresses were investigated during axial turning of an advanced wrought disk alloy (AD730®). It was found that both HPC application and cutting speed influence residual stress depth profiles, which are known to enhance fatigue life and can therefore be optimised to improve in-service performance. Consequently, this research has significant implications for industries such as aerospace and power generation.

Introduction

Due to the high energy demands required to deform Ni-based superalloys, machining processes tend to result in large thermomechanical loads that impact the surface integrity. Various alterations can occur to the microstructure, which may include grain refinement, phase transformation, slip banding, twinning, and swept grains, all of which affect the surface residual stresses (RS). In conventional machining, it is widely understood that thermal effects largely control the surface and near-subsurface RS, which are often tensile, whereas mechanical effects extend deeper into the bulk and tend to induce compressive RS. In reality, these effects are not easily decoupled.

Within industrial processes, many other considerations such as tool wear and coolant application methods must be taken into account. In this way, the surface integrity is continuously evolving throughout machining processes. Therefore, limits must be set on the depth of microstructural alterations in the component caused by machining that are understood to be detrimental to its functional performance. Understanding the impact of changing processing parameters on surface integrity can therefore be extremely costly, since it requires substantial experimental work using destructive methods. This is a major restriction to the implementation of new materials, especially in safety-critical applications involving complex loading scenarios, such as those encountered within gas turbines.

AD730® is a wrought alloy developed as a high-performance alternative to traditional alloys such as IN718. Due to its relatively low cost, high microstructural stability, forgeability, and good in-service mechanical properties [1], there is growing industrial interest in expanding its current usage. To date, several studies have been conducted to investigate various aspects of AD730® machinability when turned using CBN (cubic boron nitride) tools [2-5]. CBN tools are typically used for turning the hardest Ni-based materials (>50 HRC), offering a substantial performance advantage over coated carbide tools, which are better suited for machining softer alloys. The high cost of CBN tooling can be offset by operating at higher material removal rates, thereby improving process efficiency. However, to realise these potential gains, any changes to the processing parameters, e.g. turning at higher speeds, must not compromise the surface integrity requirements.

Zhou et al. [4] compared the machinability of IN718 and AD730® when turned with CBN tools, observing a 10% reduction in cutting forces when machining AD730®. When analysing the cross-sections of the machined surface, slip band formation and material deformation were detected. On the machined surface, various defects were observed, such as cracks, side flow, smearing, and plastic flow. In their subsequent work [2], surface RS were found to be comparable in the feed direction for both alloys. However, greater tensile stresses were observed in the cutting direction when turning AD730®, along with a greater depth of plastic deformation. Similar findings have been reported by other authors regarding the increased sensitivity of higher-strength Ni-based alloys in terms of surface integrity [6,7]. In said cases, higher depth and intensity of deformation in terms of subsurface strain correlated well with increased cutting forces, as might be expected, when comparing IN718 to more advanced wrought alloys such as 720Li. The fact that this was not observed when comparing IN718 and AD730®, suggests that there is still much to learn about the interactions between tool and workpiece during the machining of AD730®.

The RS profiles of machined Ni-based superalloy components are inextricably linked to processing parameters [8], tool material [9], and tool wear level [10]. One area that has received less attention is the relationship between high-pressure coolant (HPC) and RS generation in machined Ni-based superalloy components. Due to the high temperatures that develop during machining, HPC is typically used to prevent premature tool failure caused by overheating. Devillez et al. [11] compared the RS generated when turning IN718 under dry and low-pressure cooling conditions, (pressure = 5 bar), at moderate cutting speeds, (40–80 m/min). They observed that at the lowest cutting speed of 40 m/min, tensile RS in the first 20 μm of the subsurface in both cutting and feed directions were reduced by a factor of 2 to 5 with the application of low-pressure cutting fluid. At the same cutting speed compressive RS were observed to extend half the distance into the bulk when using cutting fluid, and the maximum compressive RS was also observed to reduce by approximately 200 MPa. In contrast, when turning at 80 m/min, similar RS profiles were obtained in both the cutting and feed directions regardless of coolant application. When comparing conventional cooling with dry cutting using CBN tools to turn IN718 at cutting speeds ranging from 150-375 m/min, it has also been observed that the application of coolant reduces surface RS from tensile to compressive; no information was supplied by the authors regarding the coolant delivery system, however, it is likely they are referring to flood cooling [12].

Based on the preceding discussion, there appears to be a clear research gap regarding the impact on surface integrity, in terms of surface damage and induced RS, when turning AD730® using HPC. The following work aims to address this gap, providing valuable insights that can assist manufacturers in producing components with improved performance under conditions optimised to balance both production rate and part quality.

Experimental Method and Materials

Single-point axial turning was employed on cylindrical bar stock using HPC Jetstream cooling (Holder: DCLNL3232X12JETI); jets were applied to both the rake and flank faces of the tool simultaneously (Figure 1). CBN tools (grade: CBN170, geometry CNGA120408E25) were used with the following cutting data: feed rate (f_n) = 0.1 mm/rev, depth of cut (a_p) = 0.25 mm. Both cutting speed (v_c) and HPC pressure were varied to produce five separate surfaces; the test matrix is displayed in Table 1. X-ray diffraction (XRD) was carried out in accordance with SS-EN 15305:2008 using a Stresstech Xstress G2R, featuring an Mn-tube X-ray source. To measure RS, the $\sin^2\psi$ method in modified χ mode was employed at tilt angles between -45° and 45°. The (311) plane was used to measure d-spacing in the cutting and feed directions; the following constants were used for the stress calculation (E = 199.9 GPa, ν = 0.294). To produce RS depth profiles, samples were electropolished using a Struers Movipol-5 and A2 electrolyte up to a depth of approximately 160 μm.

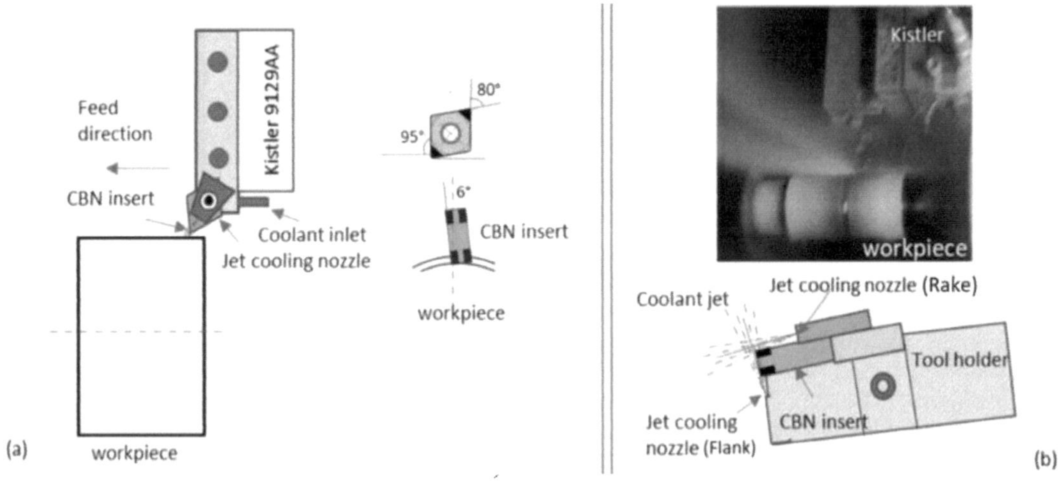

Figure 1: Experimental setup used during turning trials [2].

Table 1: Test matrix for experimental machining trial, showing variation in v_c and HPC cooling pressure used.

Sample Code	v_c (m/min)	HPC pressure (bar)
A-1	200	8
A-2	300	8
B-1	300	50
B-2	300	100
B-3	300	Dry

The material utilised in the trial was AD730®, a modern wrought Ni-based superalloy manufactured by Aubert & Duval. The composition is given in Table 2, which was supplied by the manufacturer. The microstructure of the billet material is shown in Figure 2. Given the subsolvus maximum temperature used for solution heat treatment, primary γ' precipitates can be observed throughout the microstructure; these limit grain growth during the solution heat treatment. Other features can also be observed, such as inclusions (likely σ [2], [4]), and annealing twins.

Table 2: Composition in wt.% for Aubert & Duval wrought alloy AD730®.

Fe	Cr	Co	Mo	W	Al	Ti	Nb	Ni
4	16	9	3	3	2	4	1	Base

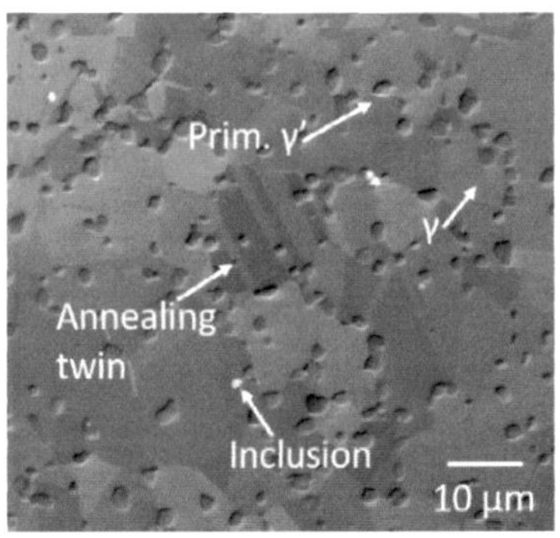

Figure 2: Back scattered electron (BS) SEM image showing bulk microstructure and key microstructural features.

Results and Discussion

XRD RS measurements presented in Figure 3a demonstrate limited variation in RS in the feed direction across the test matrix. There is some variation in the position of the compressive minimum RS, whereby for A-1, A-2, and B-2 it is shifted closer to the surface by approximately 30 μm. In contrast, when comparing the RS profiles obtained in the cutting direction (Figure 3b), significant variation is observed between 0-10 μm.

Considering the effect of coolant, there is a clear positive correlation showing that higher HPC pressure leads to increased tensile RS at the surface in the cutting direction. It is hypothesised that this is related to an increased cooling rate, combined with more aggressive chip deflection due to the rake face coolant jet [13]. In addition, it appears that machining dry leads to the transition to compressive RS occurring deeper into the bulk, similar observations have been previously reported during low speed Ni-based turning at 40-60 m/min [11]. Although dry turning results in lower tensile surface RS, it also appears to reduce the minimum compressive RS in the cutting direction. As such, it is unclear which of these effects will have the greater impact on the part's functional performance in terms of fatigue life.

When comparing the effect of cutting speed (A-1 and A-2), a marginally greater compressive RS (≈ 50 MPa) was recorded in the feed and cutting directions at the lower speed of 200 m/min (A-1); a higher tensile surface RS in the cutting direction was also measured at the lower speed (200 MPa). FWHM is known to correlate with strain in terms of swept grain formation in Ni-based superalloys, and the overall defect density [14]. When comparing the effect of coolant in Figure 4 on average FWHM, no major variations were observed in the deformation profiles for cuts performed at 300 m/min. At the lower cutting speed of 200 m/min, some reduction in FWHM values were measured, indicating lower overall levels of sub-surface strain, due to the lower overall strain rate. Considering these findings, higher tensile RS may develop at the lower speed due to an increased contact time, which facilitates heat diffusion into the workpiece surface.

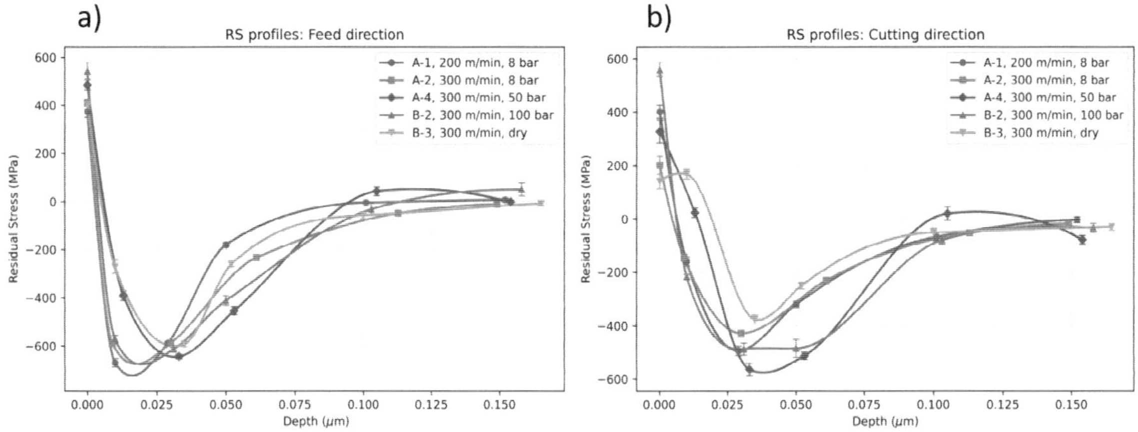

Figure 3: RS depth profiles for machined samples from surface into bulk: a) feed direction, b) cutting direction.

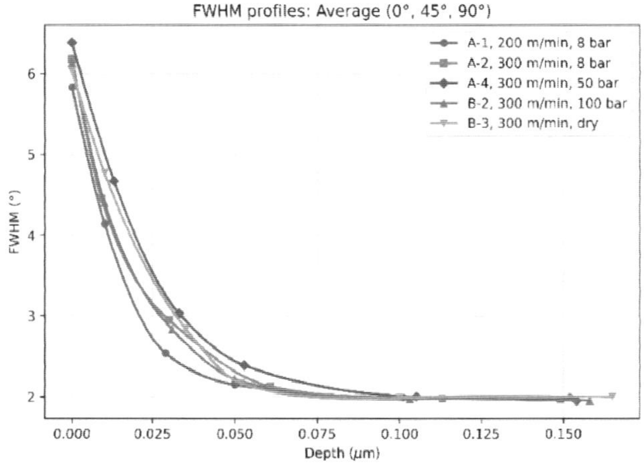

Figure 4: FWHM profiles from surface into bulk for machined samples average across 3 measurement directions (0°, 45°, 90°).

Figure 3b shows that the transition from tensile to compressive RS occurs within the first 5–20 μm when measured in the cutting direction, depending on the cutting conditions. Observing the subsurface microstructure in Figure 5, a strong deformation gradient is evident, indicated by microstructural alterations. Near the surface (0–3 μm), where the strain is highest, precipitate shearing is observed and the grain structure is no longer discernible; this region is typically referred to as a severely plastically deformed (SPD) layer. Deeper into the bulk (5–15 μm), slip bands become more apparent, with some evidence of cross slip. At greater depths (>15 μm), where the strain decreases further, deformation becomes more diffuse and only intragranular misorientation contrast is detected.

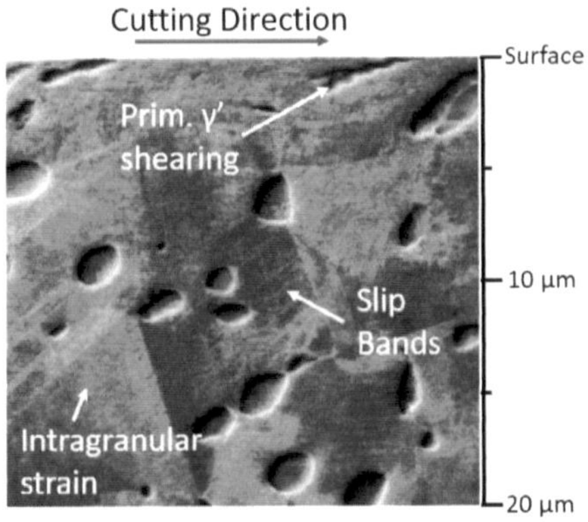

Figure 5: BS SEM micrograph showing cross-section of machined surface and sub-surface viewed normal to the cutting direction for sample B-2 (300 m/min, 100 bar).

Comparing the EBSD scans of the subsurfaces for samples B-2 and B-3 in Figure 6 no clear differences are observed. Kernel average misorientation (KAM) mapping was performed using MTEX by averaging the local misorientation between each pixel and its nearest neighbours, whilst excluding high-angle boundaries which correspond to grain boundaries and therefore are not related to the machining induced strain. This approach shows the variations in lattice distortions associated with plastic strain. Compared to Figure 4, the KAM method (Figure 7) appears to predict a shallower profile, with no deformation evident beyond a depth of 40 μm. Nevertheless, in both methods, no clear variation in deformation profile shape was observed. This suggests that the variations in RS seen in Figure 3b are more likely related to cooling rate, whereby high tensile stresses occur in the cutting direction when using HPC due to the quenching effect. It should be noted that grain size variations between the two locations shown in Figure 6 are also evident, presumably related to previous forming steps. Typically, an increase in grain size leads to reduced strengthening, as described by the Hall-Petch relation, resulting in more diffuse deformation extending into the bulk and greater variation in severe plastic deformation (SPD) layer thickness when turning Ni-based superalloys, which could have influenced the results [15], although more investigation would be required to confirm this.

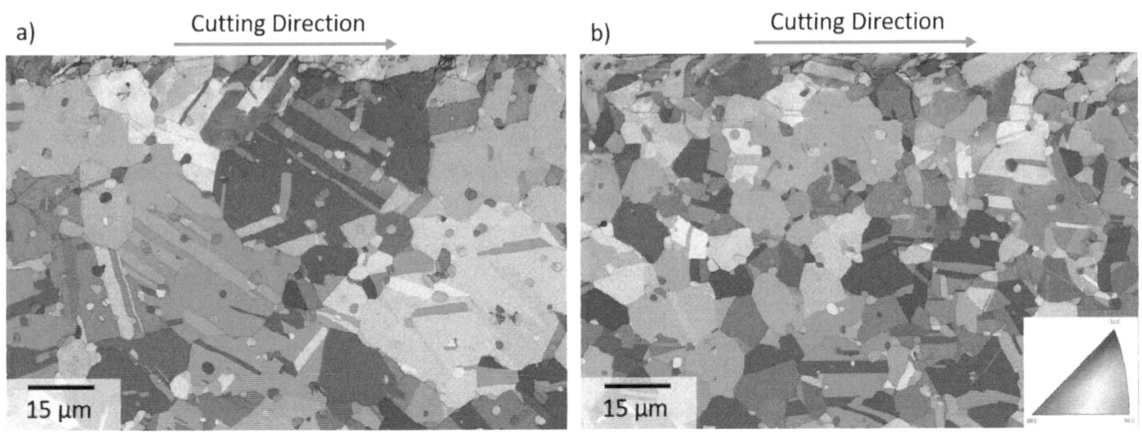

Figure 6: EBSD IPF-Z map showing cross-section of machined sub-surface viewed normal to the cutting direction: a) B-2 (300 m/min, 100 bar), b) B-3 (300 m/min, dry).

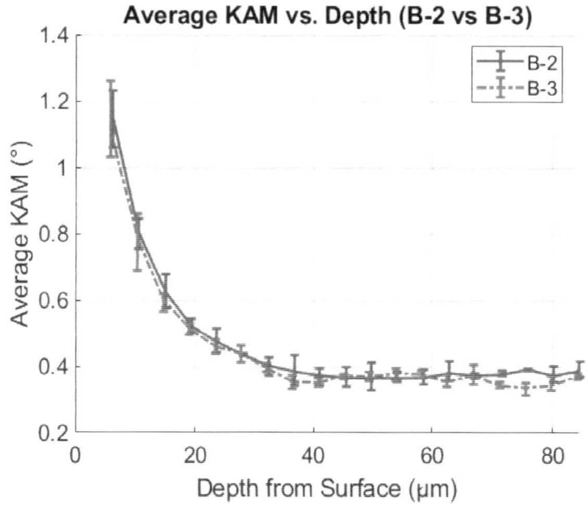

Figure 7: Average KAM against depth from surface for B-2 and B-3. EBSD maps were divided into 3 vertical bands; error bars show horizontal experimental spread.

Observing the same surfaces (B-2 and B-3) from above using secondary electron (SE) SEM (Figure 8), more surface defects—such as material smearing—are detected when machining dry. This is likely related to the lack of cooling, which leads to higher workpiece surface temperatures, making the surface more susceptible to material adhesion and damage. The influence of chip formation and chip control may also be a contributing factor, as the HPC coolant jet can assist in chip removal. Surface roughness measurements shown in Table 3 also confirm that increased R_z occurs when turning dry, which is indicative of surface defects. The lowest R_a values were measured when turning using an HPC pressure of 8 bar, and the highest when the coolant pressure was 50 bar. This highlights the need to control HPC pressure in order to achieve optimal surface quality.

Table 3 Surface roughness measurements performed on machined samples taken normal to the feed direction.

Sample	R_a	R_z	v_c (m/min)	Pres. (bar)
A-1	0.453	2.316	200	8
A-2	0.449	2.353	300	8
A-3	0.751	3.099	200	50
A-4	0.669	3.649	300	50
B-1	0.569	3.258	200	100
B-2	0.516	2.595	300	100
B-3	0.560	3.778	300	dry

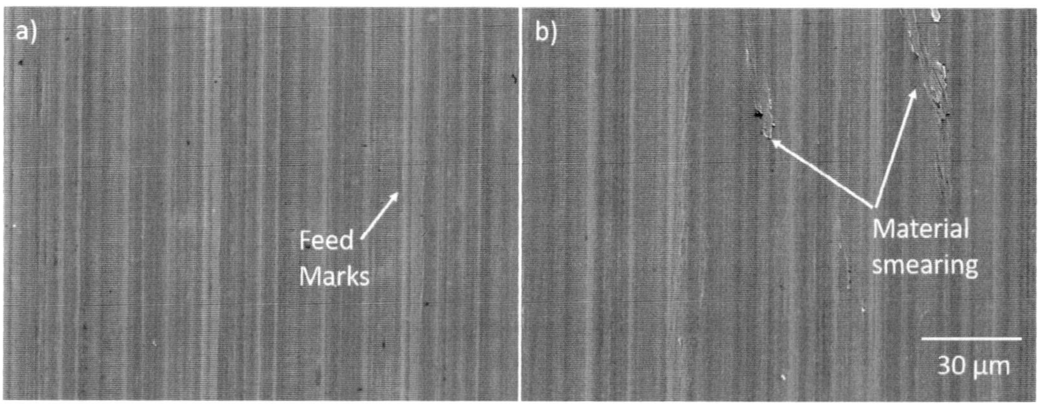

Figure 8: SE SEM micrograph showing surfaces produced: a) B-2 (300 m/min, 100 bar), b) B-3 (300 m/min, dry).

Conclusions

From the work presented, the following conclusions can be drawn:

- Higher HPC pressure appears to induce more tensile surface RS in the cutting direction. In addition, high coolant pressures (>50 bar) appear to induce a deeper compressive RS in the cutting direction when compared to lower pressure or dry conditions.
- Dry machining promotes material smearing but does not significantly affect the degree of sub-surface deformation. This suggests that the variations observed in surface tensile RS in the cutting direction under HPC are related to the quenching effect.
- Turning at the lower cutting speed of 200 m/min at 8 bar resulted in higher tensile RS in the cutting direction compared to 300 m/min at 8 bar. Conversely, the FWHM profiles indicate that higher levels of plastic strain were induced at the higher cutting speed. This shows that although higher cutting speeds promote more severe deformation and a greater defect density, the development of RS is also influenced by the overall contact time, which is longer at lower cutting speeds.
- HPC and cutting speed appear to have minimal impact on surface or subsurface RS in the feed direction.
- Optimal surface roughness was achieved when turning with a HPC pressure of 8 bar.

Further work is now required to ascertain the impact of these effects on functional performance in terms of fatigue life, to investigate how grain size influences workpiece deformation, and to perform more in-depth studies on SPD layers using higher-resolution microscopy techniques.

Acknowledgments

The authors would like to thank the NEXT competence centre (Vinnova 2023-00553) for financial support, as well as Lund University for performing all experiments, and Siemens Energy and SECO Tools for their technical support.

References

[1] A. Devaux, B. Picque, M. F. Gervais, E. Georges, T. Poulain, and P. Heritier, *AD730 TM - A New Nickel-Based Superalloy for High Temperature Engine Rotative Parts*. 2012, p. 919. doi: 10.7449/2012/Superalloys_2012_911_919.

[2] J. Zhou, Z. Chen, H. Persson, R. L. Peng, R. M'Saoubi, and D. Gustasson, 'Comparative assessment of the surface integrity of AD730 and IN718 superalloys in high-speed turning with a CBN tool', *J. Manuf. Mater. Process.*, vol. 3, no. 3, Sept. 2019, doi: 10.3390/JMMP3030073.

[3] H. Persson, V. Bushlya, L. Franca, J. Zhou, J. E. Ståhl, and F. Lenrick, 'Performance and wear mechanisms of different PcBN tools when machining superalloy AD730', *Ceram. Int.*, vol. 48, no. 16, pp. 22733–22742, Aug. 2022, doi: 10.1016/J.CERAMINT.2022.04.042.

[4] J. M. Zhou *et al.*, 'Surface Characterization of AD730TM Part Produced in High Speed Turning with CBN tool', *Procedia CIRP*, vol. 71, pp. 215–220, Jan. 2018, doi: 10.1016/j.procir.2018.05.068.

[5] H. Persson *et al.*, 'Modelling tool life in high speed machining of AD730®', *Procedia Manuf.*, vol. 25, pp. 316–321, Jan. 2018, doi: 10.1016/j.promfg.2018.06.089.

[6] R. M'Saoubi, D. Axinte, C. Herbert, M. Hardy, and P. Salmon, 'Surface integrity of nickel-based alloys subjected to severe plastic deformation by abusive drilling', *CIRP Ann.*, vol. 63, no. 1, pp. 61–64, Jan. 2014, doi: 10.1016/J.CIRP.2014.03.067.

[7] R. M'Saoubi *et al.*, 'Surface integrity analysis of machined Inconel 718 over multiple length scales', *CIRP Ann. - Manuf. Technol.*, vol. 61, no. 1, Art. no. 1, Jan. 2012, doi: 10.1016/j.cirp.2012.03.058.

[8] F. Jafarian, H. Amirabadi, and J. Sadri, 'Experimental measurement and optimization of tensile residual stress in turning process of Inconel718 superalloy', *Measurement*, vol. 63, pp. 1–10, Mar. 2015, doi: 10.1016/j.measurement.2014.11.021.

[9] J. C. Outeiro, J. C. Pina, R. M'Saoubi, F. Pusavec, and I. S. Jawahir, 'Analysis of residual stresses induced by dry turning of difficult-to-machine materials', *CIRP Ann.*, vol. 57, no. 1, pp. 77–80, 2008, doi: 10.1016/j.cirp.2008.03.076.

[10] A. R. C. Sharman, J. I. Hughes, and K. Ridgway, 'An analysis of the residual stresses generated in Inconel 718^{TM} when turning', *J. Mater. Process. Technol.*, vol. 173, no. 3, pp. 359–367, Apr. 2006, doi: 10.1016/j.jmatprotec.2005.12.007.

[11] A. Devillez, G. Le Coz, S. Dominiak, and D. Dudzinski, 'Dry machining of Inconel 718, workpiece surface integrity', *J. Mater. Process. Technol.*, vol. 211, no. 10, pp. 1590–1598, Oct. 2011, doi: 10.1016/j.jmatprotec.2011.04.011.

[12] R. M. Arunachalam, M. A. Mannan, and A. C. Spowage, 'Residual stress and surface roughness when facing age hardened Inconel 718 with CBN and ceramic cutting tools', *Int. J. Mach. Tools Manuf.*, vol. 44, no. 9, pp. 879–887, July 2004, doi: 10.1016/j.ijmachtools.2004.02.016.
[13] B. Denkena, A. Krödel, and L. Ellersiek, 'Influence of metal working fluid on chip formation and mechanical loads in orthogonal cutting', *Int. J. Adv. Manuf. Technol.*, vol. 118, no. 9, Art. no. 9, Feb. 2022, doi: 10.1007/S00170-021-08164-2.
[14] M. Brown, P. Crawforth, and D. Curtis, 'Rapid non-destructive sizing of microstructural surface integrity features using x-ray diffraction', *NDT E Int.*, vol. 131, p. 102682, Oct. 2022, doi: 10.1016/j.ndteint.2022.102682.
[15] H. R. Boyle, 'Tool Wear and Workpiece Surface Integrity during Ni-Based Superalloy Machining with Multi-layer CVD Alumina Coated Carbides', phd, University of Sheffield, 2025. Accessed: June 25, 2025. [Online]. Available: https://etheses.whiterose.ac.uk/id/eprint/36994/

ICRS 12: Proceedings from the 12th International Conference on Residual Stresses, 2025
October 20–23, 2025; Detroit, Michigan, USA
https://doi.org/10.31399/asm.cp.icrs2025p0065

Investigation on the residual stresses in a welded longitudinal gusset joint using computational weld mechanics and qualitative Barkhausen noise analysis

Riski, Jani, Pesonen, Tero, Ahola, Antti, Skriko, Tuomas, Björk, Timo
Steel Structures, LUT University, Lappeenranta, Finland
Responsible author: jani.riski@lut.fi

Skriko, Tuomas
Welding Technology, LUT University, Lappeenranta, Finland

Abstract

The effect of residual stresses on the fatigue performance of metallic materials has been recognised for decades, and they can be considered in fatigue assessments, for example, by conducting mean stress corrections. However, based on simulations and in-situ field measurements, there might be considerable variation in the residual stresses after fabrication, introducing uncertainty in the determination of fatigue life assessments. To fully utilise the fatigue capacity of welded structures, the uncertainty related to the residual stress state must be addressed. In specific fatigue-critical areas, unfavourable high tensile residual stresses can be diminished by using local post-treatments, such as thermal stress relief techniques and/or mechanical treatments. However, in larger, highly stressed weldments, post-weld treatments are neither practical nor cost-effective, creating a need for more accurate prediction or determination of residual stresses. In this study, the welding of a longitudinal gusset joint was simulated using finite element modelling, considering different mechanical boundary conditions and the geometrical stiffness of the joint. Both factors were found to affect the resulting residual stress distribution. Additionally, the suitability of the Barkhausen noise method for qualitatively verifying the residual stress state in welded structures was investigated.

Introduction

In the design of steel structures, fatigue is often a key design criterion. Several approaches can be used for fatigue assessment, but to ensure reliable and widely accepted outcomes, internationally validated methods supported by scientific societies are typically preferred. Such methods are provided, for example, by ISO and ASM standards, as well as IIW Recommendations[1, 2]. However, these standards and guidelines often allow designers to freely select an applicable method, with the choice depending on the specific detail under consideration and the desired level of accuracy. Generally, the simpler the structural detail, the simpler the applicable design methods can be applied for fatigue assessments.

For modern and next-generation structures, clients often impose requirements that did not exist when the traditional design guidelines were developed. These include, for instance, high load-carrying capacity, low self-weight, and large open spaces within the structure. In response, new materials, advanced manufacturing techniques, and highly developed design, analysis, and simulation tools have been introduced to the market. These developments enable more efficient and environmentally sustainable use of materials and structures throughout their entire life cycle, from manufacturing to recycling. However, as a consequence, the structural details to be designed have become increasingly complex, and traditional simplified methods can no longer provide sufficient accuracy. This has led to the emergence of new fatigue assessment methods, such as the 4R method [3–5], developed to meet the evolving needs of the industry.

This study focuses on residual stress, a fundamental parameter in the fatigue assessment of welded structural details. The existence of residual stresses and their contribution to fatigue have been recognised for a long time. Their influence has traditionally been accounted for in fatigue design, for example, by correcting the mean stress level.

However, increasing demands for improved fatigue performance of individual details have also highlighted the importance of considering individual parameters in greater depth. For instance, weld toe geometry and the residual stress state can be specified as manufacturing requirements, thereby avoiding the need for additional post-treatment operations [6].

Due to the limitations of residual stress measurement techniques, verifying the residual stress state in manufactured engineering components remains a challenging task. Typically, verification is based on characterised reference measurements, indirect methods such as strip test or deformation analysis, or on random control measurements. However, measurements performed on actual ship structures as part of this study demonstrated that residual stress can vary by nearly twice the material's yield strength. Moreover, simple welding simulations conducted in the study, where only the support conditions during welding were modified, showed that significant variations in the residual stress field can be induced with minimal changes to boundary conditions. From a fatigue assessment perspective, such variation in residual stress is significant, leading to a situation where beneficial residual stress states cannot be fully utilised due to uncertainty. Based on the findings of this study and previously made research[7, 8], the Barkhausen noise method shows promise as a qualitative measurement technique for evaluating residual stress. The suitability of the method for monitoring residual stress quality, similarly to weld toe geometry or welding parameters, should be further investigated in future research.

Fatigue assessment applying finite element analysis

4R method

The 4R method is an effective stress method for fatigue assessment of structural components. It was developed at LUT University around the early 2010s by Dr. Timo Nykänen and Professor Timo Björk. The method was originally designed to address the specific fatigue behaviour observed in welded high-strength steel joints failing at the weld toe. In those joints, high fatigue performance was observed, and the conventional effective stress methods failed to capture increased fatigue strength adequately [3, 4]. Shortly after its development, the 4R method was also found to apply to components with cut edges [9]. Since then, significant developments have been conducted around the method, including investigations into its applicability to cut edges, additively manufactured components and root side failure, as well as its generic applicability across various cases and sensitivity analysis, for example.

In addition to the research conducted at LUT, the 4R method has attracted international interest. Results obtained using this method have been frequently compared with those of established approaches, often with promising outcomes. For instance, Gkatzogiannis et al. [10] applied the method in combination with simulated HFMI treatment, obtaining results that were in good agreement with experimental findings. Furthermore, Brömer et al. [11] applied the method using an actual scanned geometry. Their results indicated that, with minor adaptations, the method can be effectively applied to scanned geometries, while also allowing for a reliable consideration of different residual stress states in the analysis. Shojai et al. [12] have also mentioned the 4R method in the conclusions of their research as one possibility for investigating structures damaged by corrosion.

The fundamental principle of the 4R method is to evaluate elastic-plastic material behaviour at the notch root. The method is based on estimating the local elastic stress range induced by external loading, which can be calculated using approaches such as the effective notch stress (ENS) method. This elastic stress range is then converted into the corresponding local elastic-plastic material response using the Ramberg–Osgood material model with the kinematic hardening rule, in combination with Neuber's rule (i.e., Neuber's hyperbola). From maximum and minimum stresses, the local stress ratio (R_{local}) can be defined. The fatigue life is finally assessed using the 4R master *S–N* curve, which has been determined using a large set of experimental data. However, unlike the often-applied mean stress calculations by the Smith–Watson–Topper (SWT) approach, which assumes that $\sigma_{max}\varepsilon_a$ = constant, the 4R method applies the material's response under the calculated local stress ratio R_{local}, as shown in the equation given in Figure 1. [4]

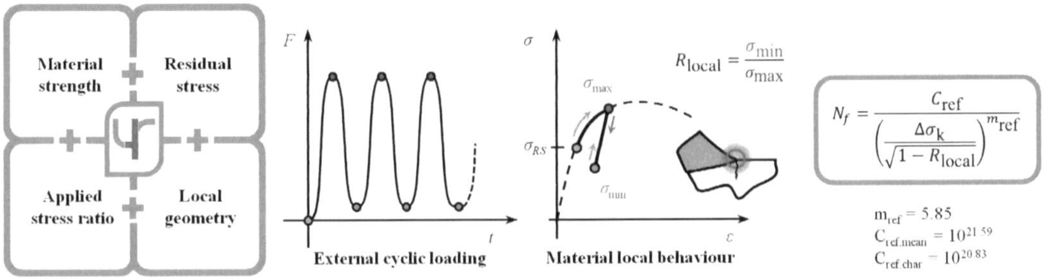

Figure 1: Basic calculation principle of the 4R method. (Modified from LUT Steel Structures group presentation)

In addition to considering stresses due to external loading, the 4R method also accounts for the effect of residual stresses within the component. However, the residual stress state in real-life structures is not always known. In such cases, the analysis can utilise estimated residual stress values for specific detail types, which have been shown to yield realistic fatigue life predictions [5]. As illustrated in Figure 2, the magnitude of residual stress significantly affects the local material behaviour according to the principles of the 4R method. This difference in behaviour has a direct computational impact on the predicted fatigue life. Nevertheless, under variable amplitude (VA) loading, occasional overloads have been observed to improve fatigue performance. This effect can likewise be explained through the material's local response characteristics, as captured by the 4R methodology. [13]

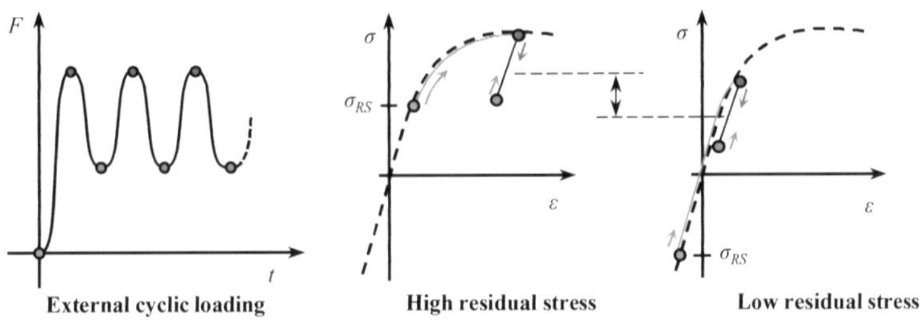

Figure 2: A schematic representation of the effect of residual stress state on the local behaviour of a material.

Welding simulation

The effect of varying boundary conditions on the resulting residual stress field was investigated by simulating the welding of a longitudinal stiffener using Simufact Welding 2023.1 software. Sequentially coupled analysis, utilising a moving Goldak heat source and temperature-dependent material parameters, was employed. As a base material, S690QL steel was used, and the stiffener was modelled using S355J2G3 steel, with thicknesses of 15 mm and 10 mm, respectively. Material parameters for both steels were obtained from the software's internal material library. In the simulation, a 100 mm long and 50 mm high stiffener was welded at the centre of a 70 mm wide base plate. The weld started and ended at the midpoint of the stiffener's long edge. A schematic of the welded configuration is presented in Figure 3.

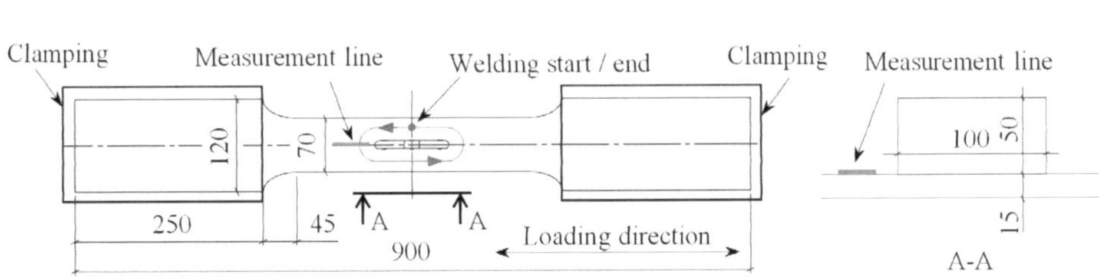

Figure 3: Dimensions of the specimen used in welding simulation. In the experimental tests, a gusset was also added on the opposite side of the base plate.

During welding, the base plate was supported at both gusset ends using four different boundary conditions: entirely free, clamped between plates, fully fixed, and pre-stressed with an axial membrane stress of 200 MPa. In the constrained boundary condition cases, the supports were released once the material had cooled to approximately 30 °C. Residual stresses were extracted from the surface of the base plate, at a distance of 1 mm in front of the stiffener weld toe. Based on the results (Figure 4), it can be observed that the increased stiffness (free – clamped – fully fixed – pre-stressed) of the boundary conditions during welding reduces the magnitude of the residual stress peak at the front of the stiffener. Conversely, this reduction in the tensile peak leads to lower compressive residual stresses near the plate edges, as required by the equilibrium of residual stresses across the plate cross-section.

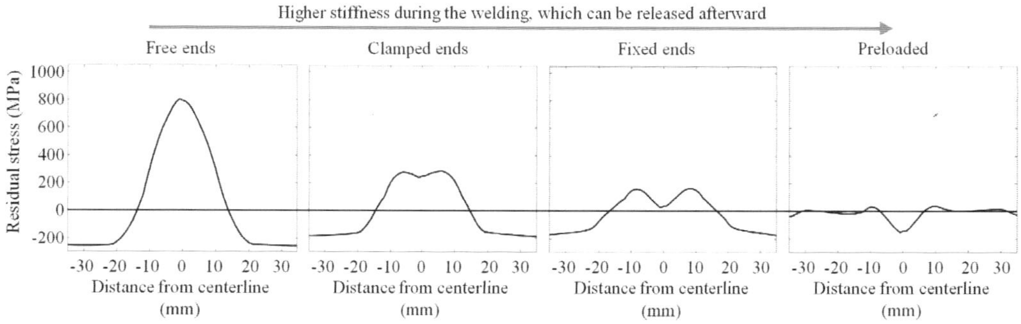

Figure 4: Residual stress distribution determined by welding simulation. Stresses are read in the transverse direction from the surface of the base plate, at a distance of 1 mm in front of the stiffener weld toe.

Experimental tests

Barkhausen measurements under laboratory conditions

As part of the study, Barkhausen noise measurements were conducted under laboratory conditions. In the measurements, a double-symmetric gusset joint geometry, previously used in welding simulations, was employed. Three separate test specimens were prepared, each designed to produce a distinct residual stress state. The first specimen was left in the as-welded condition (AW). In the second specimen, the weld toes were treated using high-frequency mechanical impact (HFMI) with a HiFIT device equipped, applying a 4 mm diameter treatment tool. In the third specimen, tensile pre-stressing (PS) of the base plate was used during welding by supporting the specimen at its centre and bending both ends downward. After welding, the specimen was released from the fixtures.

The measurements were carried out manually using a Rollscan 350 signal analyser in combination with a dual-coil custom sensor (Figure 5), in which the same coils were used for both magnetising the material and detecting noise. The primary advantage of this specialised sensor configuration was the ability to position the sensor and obtain measurements as near the weld toe as possible.

The measurements were performed by placing the sensor contact surfaces both longitudinally and transversely on the specimen surface, aligned with the direction of the load (Figure 1). All measurements were performed using consistent settings, based on values recommended by the equipment manufacturer. The influence of device parameters on the results was not investigated in this context. In addition to Barkhausen noise measurements, residual stresses were determined from the surface of all test specimens in the loading direction. The stresses were determined using a Stresstech Xstress G3 diffractometer in combination with the Xstress 3000 main unit. The Xtronic software was employed for both instrument control and data processing.

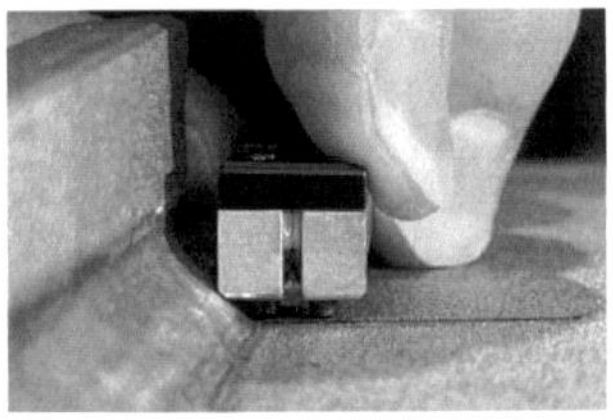

Figure 5: The specialised two-coil Barkhausen noise sensor positioned as close as possible to the weld toe. The distance between coils is about 3 mm, and the measurement direction in the picture is longitudinal.

Before the measurements, the gusset tip areas were prepared by verifying the flatness of the plate surface and removing any weld spatter that could interfere with the measurement. This procedure was intended to ensure proper contact between the Barkhausen noise sensor and the specimen surface. Both X-ray diffraction and Barkhausen noise measurements were conducted up to a distance of 25 mm from the weld toe, as illustrated by the distributions presented in Figure 6.

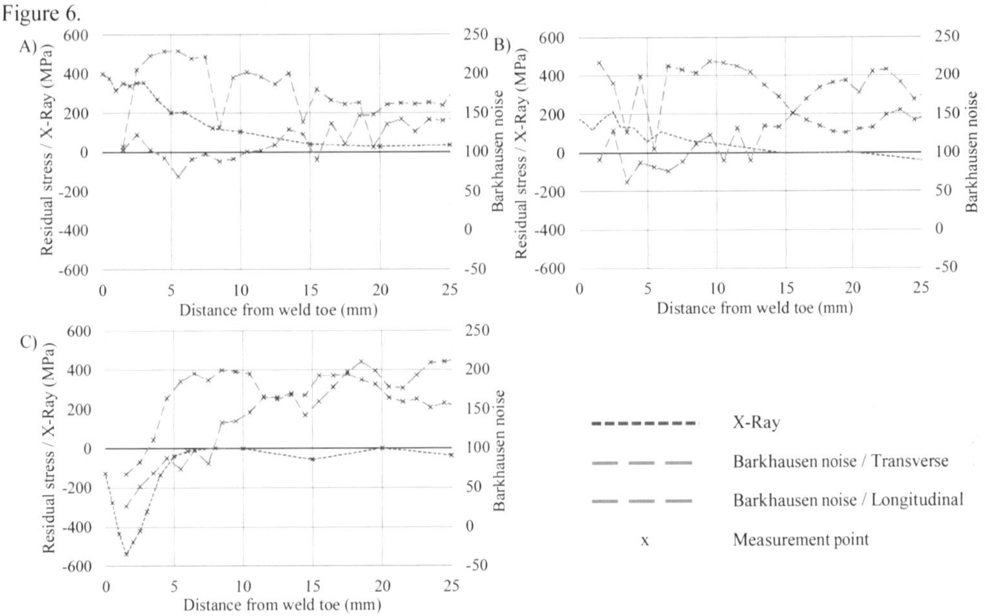

Figure 6: Barkhausen noise measurement for the tip of the gusset. Specimen in A) As-welded, B) HFMI treated, C) Pre-tensioned condition.

The X-ray diffraction measurements revealed differences in the residual stress states near the weld toe between the three specimens. In the AW condition, the residual stresses at the weld toe reach approximately 400 MPa in tension. In the pre-stressed specimen, the residual stresses are lower, with values reaching a maximum of roughly 200 MPa in tension. This difference persists up to around 5 mm from the weld toe. In the HFMI-treated specimen, compressive residual stresses are defined near the weld toe, extending approximately 5 mm, after which the stress gradually levels off to the level of the base material. In contrast, in the as-welded specimen without treatment, the stress returns to the base material level only at approximately 15 mm from the weld toe.

The results of the Barkhausen noise measurements demonstrate a clear correlation with the longitudinal residual stress values determined by X-ray diffraction: higher residual stresses are associated with increased Barkhausen responses, while lower stresses are associated with lower readings. However, in both the AW and pre-stressed specimens, the measurements exhibit noticeable fluctuation between some adjacent points. In contrast, the HFMI-treated specimen displays relatively consistent values throughout the measured region.

The results of the transverse Barkhausen measurements do not appear to correlate with the longitudinal residual stresses measured by X-ray diffraction. Nevertheless, a distinct change in the Barkhausen response is observed in all specimens at approximately 15 mm from the weld toe. Due to the specimen's geometry and accessibility, transverse residual stresses could not be determined using X-ray diffraction. However, the Barkhausen results appear to show a rough correlation with the transverse residual stress distribution predicted by numerical simulation.

As part of the Barkhausen measurements, an experiment was conducted in which the base material surface was measured using a Barkhausen noise sensor at a distance of 50 mm from the gusset, aligned in the longitudinal direction. During the measurement, the nominal stress was gradually increased from the unloaded condition to 190 MPa. The results of this experiment are presented in Figure 7. A linear correlation between the Barkhausen response and the applied stress level can be observed. However, it is notable that up to 100 MPa, the Barkhausen signal increases by approximately 50 units per 100 MPa, while in the range of 100–190 MPa, the rate of increase is reduced to around 30 units per 100 MPa.

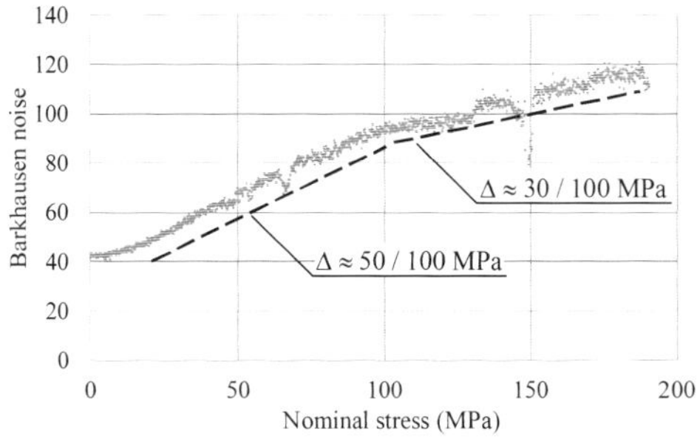

Figure 7: Barkhausen noise measurement from the base material in the longitudinal direction during the tensile loading cycle from 0 to 190 MPa.

X-ray measurements at the shipyard

As part of the study, in-situ residual stress measurements were carried out at a shipyard. The measurements focused on the pre-selected critical details on the plate surface, specifically the weld toe and cut edge, which were identified in advance using finite element analysis. These details were fabricated from either NV A36 or NV D690QT steel plates, corresponding to the yield strength classes of 355 MPa and 690 MPa, respectively. In selecting the measurement locations, particular attention was given to the ability to repeat the measurements at different stages of the production process: the bulkhead, block, and grand block stages. No local post-weld treatments, such as grinding or sandblasting, that could influence the residual stresses locally were applied to the welds or details between the stages. Therefore, any changes in the residual stress state could be attributed purely to global deformations and stress variations induced by the successive production phases. The measurements were performed using the same X-ray equipment as for the Barkhausen measurements.

The results of residual stress measurement for a cut edge and a butt joint are presented in Figure 8. The materials correspond to steel grades with yield strengths of 690 MPa and 355 MPa and material thicknesses of 25 mm and 16 mm, respectively. The results indicate that the residual stress state undergoes significant changes during the production process. In both cases, the stresses measured during the first and second stages (bulkhead and block phases) remain relatively constant, either in tension or compression. However, during the third measurement stage (grand block phase), the stress state shifts noticeably, from compression to tension or vice versa.

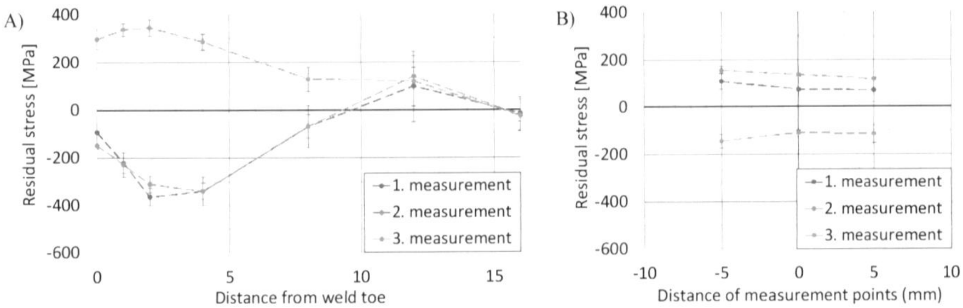

Figure 8: Results of residual stress measurements at the shipyard at different stages of production (measurements 1-3). Locations: A) Butt joint in the plate field. B) Plate surface at the 5 mm tangential distance from the cut edge.

For the cut edge, the change in residual stress is approximately 300 MPa, whereas for the butt joint, the variation reaches up to 650 MPa. In the butt joint, this stress transition is distinctly observable up to about 12 mm distance from the weld, corresponding 0.75 times the material thickness.

Due to production-related constraints, the number of joints accessible for residual stress measurements remained limited and was primarily restricted to the same welded details monitored throughout the manufacturing process. However, during the grand block phase, additional measurements were conducted on comparable butt joint details, some of which had also been included in earlier measurement stages. The results are presented in Figure 9.

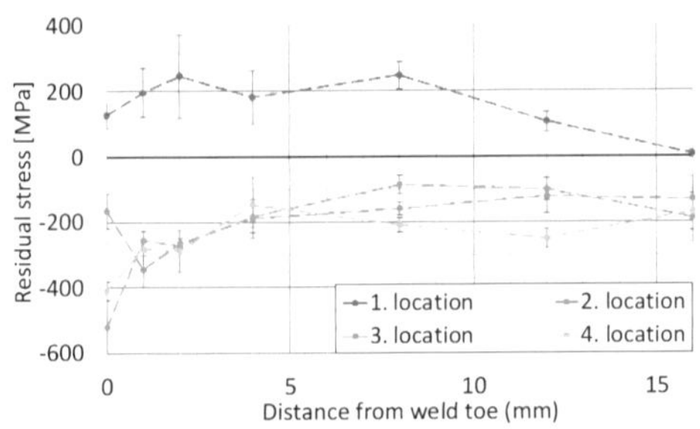

Figure 9: Measurement results of residual stress measurements at the yard at different stages of production.

The data indicate that the residual stress distribution is not uniform throughout the structure but varies according to the spatial location of the detail. The magnitude of this variation is comparable to that observed between different production phases, as shown in Figure 7.

Discussion

Since the development of the 4R method, its applicability and robustness have been extensively studied from various perspectives, thereby confirming and expanding its application domain. In terms of residual stresses, accurate input values, either measured experimentally or derived from calculations, can be employed in fatigue life assessments. However, in real structures, residual stresses can vary significantly, and measuring a single detail does not necessarily provide an unambiguous representation of the overall stress state. This was demonstrated in the measurements performed at the shipyard, where residual stresses were found to vary both within the same structural detail at different stages of production and between similar details produced under nominally identical conditions. In all cases, the magnitude of stress variation was nearly twice the yield strength of the material, which has a notable impact on fatigue life predictions. Compressive residual stresses are generally considered highly beneficial for fatigue life, whereas tensile residual stresses represent a less favourable condition. However, when the actual service loading is taken into account, an initial tensile stress state does not necessarily lead to significantly lower fatigue performance [13].

In fatigue life assessment, residual stress is often considered by assuming or predefining a representative residual stress level for a given detail. For welded joints in the as-welded condition, this typically involves assuming a high level of tensile residual stress, which is already accounted for in conventional fatigue design curves (e.g., IIW Recommendations [1]). However, the fatigue performance of a joint can be improved when necessary by employing various fabrication-related techniques that directly or indirectly influence the residual stress state. Such methods include the use of low transformation temperature (LTT) welding consumables [14], weld toe treatments [15], and post-weld treatments applied to the entire joint region, such as peening [16, 17]. A common feature of these methods is that their beneficial influence on the residual stress state does not necessarily need to be verified explicitly, as a properly executed treatment is considered sufficient. Nevertheless, to ensure proper execution, treatments must be performed following specified guidelines. In addition, verification may involve visual inspections, calibration procedures, and test runs such as the Almen strip tests. [6, 15, 16]

Through the simplified welding simulations conducted in this study, it was demonstrated that the residual stress level can vary by more than the yield strength solely as a result of changes in boundary conditions during welding. These findings are consistent with the residual stress measurements performed on site at the shipyard, which showed that a similar level of scatter exists in real structures. From a production quality perspective, however, considerable variability in any parameter is undesirable, as it can lead to overdesign, unnecessary post-treatments, and increased manufacturing costs (ref.). With X-ray, the numerical verification of residual stress states is possible as part of the production process. However, the method is neither straightforward, rapid, nor easily integrated into manufacturing workflows [18], meaning that practical solutions are often case-specific and tailored to the situation. Additionally, such verification typically requires an extra production step, resulting in increased time and cost.

Based on the Barkhausen noise measurements conducted in this study, as well as prior research, the method has demonstrated potential for determining residual stresses in welded joints, while also offering a high degree of flexibility in terms of availability, portability, and workability. However, compared to the X-ray, BN has proven challenging to extract accurate numerical stress values due to the large number of influencing parameters [18], although with careful calibration to the measurement conditions, the Barkhausen noise method has also been applied to obtain numerical values, observing a good correlation between the Barkhausen noise signal and residual stress levels in welded structures [8, 19]. In the Barkhausen noise measurements carried out as part of the study, the instrument was not calibrated separately for the material being measured. Nevertheless, the measurement results can be found to correlate quite well with the X-ray results, indicating the suitability of the method for qualitative use even without calibration.

Conclusions

This study focused on evaluating the applicability of the Barkhausen noise method for the qualitative assessment of residual stress fields. The experimental investigations focused on a welded longitudinally double-symmetric gusset joint prepared under laboratory conditions, where the residual stress state at the gusset end toe was modified using different fabrication techniques. The resulting stress states were examined using both X-ray diffraction and Barkhausen noise methods. Additionally, welding simulations were conducted to investigate the effect of fixture stiffness on the residual stresses in the test specimen. As part of the study, X-ray diffraction measurements were also carried out on site at the Meyer shipyard in Turku.

Based on the results of the present study, the following conclusions can be drawn:

- Existing computational tools enable the prediction of residual stress fields in a structure before fabrication. However, the use of calculation and simulation as part of production often relies on considering deformation during the manufacturing process.
- The actual residual stress state is established only after the final manufacturing step and can vary significantly. A reliable understanding of the real stress state can only be achieved through on-site measurement.
- In field conditions, the use of X-ray diffraction for stress measurements is often laborious, time-consuming, and not always feasible due to accessibility limitations. Barkhausen noise measurement offers notable advantages in terms of accessibility and speed.
- Based on both the literature review and laboratory measurements, the Barkhausen noise method is suitable for assessing residual stress fields. Nevertheless, its application is limited by technical and material-specific limitations that must be considered during both measurement and interpretation of results.

- The Barkhausen noise method shows strong potential, particularly for qualitative identification of residual stress states. Such a qualitative assessment may serve as a valuable tool for production quality control, especially when residual stresses are expected to be in a specific range after fabrication.

Acknowledgments

The research is funded by the Business Finland project: Carbon-neutral lightweight ship structures using advanced design, production, and life-cycle services (CaNeLis). The authors express their gratitude to Business Finland for the funding. The researchers also thank Stresstech Oy for the cooperation in the measurements.

Author contribution

J. Riski: conceptualisation, investigation, formal analysis, methodology, validation and writing – original draft. T. Pesonen: investigation, methodology, validation, and visualisation. A. Ahola: conceptualisation, writing, review and editing. T. Skriko: conceptualisation, review, editing, and supervision. T. Björk: conceptualisation, supervision, and funding acquisition.

Declaration of generative AI in scientific writing

Copilot, ChatGPT, DeepL, and Grammarly were used to improve the readability and language of the work. All the edits were reviewed and made by the authors as necessary. Authors take full responsibility for the content of the publication.

Declaration of competing interest

The authors are not aware of any competing financial interests or personal relationships that could have influenced the work reported in this paper.

References

[1] Hobbacher AF, Baumgartner J (2024) Recommendations for Fatigue Design of Welded Joints and Components. https://doi.org/10.1007/978-3-031-57667-6

[2] European committee for standardization (2025) EN 1993-1-9:2025, Eurocode 3: Design of steel structures. Part 1-9: Fatigue

[3] Nykänen T, Björk T (2015) Assessment of fatigue strength of steel butt-welded joints in as-welded condition – Alternative approaches for curve fitting and mean stress effect analysis. Marine Structures 44:288–310. https://doi.org/10.1016/J.MARSTRUC.2015.09.005

[4] Nykänen T, Björk T (2016) A new proposal for assessment of the fatigue strength of steel butt-welded joints improved by peening (HFMI) under constant amplitude tensile loading. Fatigue Fract Eng Mater Struct 39:566–582. https://doi.org/10.1111/FFE.12377

[5] Nykänen T, Mettänen H, Björk T, Ahola A (2016) Fatigue assessment of welded joints under variable amplitude loading using a novel notch stress approach. https://doi.org/10.1016/j.ijfatigue.2016.12.031

[6] Marquis GB, Barsoum Z, Leitner M (2025) New developments and guideline updates for HFMI treatment for improving the fatigue strength of welded joints. Welding in the World. https://doi.org/10.1007/S40194-024-01882-7

[7] Nitschke-Pagel T (2005) A strategy for life prediction of fatigue loaded welded constructions by use of continuously measured micromagnetic parameters. Materials Science Forum 490–491:409–417. https://doi.org/10.4028/0-87849-969-5.409

[8] Nitschke-Pagel T, Dilger K, Eslami-Chalandar H (2011) Determination of residual stresses in welded structural steels with help of micromagnetic measurements. Materials Science Forum 681:194–201. https://doi.org/10.4028/WWW.SCIENTIFIC.NET/MSF.681.194

[9] Peippo J, Björk T, Nykänen T (2018) A novel method for fatigue assessment of steel plates with thermally cut edges. Welding in the World 62:105–115. https://doi.org/10.1007/s40194-017-0529-7

[10] Gkatzogiannis S, Schubnell J, Knoedel P, et al (2024) Estimating the fatigue strength of HFMI-treated weldments by means of numerical analyses. J Constr Steel Res 218:108675. https://doi.org/10.1016/j.jcsr.2024.108675

[11] Brömer T, Widerspan V, Shojai S, Ghafoori E (2025) Enhanced local fatigue approach for welded tubular joints using 3D digital scans and implicit gradient model. Int J Fatigue 199:109060. https://doi.org/10.1016/J.IJFATIGUE.2025.109060

[12] Shojai S, Schönamsgruber F, Köhler M, Ghafoori E (2025) Impact of accelerated corrosion on weld geometry, hardness and residual stresses of offshore steel joints over time. Mater Des 251:113578. https://doi.org/10.1016/J.MATDES.2024.113578

[13] Grönlund K, Ahola · Antti, Riski J, et al (2024) Overload and variable amplitude load effects on the fatigue strength of welded joints. Welding in the World 68:411–425. https://doi.org/10.1007/s40194-023-01642-z

[14] Hübner M, Dittmann F, Kromm A, et al (2025) Residual stress reduction using a low transformation temperature welding consumable with focus on the weld geometry. Welding in the World. https://doi.org/10.1007/S40194-025-02094-3

[15] Marquis GB, Barsoum Z (2016) IIW Recommendations for the HFMI Treatment. Springer Singapore, Singapore

[16] Hensel J, Eslami H, Nitschke-Pagel T, Dilger K (2019) Fatigue Strength Enhancement of Butt Welds by Means of Shot Peening and Clean Blasting. Metals 2019, Vol 9, Page 744 9:744. https://doi.org/10.3390/MET9070744

[17] Ahola A, Lipiäinen K, Riski J, et al (2023) Fatigue strength of shot-peened as-welded joints and post-weld-treated and subsequently clean-blasted fillet weld joints. Welding in the World 67:1559–1574. https://doi.org/10.1007/S40194-023-01498-3/TABLES/8

[18] Lachmann C, Nitschke-Pagel T, Wohlfahrt H (2000) Characterization of residual stress relaxation in fatigue loaded welded joints by X-ray diffraction and Barkhausen noise method. Materials Science Forum 347:374–379. https://doi.org/10.4028/www.scientific.net/msf.347-349.374

[19] Górka J, Przybyła M (2024) Research on the Influence of HMFI and PWHT Treatments on the Properties and Stress States of MAG-Welded S690QL Steel Joints. Materials 2024, Vol 17, Page 3560 17:3560. https://doi.org/10.3390/MA17143560

ICRS 12: Proceedings from the 12th International Conference on Residual Stresses, 2025
October 20–23, 2025; Detroit, Michigan, USA
https://doi.org/10.31399/asm.cp.icrs2025p0075

Residual Stress Development in Cold Spray Coatings: Insights from Curvature Measurements and Finite Element Analysis

Alfredo Valarezo, Westly Castro
Universidad San Francisco de Quito, Quito, Pichincha, Ecuador
avalarezo@usfq.edu.ec

Sanjay Sampath
Stony Brook University, Stony Brook, New York, USA
sanjay.sampath@stonybrook.edu.ec

Abstract

Cold spray is a solid-state deposition technique in which particles adhere to a substrate through severe plastic deformation at relatively low temperatures, typically resulting in compressive residual stresses. However, recent in-situ curvature measurements have revealed unexpected instances of tensile residual stress development in cold-sprayed coatings. This study aims to quantify and elucidate the evolution of residual stresses in cold spray deposits through finite element modeling. Single-particle impacts of copper and AISI 316 stainless steel onto similar substrate materials are simulated under comparable process conditions to systematically investigate the interplay between impact dynamics, layer buildup, and post-impact thermal contraction. The comparison of materials with differing mechanical behavior highlights the influence of strength, strain-hardening characteristics, and strain-rate sensitivity on peening-induced compressive stresses, as well as the role of thermal mismatch in generating quenching-induced tensile stresses. Microstructural analyses support the numerical predictions, providing additional insights into the mechanisms driving stress evolution. By drawing parallels with residual stress profiles observed in conventional thermal spray processes, this work proposes a set of hypotheses to explain the emergence of tensile stresses in cold spray coatings.

Introduction

Cold gas spray is a solid-state coating process used to deposit metal powders by impact of particles at high velocities without melting. The coating material in powder form is injected into a high pressure and medium temperature supersonic gas jet and directed onto a substrate. Since the coating particles are propelled at relatively low temperatures (significantly lower than the melting point), the bond process with the substrate is achieved by plastic deformation upon impact. Depending on the manufacturing parameters, the resulting coating can exhibit a dense well-adhered distribution [1], [2]. The conventional thermal spray process relies on the high temperature reached by the coating particles; however, this fact leads to problems associated with oxidation, recrystallization, phase transformations and residual stresses. On the other hand, cold spray emerges as a solution to the mentioned drawbacks based on its low-temperature coating particle operation yielding a coating on compressive residual stresses due to the peening effect[2], [3].

In high-temperature thermal spray process (plasma spraying, arc spraying), the coating particles are heated above melting point before being propelled onto the substrate. When the particles reach the substrate, the solidification and subsequent cooling reduces the particle volume. This contraction is limited by the adherence with the substrate resulting in the generation of quenching stresses that are tensile in nature. In contrast, high-velocity thermal spray systems (High-Velocity Oxy-Fuel, High-Velocity Air-fuel, Cold spray) introduce compressive stresses (known as peening stresses) generated by the high kinetic energy of the particle at the onset of the impact. After the coating as been deposited, the bonded material pair (substrate and deposited layers) undergoes cooling to room temperature which gives rise to the appearance of thermal stresses based in the mismatch between the coefficient of thermal expansion of the substrate and the coating (ΔCTE) [2], [4]. These thermal stresses could be tensile, compressive or even neutral in nature depending on the magnitude of ΔCTE, and the temperature gradient (ΔT). These three kinds of stresses are always present in thermal spray coatings; however, the magnitude of each one is variable and strongly related to the selection of manufacturing parameters and thermal spray process.

In the cold spraying technique, the coating particles are accelerated by a gas stream until they reach velocities between 500-1000 m/s. The high velocity of the particles produces plastic deformation onto the substrate during the impact due to shear adiabatic instabilities forming a bond between them. The nature of the impact generates a localized temperature increase due to the conversion of kinetic energy from the particle into plastic deformation (particle and substrate) and heat. It is expected that the temperature increase does not significantly affect the substrate in zones far from the impact; thus, with the peening effect being the primary source of residual stresses [5]. This compressive stress presents significant advantages in the coating performance, especially those related to fatigue life, crack initiation and adhesion strength [2].

Although cold spray coatings are expected to exhibit dominant compressive residual stress due to the peening effect, recent in-situ curvature measurements (In situ Coating Properties sensor, ICP) have revealed cases where unexpected tensile residual stresses appeared through the coating. In order to understand the complex interactions that could originate these unanticipated tensile stresses, this study investigates residual stress evolution in cold spray coatings using a finite element modeling approach with single and multiple particles impact using Cu, and SS316L materials for reference [3], [5], [6].

Finite element models have been employed to investigate the development of layer growth and residual stress evolution of cold sprayed coatings under a broad range of complex conditions: different materials, contact interaction, as-sprayed stress profiles, residual post cooling stresses, temperatures profiles, energy transformation on impact events and the strain-rate behavior of materials. For the current work, an explicit-implicit FE approach is used [7]. Cu and SS316L particles are propelled into different substrates under varying process parameters using an explicit FE analysis to obtain insights about the correlation between impact dynamics and layer buildup. The resulting stresses, temperature and deformation profiles throughout the coating-substrate configuration are used as input in an implicit FE analysis to correlate the information of post-deposition cooling and the development of residual stresses. Moreover, microstructural evaluations are used to corroborate the observed stress states and modeling [3],[8]. With these comparisons, the study presents well supported hypotheses regarding the origin of unusual tensile residual stresses in cold spray coatings. The findings contribute to a deeper understanding of residual stress development in high-velocity deposition methods and offer guidance for optimizing process parameters to achieve desired stress states.

Materials and Methods

Finite Element (FE) Model

The final residual stress state of a cold spray coating is the result of the overlapping of highly dynamic effects. The peening stresses generated from the high velocity impact of coating particles onto the substrate causing plastic deformations, the quenching stresses that arise from the solidification and cooling of individual splats and the thermal stresses that are induced by thermal expansion mismatch between coating and substrate during the cooling down process after deposition. Additionally, the repeated deposition of coating layers also contributes to the final stress profile by superposition.

To investigate the correlation of these complex dynamic events, a hybrid explicit-implicit FE model was implemented using Abaqus, as similarly conducted previously by [3], [7]. Initially, a high-velocity single particle impact over a substrate was simulated using an explicit approach. Abaqus/Explicit module performs a direct integration of motion equations (position, velocity, forces) in the current time to obtain the kinematic state of the next small-time increment. The direct calculation scheme allows Abaqus/Explicit to avoid equilibrium iterations and makes it well-suited for highly dynamic nonlinear problems in small periods of time [9]. In the present study, the explicit simulations yield the results of peening stresses, temperature and deformation states just after the particle impact. The nature of the explicit FE formulation made it adequate for short-time-scale phenomena; however, it is not recommended for processes that involve long-time spans like effective heat transfer; for this reason, the resulting information from the impact models are used as input for subsequent implicit FE models. Abaqus/Implicit offers capabilities to solve static or slow evolving problems by using an iterative approach that involves the equilibrium of the present and future state of a system [9] . An implicit FE analysis is used to quantify the evolution of stresses (quenching and thermal stresses) after an effective heat transfer phenomenon has taken place defining the final residual stress profile of the particle–substrate system.

Explicit Impact Analysis (First and Second impact)

The model consists of a two-dimensional axisymmetric spherical particle that impacts onto a cylindrical substrate. The entire impact event occurs in a time span of 400 ns. This time is assumed to be small enough to avoid the interaction between particles, making an independent single particle impact statistically representative[3]. Two models were simulated: a stainless steel (SS316) particle was impacted onto the SS316 substrate; on the other hand, a copper particle was propelled into a Cu substrate, independently. The cylindrical substrate has a radius and height of 500 μm while particles were used with a diameter that varied depending on the material. The particle diameter was fixed at 40 μm and 20 μmor SS316 and copper, respectively. The substrate dimensions were chosen in order to avoid edge effects that could interact with the development of stresses due to impact. Nodes located at the bottom of the substrate were fixed under the ENCASTRE boundary conditions that disabled their movement and rotation in every direction. Particles and substrates were meshed using quadrangular linear elements.

The velocity with which particles approached perpendicular to the substrate was 800 m/s. The substrate temperature, in all simulations, was defined by a predefined field with a value of 398K just before the impact. With respect to the particle, the temperature was fixed at a value of 698K; this value is below the melting point of all the materials involved and ensured that the impact and deformation process of the particle occurred in solid state.

The inelastic heat fraction is one of the key parameters that define the contact interaction between particle and substrate after the impact. This value represents the amount of plastic work (inelastic dissipation) that is converted into heat during plastic deformation; it is important in simulations with high-speed and large amounts of inelastic strain scenarios where the properties of materials are temperature dependent. In this study, this value was set at 0.9 in all simulations meaning that 90% of the plastic work was converted into heat. Other important parameters of contact interaction are normal and tangential behavior. In the simulations, the normal behavior was defined as a "hard-contact" that involved a no-penetration condition between surfaces (contact pressure could rise to any value to reinforce this constraint); additionally, it was not allowed separation of contact surfaces to avoid debonding phenomena. The tangential behavior was formulated with a friction penalty function with a friction coefficient of 0.3 that is a common value between metals. Finally, the simulations considered that all frictional work was transformed into heat and this heat was distributed in the particle and the substrate. All of the above mentioned parameters ensured that the bonding process occurred through high plastic deformation and heat generation by the impact, which are the fundamentals of adiabatic shear instability [3], [4], [7].

The material properties for particle and substrate were defined by the Johnson-Cook (JC) constitutive model to ensure a correct response to high strain, strain rate, and temperature generated by the impact. The stress under plastic deformation is defined by equation 1:

$$\sigma=\left[A+B(\varepsilon_p)^n\right]\left[1+C\ln\left(\frac{\dot{\varepsilon}_p}{\dot{\varepsilon}_{p0}}\right)\right]\left[1-(\hat{T})^m\right]$$

Equation 1

Where ε_p is the equivalent plastic strain, $\dot{\varepsilon}_p$ is the equivalent plastic strain and $\dot{\varepsilon}_{p0}$ is the reference plastic strain rate. A,B,C,n,m are material parameters and represent: yield stress (MPa), hardening modulus (MPa), strain rate sensitivity coefficient, strain hardening exponent and thermal softening exponent, respectively. $\hat{T}$ is non-dimensional temperature defined by:

$$\hat{T}=\begin{cases} 0 & T<T_0 \\ \dfrac{T-T_0}{T_{melt}-T_0} & T_0\leq T\leq T_{melt} \\ 1 & T>T_{melt} \end{cases}$$

Where T is the temperature, T_{melt} is the melting temperature of the material and T_0 is the transition temperature. The material properties and JC parameters of the involved materials are shown in table 1. [3], [10], [11], [12]. The impact model assumed an adiabatic process, meaning that heat transfer between elements was disabled and the temperature changes observed during the simulation resulted solely from the conversion of energy. This assumption was justified based on the thermal diffusivities of the materials, the small dimensions of the model and the short duration of the impact [4]. These parameters were used in all impact simulations for the mentioned material combinations (first impact). The results include the combined effects of impact, plastic deformation, and the shockwave. The shockwave is responsible for a tensile gradient in residual stress on the order of tenths of megapascals.

The resulting stresses after the first impact, temperature and deformation states were then imported as initial conditions into subsequent explicit simulations. In these, a second particle, made of the same material and impacted under the same conditions, was impacted just above the deformed geometry of the first impact (particle-substrate system). The resulting stresses of all simulations (first and second impacts) are represented as $\sigma_i(z)$ function, where z is a depth along the symmetry axis of the simulation (particle-substrate) and σ_i could be σ_{11}, or σ_{22} stresses from the element at that depth. σ_{11} and σ_{22} are equivalent to S11 and S22 respectively, used by the finite element solver of Abaqus.

Implicit Coupled Temperature-Displacement Analysis

The residual stress evolution in cold spray coatings is composed by an overlapping of peening, quenching and thermal stresses. In this study, the explicit FE models provided insights about the quenching stresses, temperature and deformations immediately after particle impact (first and second impact); however, explicit approach is only well suited for short highly dynamic processes but it is computational inefficient for processes that involve longer time spans like cooling [3], [9] . Additionally, the explicit model parameters were set to disable the heat transfer between elements (adiabatic process). Therefore, subsequent implicit coupled temperature-displacement FE analyses were performed to evaluate the evolution of thermal stresses during cooling.

Temperature and deformations from explicit models were used as initial conditions in implicit analyses where heat transfer was allowed and the simulations were run until thermal equilibrium was reached. Some modifications were made compared to the explicit analyses. Since no-debonding assumption was made, the contact interactions in explicit models were replaced with tie constraints in nodes between particle and substrate. Moreover, the strain-rate dependency of the materials in the Johnson–Cook model was disabled $C=0$. Given the long duration of the cooling process (compared with the impact time), the plastic response of the materials could be assumed to be strain-rate independent [7]. The mesh was also modified to change the element type to one that supports temperature-displacement analysis, while the boundary conditions remained unchanged.

The resulting thermal stresses obtained after cooling process are reported in the same manner as those from the explicit simulation and are represented as $\sigma_i^{thermal}(z)$. The final residual stresses (after impact and cooling) are represented by $\sigma_{res}(z) = \sigma_i(z) + \sigma_i^{thermal}(z)$.

Tables 1 through 3 summarize the fundamental physical and mechanical properties of the materials used in this study—AISI 316 stainless steel and pure copper—which serve as input parameters for the numerical models. Table 1 presents the thermophysical properties, including thermal conductivity, coefficient of thermal expansion (CTE), density, and specific heat capacity. These properties are essential for accurately simulating thermal and stress evolution during the cold spray process. Table 2 lists the material-specific indices for the Johnson–Cook constitutive model, which was employed to capture the strain rate and temperature-dependent plastic deformation behavior observed during high-velocity impact. The melting temperature of each material is also included in Table 2 to provide reference limits for model applicability. Finally, Table 3 details the elastic modulus as a function of temperature for both materials, recognizing that temperature-dependent mechanical properties are critical to capturing stress development under thermal loading and cooling cycles.

Table 1. Thermophysical properties of the materials used in the model. Includes thermal conductivity, coefficient of thermal expansion (CTE), density, and specific heat capacity for AISI 316 stainless steel and copper.

Material	Thermal Conductivity (W/mK)	CTE (K^{-1})	Density (kg/m^3)	Specific heat (J/kgK)
SS316	16	1.6E-05	8030	457
Cu	400	1.65E-05	8960	385

Table 2. Johnson–Cook material model parameters for SS316 and Cu. The table also includes the melting temperatures for both materials, which serve as a reference for model validity in high-temperature simulations.

Material	*A (MPa)*	*B (MPa)*	*C*	*n*	*m*	**Melting Temp (K)**
SS316	316	1000	0.012	0.45	1.03	1673
Cu	90	292	0.025	0.31	1.09	1356

Table 3. Temperature-dependent elastic modulus for SS316 and Cu. Values are reported at multiple temperature points to account for the influence of thermal gradients during cold spray deposition.

Material	**SS316**						**Cu**					
Temper. (K)	298	473	873	973	1500	1643	298	473	873	973	1500	1643
Elastic (GPa)	193	184	154	148	100	75	110	109.23	107.47	107.03	104.71	104.08

Figure 1 outlines the sequential modeling strategy employed to simulate residual stress development from multiple particle impacts. The schematic illustrates the computational steps (Fig 1a) and the representative finite element model, including mesh refinement and applied boundary conditions (Fig 1b). The workflow begins with the definition of process parameters for a single-particle impact, from which the resulting stress and temperature fields are calculated. These fields are then subjected to a cooling phase to determine the residual stress distribution. This residual state serves as the initial condition for the subsequent particle impact. In each step, new stress and thermal distributions are computed and followed by cooling, allowing the accumulation of residual stress profiles over multiple impacts. This progressive modeling approach captures the interplay between impact-induced plastic deformation and thermally driven stress evolution, enabling a detailed prediction of stress buildup in both the coating and substrate.

Fig 1. A) Modeling framework for predicting residual stress evolution from sequential cold spray impacts. B) Representative finite element model showing mesh resolution and boundary conditions applied during the simulations.

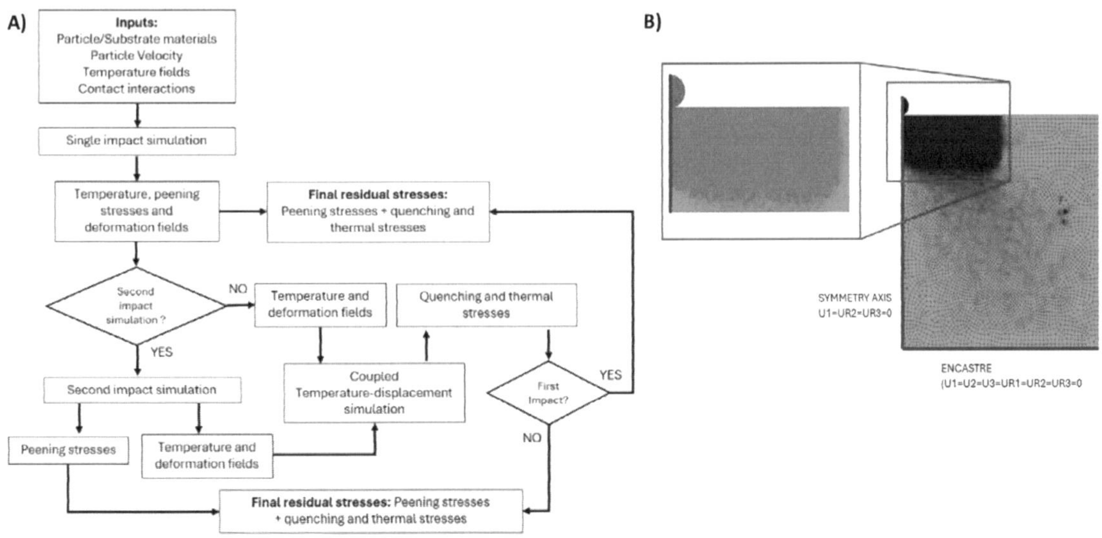

RESULTS AND DISCUSSION

Residual Stress Trends from Curvature Measurements

Experimental observations by the authors, supported by prior literature, indicate that tensile residual stresses can develop during cold spray deposition, particularly when spraying high-strength materials such as Ti [1] and Inconel 718 under conditions of high deposition rate (e.g., low raster speeds) [2]. In the present study, curvature measurements were employed to monitor the stress evolution during successive layer deposition. When Cu particles were deposited onto a Cu substrate, a negative curvature was recorded, consistent with compressive residual stresses typically expected from the peening effect of cold spray. In contrast, deposition of SS316 resulted in a measurable positive curvature, signaling the development of tensile residual stresses—contrary to the generally accepted expectation of predominantly compressive stress states in cold spray coatings. This finding aligns with a growing, though still limited, body of evidence reporting tensile stress formation in high-strength systems. For example, tensile residual stresses have been documented in Ti [1] and Inconel 718 [2] coatings. Additionally, our earlier work [3] predicted tensile stresses in SS316 coatings through finite element analysis, and further experimental confirmation will be reported in forthcoming publications by the authors [13].

Figure 2 illustrates the curvature evolution for both material systems across multiple passes, with Figure 1B emphasizing the abrupt increase in curvature after the initial pass for the SS316 case, corresponding to a high level of tensile residual stress. These findings underscore the material-dependent nature of stress development during cold spray and motivate further micro-scale analysis through single-particle impact simulations. The modeling results presented in the following sections aim to elucidate the mechanisms that drive compressive stresses in Cu and tensile stresses in SS316 by examining localized impact-induced and thermally induced stress distributions.

Fig. 2: Curvature versus temperature evolution during cold spray deposition on thin metallic substrates using ICP Instrument. (A) Copper particles sprayed onto a copper substrate, showing a negative curvature trend indicative of compressive residual stress development. (B) Stainless steel 316 deposited onto a SS316 substrate, where the observed positive curvature reflects the buildup of tensile residual stresses.

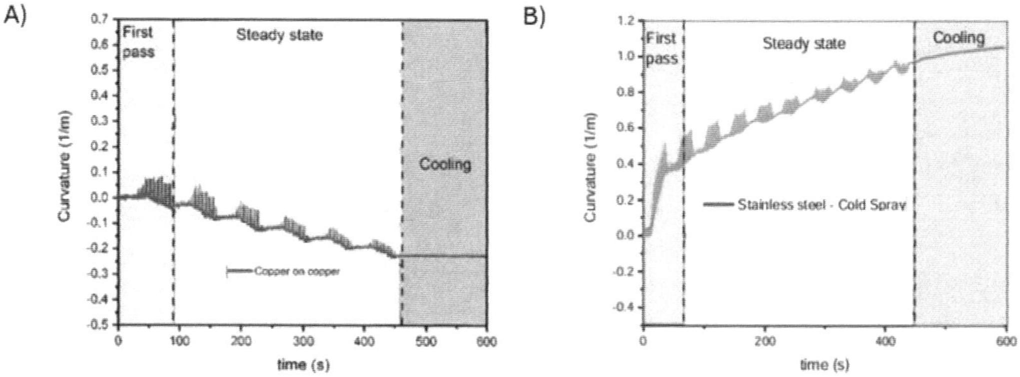

Single-Particle Impact Morphologies and Local Residual Stress and Temperature

The explicit dynamic simulations of single-particle impacts reveal distinct deformation morphologies for copper-on-copper and SS316-on-SS316 configurations, as shown in Fig. 3. In both cases, the impact generates a crater-like depression, with pronounced plastic deformation beneath the center of the contact zone. This deformation induces compressive residual stresses localized near the impact core. The particle material is displaced laterally, forming a peripheral splat with a characteristic geometry. For the Cu-case, the substrate's lower yield strength allows deeper penetration and spreading, resulting in a thinner splat and more pronounced lateral flow. In contrast, the SS316 impact exhibits more confined deformation, governed by its higher material strength. Notably, temperature elevation is observed at the periphery of both impact zones, where friction and plastic work contribute to localized heating. These thermal and mechanical fields are key precursors to the residual stress states that develop after cooling and successive layer buildup.

Fig. 3. Simulated morphologies of single-particle impact for (A) Cu-on-Cu, and (B) SS316-on-SS316 systems showing the S22 stress-fields. Both impacts produce crater formation and plastic deformation beneath the particle center, leading to compressive residual stress. Elevated temperatures are observed at the outer boundaries, where frictional effects and plastic work are most intense, for (C) Cu-on-Cu, and (D) SS316-on-SS316 systems. Stresses are in MPa, and temperature in K. The scales are selected to include the most representative stress and temperature fields.

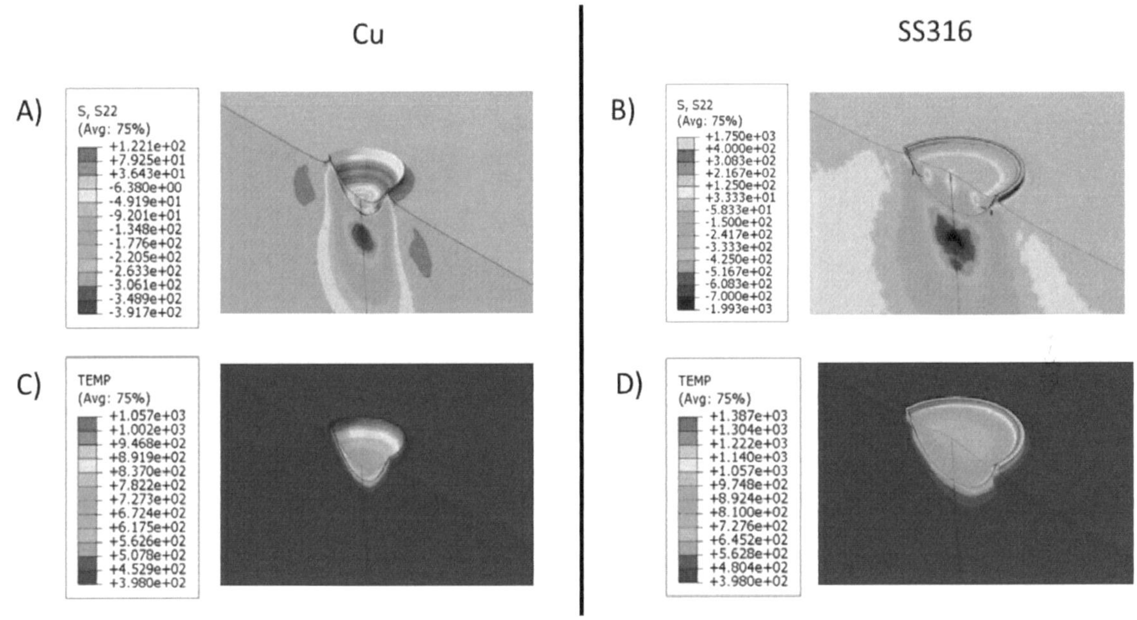

Stress Distribution beneath Single-Particle Impacts

Figures 4 and 5 present the computed stress profiles beneath the center of impact for the Cu-on-Cu and SS316-on-SS316 systems, respectively. The figures illustrate both in-plane normal stress (S11, left) and through-thickness normal stress (S22, right), providing insight into the stress state distribution generated by the crater formed during impact. Due to the pronounced plastic deformation associated with crater formation, highly compressive residual stresses develop along the through-thickness direction (S22) within the substrate, particularly near the impact center. In the in-plane direction (S11), constraint from the surrounding undeformed material further amplifies the compressive stress magnitude, making S11 typically more compressive than S22. In both material systems, the stress levels in stainless steel are noticeably higher, consistent with its higher yield strength and resistance to plastic deformation.

Within the impacting particle itself, a counterbalancing tensile stress is observed primarily in the S11 direction, while the S22 component remains near neutral. Although mesh deformation during high-strain impact may influence localized stress resolution in the particle, the presence of significant tensile stresses occurs due to the cooling effect of the particle upon contacting the colder substrate. The stress profiles shown correspond to regions close to the surface, as stresses tend to relax and approach neutral values deeper into the substrate. Overall, the results demonstrate that crater formation during cold spray impact induces compressive residual stresses in both principal directions, correlating strongly with the geometry and depth of the deformation zone, and quenching stress occurs as a result of the cooling to substrate temperature.

Fig 4. Simulated residual stress profiles beneath a single Cu-on-Cu impact. Left: in-plane stress (S11); Right: through-thickness stress (S22). Stress fields adjacent to the particle–substrate interface are not shown here because highly distorted elements in this region lead to non-robust FEA results.

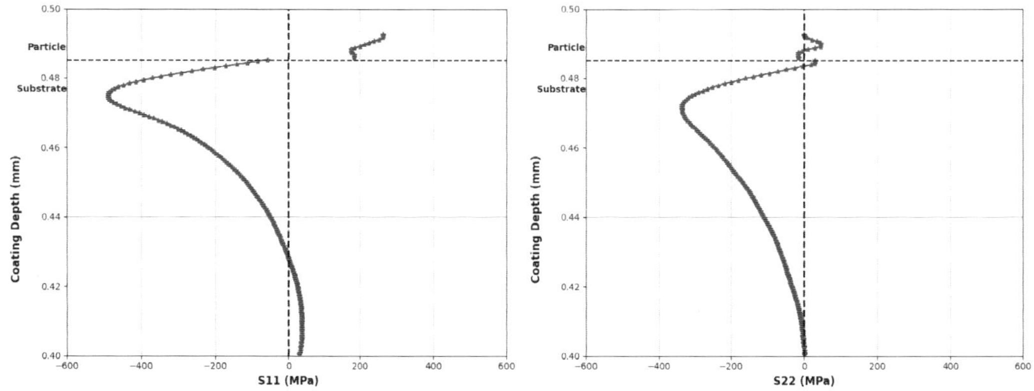

Fig. 5. Simulated residual stress profiles beneath a SS316-on-SS316 impact. Left: in-plane stress (S11); Right: through-thickness stress (S22).

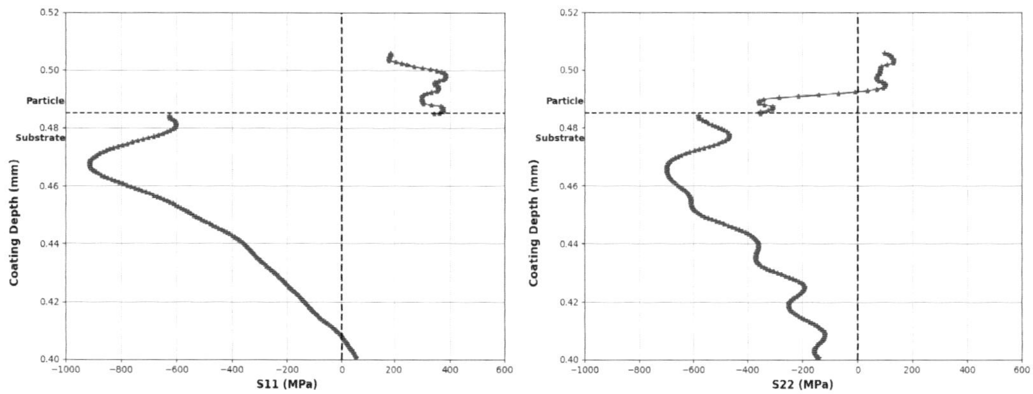

The results presented highlight the emergence of tensile stress within the particle, particularly in the S11 direction, due to post-impact cooling following the high temperatures generated by friction—especially in the case of SS316. In actual cold spray deposition, where multiple particle impacts accumulate to form a continuous layer, the combined effect of tensile-dominated particles may result in a net tensile stress state across the layer. If this tensile contribution outweighs the compressive stress induced by the peening, it can manifest as a positive curvature in curvature-based residual stress measurements. This condition is more likely for SS316, where tensile stresses in the particle can reach magnitudes approaching 400 MPa. Furthermore, the post-impact morphology of the SS316 particle tends to form a bulky, oval-shaped block that can retain significant residual stress as a single cohesive body. In contrast, the copper system produces a crescent-shaped profile, with lateral regions that more readily relax and carry much lower stress relative to the particle center. Therefore, this initial single-particle impact analysis provides critical insight into the stress distribution and orientation mechanisms that may account for the overall tensile behavior observed experimentally in curvature measurements of SS316 coatings.

Residual Stress Evolution During Second Particle Impact

Figures 6 and 7 illustrate the stress distribution in the S11 and S22 directions after the sequential impact of a second particle onto a previously deposited splat, for both the Cu and SS316 systems. The introduction of the second particle alters the stress state of the underlying layer, notably reducing the compressive stresses initially developed during the first impact. This reduction is observed in both principal directions (S11 and S22), and highlights the role of pre-existing residual stresses in limiting further plastic deformation in the first splat, despite the assumption of perfect bonding at the interface. In the case of SS316, the second particle also develops tensile stresses upon cooling, attributed primarily to the quenching effect, and appears to partially relieve the tensile stress previously present in the first splat. While the S22 component for the first particle transitions toward a slightly compressive state, the top particle maintains a predominantly tensile stress profile, especially in the S11 direction. However, due to the intense localized deformation and associated mesh distortion at the particle center, some uncertainty arises in the precise quantification of stress within the second particle. Nonetheless, the general trend supports the conclusion that the top layer continues to accumulate tensile stress as a result of rapid cooling and constrained contraction.

Fig. 6. Residual stress field for the Cu-on-Cu system, (A) in the S11 direction and (B) in the S22 direction, after the second particle impact.

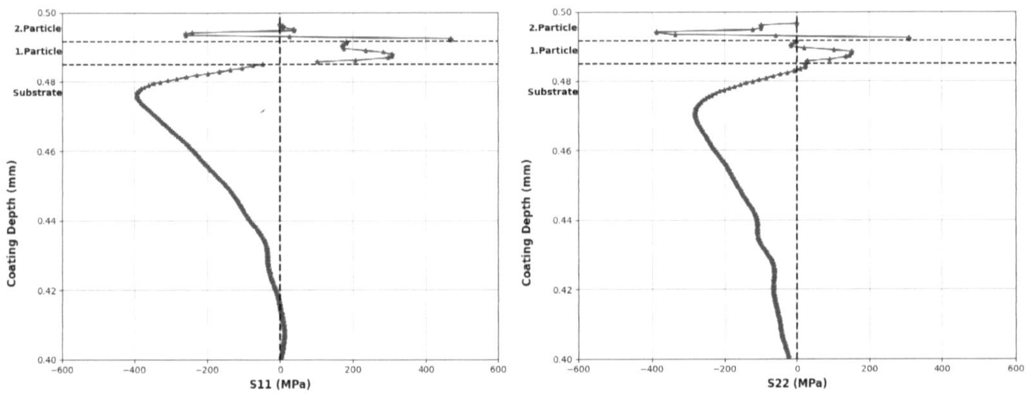

Fig. 7. Residual stress field for the SS316-on-SS316 system, (A) in the S11 direction and (B) in the S22 direction, after the second particle impact.

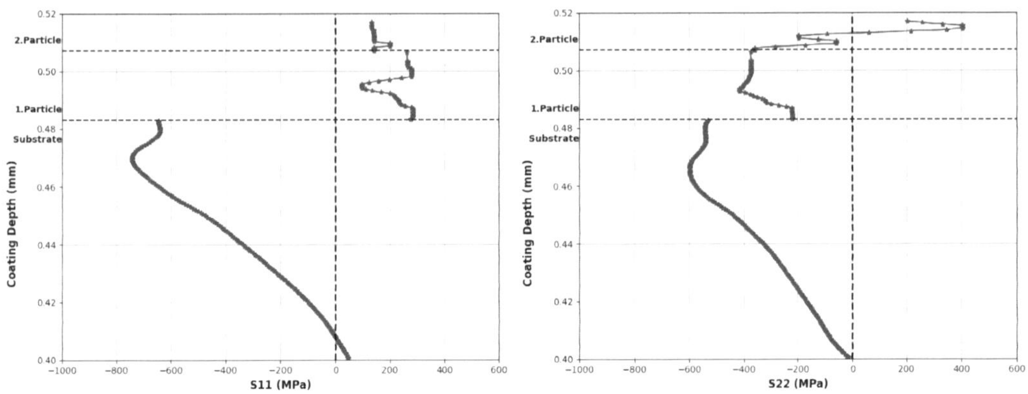

Figure 8 highlights the morphological evolution following a second particle impact, with particular focus on the effects of strain hardening in the first splat. For SS316, the strain-hardened condition of the underlying splat significantly limits further plastic deformation, resulting in a shallower and flatter splat geometry compared to the initial impact. This effect is similarly observed in the Cu system, although the Cu splat retains its characteristic crescent-moon shape with a deeper central depression and lateral spreading within the preformed crater. In contrast, the SS316 system forms a more compact, block-like top layer with minimal reduction in the height of the first splat. These morphological differences directly influence the distribution of residual stresses and the thermal response during quenching. While Cu still develops localized tensile stresses in the second splat due to cooling, the geometric confinement in SS316 leads to more pronounced in-plane quenching stresses. Additionally, in the both Cu and SS316 cases, the altered splat geometry reduces the extent of deformation, increasing frictional heating at the shear interfaces, resulting in higher peak temperatures relative to the first impact (as shown in Fig. 3c, and 3d). These findings demonstrate that the magnitude of the quenching effect is strongly governed by both the crater shape and the thermal conditions experienced during splat formation.

Fig. 8. Post-impact morphologies following the second particle impact for (A) Cu-on-Cu, and (B) SS316-on-SS316 systems showing the S22 stress-fields. Elevated temperatures are observed at the outer boundaries, where frictional effects and plastic work are most intense, for (C) Cu-on-Cu, and (D) SS316-on-SS316 systems. Stresses are in MPa, and temperature in K.

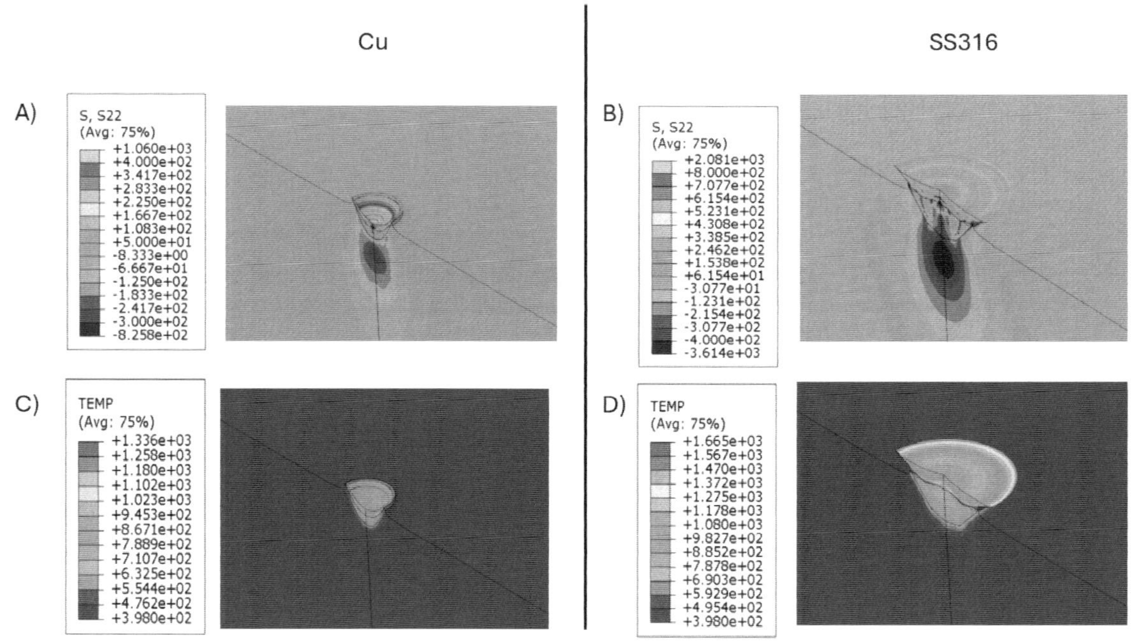

Conclusions

This study combined experimental curvature measurements with explicit and implicit finite element modeling to investigate the evolution of residual stresses in cold spray coatings, focusing on Cu and SS316 systems. While cold spray is conventionally associated with compressive residual stress due to its impact-driven peening effect, the experimental results revealed the unexpected development of tensile stresses, particularly in SS316 coatings. The curvature evolution during deposition demonstrated that the residual stress state—compressive or tensile—is highly dependent on the material's mechanical properties and layer buildup conditions. Numerical simulations of single and sequential particle impacts offered insight into the local stress distributions that underpin these macroscopic observations. In both Cu and SS316 cases, crater formation from individual impacts induced localized compressive stresses in the substrate and tensile stresses within the particle, with SS316 exhibiting higher stress magnitudes due to its higher strength and strain and strain-rate sensitivity. Quenching effects during post-impact cooling determine tensile stresses in the top layers, particularly in SS316, where increased heating, reduced deformation, and geometric confinement contributed to stress retention.

Sequential impact simulations showed that strain hardening of the first splat reduces plastic deformation during subsequent impacts and alters thermal behavior at the interface. Morphological differences between the two materials—crescent-shaped for Cu and block-like for SS316—were found to significantly influence both residual stress distribution and temperature profiles. These findings suggest that tensile residual stresses in cold spray coatings, especially for high-strength materials such as SS316, arise from a combination of limited plastic accommodation in successive layers, geometrical constraint, and thermal mismatch during rapid cooling (large quenching effect). A deeper understanding of these mechanisms is essential for improving predictive capabilities and guiding the optimization of process parameters to control residual stress evolution in cold spray applications.

The results of this study indicate that residual stress buildup is strongly influenced by the interaction between impact-induced strain hardening and thermally driven stresses. For high-strength materials (e.g., stainless steel), process conditions that favor higher thermal loads—such as elevated gas temperatures or high deposition rates—can promote tensile stress development. In contrast, lower-strength materials (e.g., copper) more readily accommodate plastic deformation, and thus tend to favor compressive stress buildup through peening. Although a direct correlation between residual stress profiles, microstructure, and properties remains under development, curvature measurements provide a promising direct in-situ metric for monitoring stress evolution. This approach may serve as a useful guideline for practitioners in tailoring process parameters to achieve desired stress states.

Acknowledgments

The authors acknowledge the financial support provided by Universidad San Francisco de Quito (USFQ) through PoliGrant No. 33618.

References

[1] T. Suhonen, T. Varis, S. Dosta, M. Torrell, and J. M. Guilemany, "Residual stress development in cold sprayed Al, Cu and Ti coatings," *Acta Mater.*, vol. 61, no. 17, pp. 6329–6337, Oct. 2013, doi: 10.1016/j.actamat.2013.06.033.
[2] J. Schmitt, J. Fiebig, S. Schrüfer, O. Guillon, and R. Vaßen, "Adjusting Residual Stresses During Cold Spray Deposition of IN718," *J. Therm. Spray Technol.*, vol. 33, no. 1, pp. 210–220, Feb. 2024, doi: 10.1007/s11666-023-01673-4.
[3] F. Oviedo and A. Valarezo, "Residual Stress in High-Velocity Impact Coatings: Parametric Finite Element Analysis Approach," *J. Therm. Spray Technol.*, vol. 29, no. 6, pp. 1268–1288, Aug. 2020, doi: 10.1007/s11666-020-01026-5.
[4] M. Grujicic, C. L. Zhao, W. S. DeRosset, and D. Helfritch, "Adiabatic shear instability based mechanism for particles/substrate bonding in the cold-gas dynamic-spray process," *Mater. Des.*, vol. 25, no. 8, pp. 681–688, Dec. 2004, doi: 10.1016/j.matdes.2004.03.008.
[5] H. Assadi, F. Gärtner, T. Stoltenhoff, and H. Kreye, "Bonding mechanism in cold gas spraying," *Acta Mater.*, vol. 51, no. 15, pp. 4379–4394, Sep. 2003, doi: 10.1016/S1359-6454(03)00274-X.
[6] T. Schmidt, F. Gärtner, H. Assadi, and H. Kreye, "Development of a generalized parameter window for cold spray deposition," *Acta Mater.*, vol. 54, no. 3, pp. 729–742, Feb. 2006, doi: 10.1016/j.actamat.2005.10.005.
[7] P. Bansal, P. H. Shipway, and S. B. Leen, "Residual stresses in high-velocity oxy-fuel thermally sprayed coatings – Modelling the effect of particle velocity and temperature during the spraying process," *Acta Mater.*, vol. 55, no. 15, pp. 5089–5101, Sep. 2007, doi: 10.1016/j.actamat.2007.05.031.
[8] W.-Y. Li, C. Zhang, C.-J. Li, and H. Liao, "Modeling Aspects of High Velocity Impact of Particles in Cold Spraying by Explicit Finite Element Analysis," *J. Therm. Spray Technol.*, vol. 18, no. 5, pp. 921–933, Dec. 2009, doi: 10.1007/s11666-009-9325-2.
[9] "Abaqus 2016 Documentation." Accessed: Jul. 01, 2025. [Online]. Available: http://130.149.89.49:2080/v2016/index.html
[10] S. A. Zelepugin, N. V. Pakhnutova, and R. O. Cherepanov, "Optimization of Johnson-Cook Constitutive Model Parameters Using the Nesterov Gradient-Descent Method.," *Materials*, vol. 16, Aug. 2023, doi: 10.3390/ma16155452.
[11] L. Kunčická, M. Jopek, R. Kocich, and K. Dvořák, "Determining Johnson-Cook Constitutive Equation for Low-Carbon Steel via Taylor Anvil Test," *Mater. Basel Switz.*, vol. 14, no. 17, p. 4821, Aug. 2021, doi: 10.3390/ma14174821.
[12] S. Zhang and B. L. Kinsey, "Influence of Material Properties on Interfacial Morphology during Magnetic Pulse Welding of Al1100 to Copper Alloys and Commercially Pure Titanium," *J. Manuf. Mater. Process.*, vol. 5, no. 2, Art. no. 2, Jun. 2021, doi: 10.3390/jmmp5020064.
[13] S. Sampath, F. Caliari, D. Lal, and A. Valarezo, "Stress Evolution During Cold Spray Deposition and Resultant Properties" (in-review)

ICRS 12: Proceedings from the 12th International Conference on Residual Stresses, 2025
October 20–23, 2025; Detroit, Michigan, USA
https://doi.org/10.31399/asm.cp.icrs2025p0086

A hybrid surface fitting approach in the contour method for residual stress assessments

Z. Cai[12] and M. J. Roy[12*]

[1]Department of Mechanical and Aerospace Engineering, The University of Manchester, Manchester M13 9PL, UK;

[2]Henry Royce Institute, The University of Manchester, Manchester M13 9PL, UK

**Corresponding author: matthew.roy@manchester.ac.uk*

Abstract

The contour method is gaining wider acceptance as a residual stress measurement technique during its 25 years of development. It offers a complete stress component orthogonal to a cutting plane by reconstructing the stresses based on the out-of-plane displacements using superposition theory. Recent advancements have focused on improving repeatability and quantifying uncertainty. Traditional analytical models often fail to adequately represent measured out-of-plane displacement data with complications. A hybrid surface fitting routine has been developed to address this, combining two different surface models to represent both high stress gradients at regions of interest and smoothing irregularities and noise in lower fidelity regions. This technique has been applied to a 30mm thick SA508 Gr.3 Cl.1 weldment, cut with self-constraint features. It will be shown that this hybrid method provides higher resolution in resolved stresses compared to singular fitting approaches, which can be compromised by cutting-induced plasticity. The hybrid fitting method proves effective in processing data with local irregularities, such as unintentional cutting interruptions. This is because it allows for a local application of a lower-weighted surface function at the most compromised regions, while maintaining finer details elsewhere with a secondary fitting function. This advancement represents an improvement in the reliability and precision of the contour method for residual stress analysis.

Keywords: the contour method; residual stress; surface fitting; model error

Introduction

The contour method (CM) is a destructive technique for residual stress evaluation [1]. Similar to other strain relief techniques such as hole drilling, the CM assumes elastic deformation induced by mechanically released internal stresses. The difference is that a planar cut is employed, rather than point-wise drilling along a measurement plane during a CM application. This in turn provides significantly greater spatial resolution that other strain relief methods. Typically, wire-electro-discharge-machining (WEDM) is employed such that the underlying stress state of components remains unchanged during cutting limiting the application to electrically conductive materials. However recent research has implied that abrasive diamond wire as a cutting medium [2] may extend the application of CM to non-metallic materials. Regardless of the cutting technique employed, the subsequent measurement of the cutting plane is required.

The out-of-plane displacements on the cutting plane are measured and processed into boundary conditions to support a finite element analysis (FEA) for the reconstruction of the original stress before relaxation. This is based on a variation of Bueckner's superposition principle [3], stating that the original stresses on a fully relaxed cut plane are equal to the stresses required to elastically push the deformed surface back to being flat [4]. Prior to the FEA analysis, an analytical surface needs to be fitted to the data to smooth spurious points which may be present and allows boundary condition to be queried at arbitrary locations [5]. The literature describes two main approaches to generate this analytical surface, which are Fourier series [4] and bivariate splines [6].

One benefit of Fourier series expansion is that due to its inherent property of representing frequency components in dataset, the high frequency noise and low frequency surface trends can be quantitatively differentiated, making it easier to select suitable level of smoothing. However, this cannot overcome the disadvantage of Fourier series being a single function. As a result, spline fitting has been predominantly employed [7]. A spline surface is defined by piecewise polynomials that is controlled by knot spacing and their order. These two parameters determine the size and flexibility of each piece. A smaller knot spacing with high order tend to be more sensitive to high frequency features, whereas a large spacing and lower order tends to smooth over higher frequency features. Both of the aforementioned techniques are susceptible to parameter selection. This is particularly the case when processing large datasets measured with high resolution techniques such as laser profilometry [8]. One of the factors affecting CM analyses is the limitation of analytical surface model in representing actual displacements on cut plane, or model error. For example, a high level of smoothing will result in a loss of resolution, but a low level of smoothing will introduce unwanted noise and further contributes to uncertainty in back-calculated stresses. However, while noise is typically uniformly distributed in the resulting surface displacement data, singularities or artefacts are not. As a result, the quality or resolution of a fit is compromised by the presence of surface defects as a larger smoothing factor is required to smooth the surface, and the spatial resolution is limited due to finer knot spacing that is suitable for shorter wavelength variations can result in overfitting elsewhere [9], as shown in Fig. 1. As a mitigation, a hybrid surface fit has been developed encompassing a two-step fit, allowing a more precise representation at selected regions while keeping a high level of smoothing at regions which are either noisy or contain cutting artefacts.

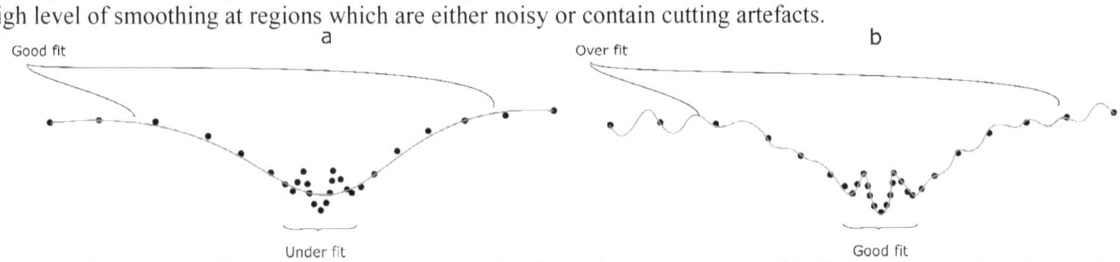

Figure 1 Illustration of how spline functions (a) with a large knot spacing suitable for longer wavelength signal can result in underfit at high frequency components and (b) small knot spacing to fit short wavelength part can result in overfit at other regions.

Methodology

The hybrid surface fitting module introduced here is based on the pyCM (Python Contour Method) package, modifying its surface fitting module. The registration, alignment and averaging steps remain unmodified, as described here [6]. It is from this point that the output of pyCM can be imported into the proposed hybrid surface fitting package, and the result returned to the pre-processing module of pyCM for FE analysis. The sequence is to first fit a NURBS function to the aligned and averaged point cloud, following a qualitative evaluation to determine the under-fitted regions for a secondary fit. In this manner, the NURBS surface can represent the longer wavelength of surface displacement that can have a large smoothing factor, and secondary spline fitting picking up more detailed local variance. With the primary and secondary surface blended together, a determinative non-uniform surface can be achieved with specific regions determined by the practitioner.

Primary surface fitting

NURBS is a robust tool for representing curves and surfaces and exhibits no apparent disadvantages. Its non-uniform property allows for the generation of a more flexible function compared to standard splines, especially when the knot spacing is large to achieve elevated smoothing. However, accurately representing certain shapes using B-splines without higher-order polynomials is challenging. For instance, representing a circle in a 2D domain requires at 7th order splines with at least 8 knots to avoid tessellation, leading to a significant waste of computing power. In the current practice of CM data processing, the fit quality still relies on the judgment of practitioners through sensitivity analysis [7]. This involves iteratively running the fitting process and comparing results obtained with different parameters, increasing the knot spacing for a given B-spline order until there is little change in the overall fit before showing the underling surface roughness features. Therefore, we have decided to use NURBS for primary surface fitting, implementing the NURBS-python (*geomdl*) library [10].

As described earlier, CM surface datasets are measured topographical displacements of cut surfaces containing noise, thus precise representation (interpolation) of measured data results in the potential for over-fitting. To interpolate, a function is generated so that every given data point can be satisfied, while in approximation, the constructed curve or surface does not necessarily satisfy the data precisely, only to satisfy certain constraints. That is, the fitting algorithm is devised to satisfy certain error boundaries or fit with certain amount of control points [12]. The NURBS fitting function uses a least squares criterion for approximation [11]. For example, to seek a function $C(u)$ from the dataset consists of m points $Q_i (i = 0,1, \dots, m)$, the following conditions need to be satisfied:

$$\begin{cases} C(0) = Q_o \ and \ C(1) = Q_m \\ \sum_{k=1}^{m-1} (Q_k - C(u_k))^2 = minimum \end{cases}$$

Which means the start/end point will satisfy the dataset with the remaining points in $\{Q_m\}$ satisfies the least square condition, where $C(u)$ is the NURBS base function. An evaluation as part of the supporting functions is designed to evaluate the NURBS surface by querying parameterised coordinates on designated locations. This feedback is used for the optimisation of fitting parameters and works as an indicator to show where larger errors exist, such that a subsidiary spline patch can be applied for the secondary fit.

The purpose of this primary fit is to capture the low frequency trend and pass the high frequency components to secondary fit. Fitting quality indicators are not applied at this stage, such as Root Mean Square Error (RSME) as it is more focused on where the error is rather than the goodness of fit. That is, the distribution of fitting residuals is more important in the primary fit than the absolute value of fitting errors. The inadequacy of localised fitting and regions of specific interest will be treated with secondary fitting so the overall fit quality can be improved. The residuals of the primary fit can be simply calculated by the difference between observed values from the averaged dataset and the fitted values.

Secondary surface fitting

The secondary surface fitting is a direct implementation of surface fitting module in pyCM but only to incorporate higher frequency residuals and potential model errors from primary fitting step. The returned surface is controlled by fixed knot spacing and fitting order, a smaller knot spacing with higher order of fit are more favourable in representing fine topographical details. The quality of secondary surface fitting is depending on subjective judgment as suggested in good practice guides for the contour method [7].

Surface blending

With both primary and secondary surface fitted to the datasets, the entire domain can be represented by:

$$z(x,y) = \begin{cases} f_{primary}(x,y) + f_{secondary}(x,y) & (x,y) \in secondary \ area \\ f_{primary}(x,y) & (x,y) \notin secondary \ area \end{cases}$$

However, as distinct boundaries exist between the primary and secondary areas, some sharp discontinuities on the blended surface are observed at the adjoining regions. This observed phenomenon arises from the *approximation* process employed in the secondary fit. In this process, the generated function invariably intersects both the initial and final data points. Consequently, the boundaries of the secondary surface exhibit precise fitting to the corresponding data, while the remainder of the area satisfy the least square criteria and is only approximated to the given data point. The abrupt reduction in fitting residual distinctly delineates this boundary from others, as shown in Fig. 2a-c.

The solution is to apply a weight function on the secondary surface which gradually change from 1 to 0 at the boundary on a defined buffer zone, with weight 1 correspond to more focused region. Thus z can be calculated by:

$$z(x,y) = \begin{cases} f_{primary}(x,y) + f_{secondary}(x,y) * weight & (x,y) \in secondary \ area \\ f_{primary}(x,y) & (x,y) \notin secondary \ area \end{cases}$$

The modified blending function produces smooth output as illustrated in Fig. 2d.

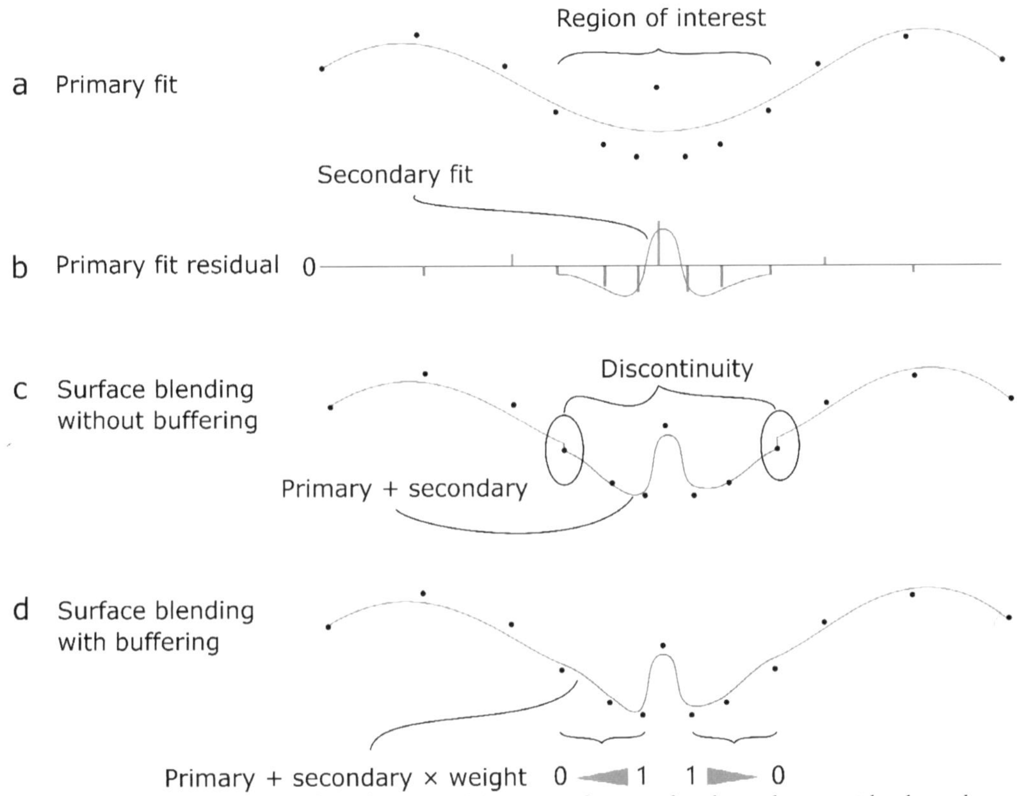

Figure 2 An illustration of hybrid surface showing a (a) primary function fitted to a dataset with a large knot spacing resulting a higher fitting residual at (b) region of interest, where the secondary function will be applied. The (c) blending of two surface functions will introduce discontinuity that can be mitigated by (d) multiplying a weight function to the secondary fit.

Illustrative example

The proposed hybrid surface fitting routine is demonstrated with a contour method dataset collected from a 30mm thick SA508 Gr.3 Cl.1 rectangular weldment previously employed in the NNUMAN project [12]. The specimen comprised a butt weld carried out with cold-wire laser welding in conduction mode, with the cross-sectional of the specimen shown in Fig. 3. Self-constraint features were applied in WEDM cutting process to reduce the effect of cutting induced plasticity [13] and relative movement between the two halves during the main cutting process. Two pilot holes positioned 12mm from the edges and two contour cuts were performed by cutting the central part first and then cut the self-constraint part open on the right side. These resulted in three segmentations in both datasets measured directly from cut surfaces and cause further disruption in aligned and averaged data (Fig. 4) due to i) out-of-plane data at two pilot holes were missing, ii) areas beyond pilot holes from two sides contains more error due to the misaligned surface resulted from localised yielding, and iii) the area closer to self-constraint features are more affected by the potential existence of very high levels of cutting introduced plasticity due to constraint.
The region closer to the central welding seam is expected to have more variations in measured displacements linked directly with processing and would therefore be of more interest and is considered a "focused area" in the demonstration. This is due to a higher expected level of residual stress magnitudes and variance at the vicinity of the weld due to concentrated heat input, in turn requiring a finer, localised surface fit. The regions closer to self-constraint features requires heavier smoothing or even interpolation/extrapolation at the missing area or closer to the edge, which can be achieved in primary surface fitting without affecting the central region of interest.

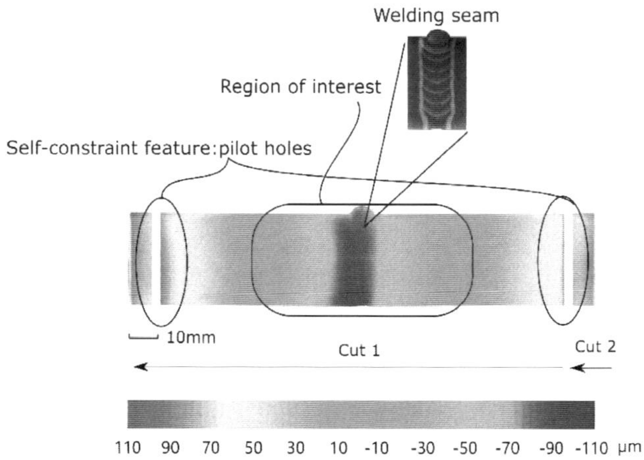

Figure 3 Displacement measurements of specimen showing self-constraint features and a region of interest (welding seam) far from these features.

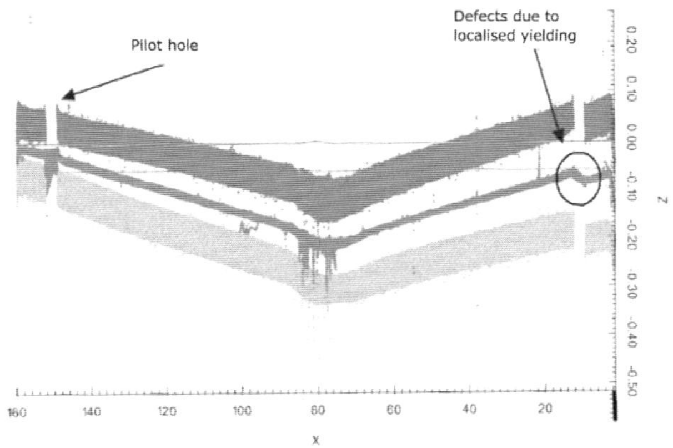

Figure 4 Defects in averaged point cloud due to presence of localised yielding near pilot holes (green: averaged point cloud; orange/yellow: reference/floating point cloud from two cut surfaces).

It has been previously reported by Javadi *et al.* [14] that the data beyond pilot holes are usually discarded and only data between two pilot holes is used for stress reconstruction. This could in theory introduce a domain error where the FEA does not represent the actual component, causing under or overestimation of the resulting stresses. For cases where a high magnitude of residual stresses that extend to the edge of the specimen, ignoring those regions is incorrect. For example, Javadi *et al.* [14] also reported that in their application of self-constraint cuts, the pilot holes have to be arranged far from edges because the internal stresses are too large and distributed across the whole specimen width. In this demonstration, an extrapolation was applied at with a sampling length of 25mm from the edge. As it is hard to evaluate the quality of extrapolation without a priori knowledge of the true stress distribution, a higher smoothing factor was applied in the initial fit to potentially reduce the extra uncertainty imposed by the extrapolation. This is achieved by interpolating the input data on a 5mm spaced grid as marked with red dots in Fig. 5. An evaluation of primary surface is shown in Fig. 6a indicating that there were remaining features unfitted in proximity to the welding seam as the topographical pattern was clearly different from far field area. Notably, the frequency of these features does not align with the underlying surface roughness. The secondary fitting that has a 1.5mm knot spacing with 3rd order spline functions in both *(x)* and *(y)* direction was applied closer to the welding seam to increase the overall fitting resolution particular at the focused area.

The primary and secondary fitted surface are plotted with the x200 magnification ratio in Fig. 6b. It is observed that the secondary surface shows little low frequency components indicates that the primary NURBS fitting successfully captured the global surface trend leaving only high frequency components for secondary fitting. The final surface blending has a buffer area of 10mm length in *(y)* direction to cover the full range in *(x)* direction was then applied to blend the initial and secondary fitting.

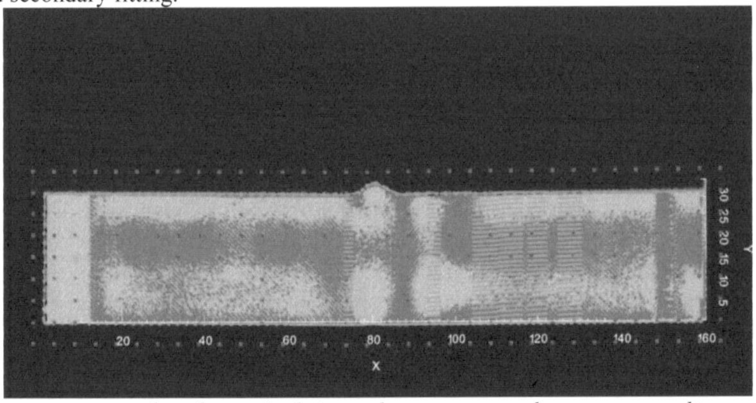

Figure 5 Input data in primary surface fitting (red: input data, grey: raw data; green: outline; yellow: primary fitted surface).

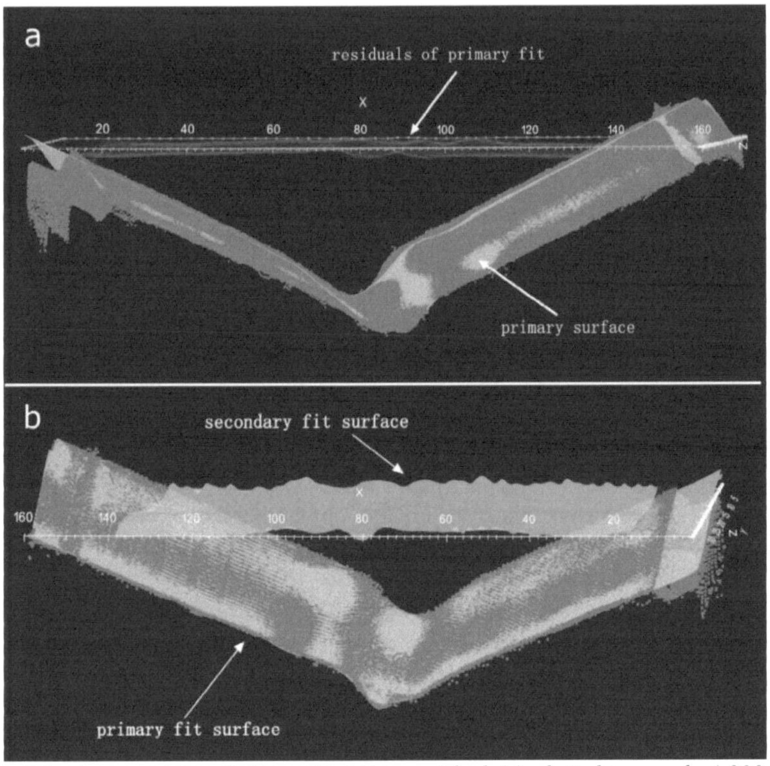

Figure 6 Evaluation of primary surface fitting with fitting residual (a) plotted in purple (x200 displacement) and secondary surface (b) fitting onto the residual at region of interest.

It is observed in Fig. 7a that compared to a standard, one-step spline fitting generated (Fig. 7c) directly from pyCM's fitting module under same fitting parameter (3rd spline with knot spacing of 1.5mm), a more uniform analytical surface can be achieved with less fitting errors closer to the edges benefiting from high level of smoothing out of interested area. The negative effect of pilot hole has been mitigated with less local stress concentration. With the hybrid surface generated, the analytical model was imported into pyCM's preprocessing analysis for stress reconstruction with the same parameters as in published routine [13]. Excessively high compressive regions exist adjacent to the pilot holes due to the aforementioned surface misalignment caused by local yielding, accompanied by localised tensile regions that are believed to be the artificial stresses to balance these compressive stresses. As the plasticity effect is sensitive to cutting direction with more plasticity tends to occur towards the end of a cut, this explains the different magnitude of artificial compressive stresses at the pilot holes. It is believed that this phenomenon also caused the asymmetrical boundary of maximum tensile region at the welding zone, which in theory should occur in a symmetrical manner considering the welding configuration as shown previously in Fig. 3. With the hybrid surface employed, the artificial stresses due to self-constraint features are greatly reduced, and the stresses at the welding zone is more symmetric while maintaining a similar magnitude, as shown in Fig. 7b. The distribution of central tensile region also agrees better with the welding process that the upper region is more tensile due to faster cooling, which is not apparent with single step spline fitting in Fig. 7d.

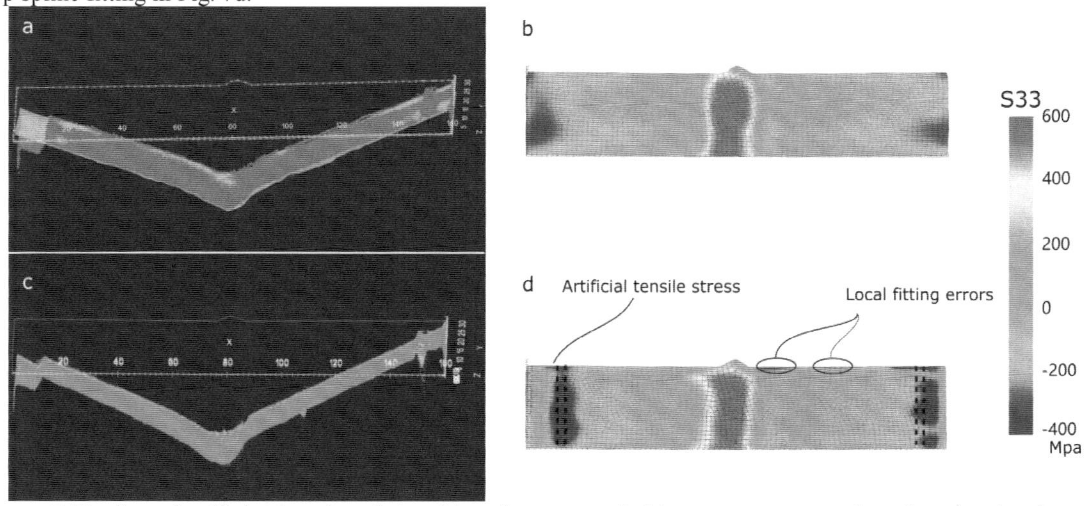

Figure 7 Final result of hybrid surface fitting (a) and stress result (b) in comparison with surface fitted with pyCM spline surface fitting module (c) and stress map(d) under same fitting parameters (1.5mm knot spacing, 3rd order splines).

Summary

The proposed hybrid surface fitting method shows its ability to process complicated dataset where different surface fitting resolution is needed. Primary surface fitting is capable of capturing low frequency topographical trends fitted with a NURBS surface, leaving the residuals to secondary spline fit. The multi-step processing of low frequency surface trends and high frequency surface details make it possible to achieve good level of smoothing while keep fine details at focused regions, thus can make the best use of high-resolution laser scanned data. The application of extrapolation and high level of smoothing can reduce the effect of local cutting discontinuity such as self-constraint features for mitigation cutting introduced plasticity. It is demonstrated that the artificial stresses due to such localised discontinuity can be reduced, which further reduces the balancing artificial components in the resolves stress results that leads to a better interpretation. Future work includes quantitative evaluation and integration into pyCM, making a more effective tool for CM analysis.

References

[1] M. B. Prime, "Cross-sectional mapping of residual stresses by measuring the surface contour after a cut," *J. Eng. Mater. Technol.*, vol. 123, no. 2, pp. 162–168, 2001.

[2] K. R. Praveen *et al.*, "Advancing the contour method for the measurement of residual stress in polymer composites," 2022, Accessed: Sep. 08, 2023. [Online]. Available: https://hal.science/hal-04058709

[3] Bueckner Hans F., "The Propagation of Cracks and the Energy of Elastic Deformation," *Trans. ASME, Ser. E*, vol. 80, no. 6, pp. 1225–1230, 1958, Accessed: Nov. 23, 2020. [Online]. Available: https://ci.nii.ac.jp/naid/10018203178

[4] A. T. Dewald, and M. R. Hill, "Residual stress in a thick steel weld determined using the contour method," Report under Los Alamos National Laboratory contract 32390-001-01 49, October, 2001.

[5] G. Johnson, "Residual stress measurements using the contour method," University of Manchester, Ph.D, p. 244, 2008.

[6] M. J. Roy, N. Stoyanov, R. J. Moat, and P. J. Withers, "pyCM: An open-source computational framework for residual stress analysis employing the Contour Method," *SoftwareX*, vol. 11, p. 100458, 2020, doi: https://doi.org/10.1016/j.softx.2020.100458.

[7] F. Hosseinzadeh, J. Kowal, and P. J. Bouchard, "Towards good practice guidelines for the contour method of residual stress measurement," *J. Eng.*, vol. 2014, no. 8, pp. 453–468, 2014, doi: 10.1049/joe.2014.0134.

[8] M. B. Prime, R. J. Sebring, J. M. Edwards, D. J. Hughes, and P. J. Webster, "Laser surface-contouring and spline data-smoothing for residual stress measurement," *Exp. Mech.*, vol. 44, no. 2, pp. 176–184, 2004, doi: 10.1007/BF02428177.

[9] N. Naveed, "Improving the Spatial Resolution of the Contour Method," The Open University, 2015. doi: 10.5860/choice.51-2973.

[10] O. R. Bingol and A. Krishnamurthy, "NURBS-Python: An open-source object-oriented NURBS modeling framework in Python," *SoftwareX*, vol. 9, pp. 85–94, Jan. 2019, doi: 10.1016/J.SOFTX.2018.12.005.

[11] L. Piegl and W. Tiller, *The NURBS Book (Monographs in Visual Communication)*, 2nd ed. New York: Springer, 1996. [Online]. Available: http://library.auditory.ru/1491/

[12] J. C. Feng *et al.*, "An evaluation of multipass narrow gap laser welding as a candidate process for the manufacture of nuclear pressure vessels," *Int. J. Press. Vessel. Pip.*, vol. 157, pp. 43–50, Nov. 2017, doi: 10.1016/j.ijpvp.2017.08.004.

[13] F. Hosseinzadeh, Y. Traore, P. J. Bouchard, and O. Muránsky, "Mitigating cutting-induced plasticity in the contour method, part 1: Experimental," *Int. J. Solids Struct.*, vol. 94–95, pp. 247–253, 2016, doi: 10.1016/j.ijsolstr.2015.12.034.

[14] Y. Javadi, J. N. Walsh, A. Elrefaey, M. J. Roy, and J. A. Francis, "Measurement of residual stresses induced by sequential weld buttering and cladding operations involving a 2.25Cr-1Mo substrate material," *Int. J. Press. Vessel. Pip.*, vol. 154, pp. 58–74, Jul. 2017, doi: 10.1016/j.ijpvp.2017.06.003.

ICRS 12: Proceedings from the 12th International Conference on Residual Stresses, 2025
October 20–23, 2025; Detroit, Michigan, USA
https://doi.org/10.31399/asm.cp.icrs2025p0094

Evaluating Residual Stress Profile in Cold Drawn 1018 Steel Rods Using Finite Element Analysis

Upama Biswas Tonny[1a], Mohamed Subair Syed Akbar Ali[2], Daniel Barnard[2], Sunil Kishore Chakrapani[1b]
[1]*Department of Electrical and Computer Engineering, Michigan State University, East Lansing, Michigan, USA*
[2]*Center for Nondestructive Evaluation, Iowa State University, Iowa, USA*

Corresponding authors: a: tonnyupa@msu.edu, b: csk@msu.edu

Abstract

This work explores a finite element (FE) method-based residual stress profile estimation technique for cold-drawn steel rods. The surface contour method is a well-known technique in which displacement normal to the lateral cut is used in the FE model to estimate residual stress. However, the lateral cut can affect residual stress, leading to errors in estimation. In this work, the samples were cut axially at different locations throughout the thickness, and the warp on the samples was measured and matched using FE analysis. Two categories of samples were prepared from the same rectangular cold-drawn steel bar: (i) samples with only top surface cut of different depths and (ii) samples with EDM cuts at different positions through the thickness. A three-step finite element analysis was carried out to model stress relief due to cutting. The deflection was compared with experiments, and several stress distribution profiles were developed until a best-matched residual stress was found. A residual stress profile of tension-compression-tension showed the best match for the surface-cut samples. The residual stress profile was further modified to no stress in the middle region to match the EDM cut deflections. This shows that the profile can be estimated using the approach presented here. The advantage of this approach is that the deflection over the length can be used for optimization, which will allow us to obtain unique solutions to the stress profile.

1. Introduction

Residual stress measurement is a key factor in structural design to accurately predict performance and product life [1]. Residual stress can be defined as self-equilibrating stress in a final product after machining and before external load application and can greatly affect material behavior in fracture, fatigue, stress corrosion, buckling, and post-buckling [2]. These stresses result from either plastic deformation caused by cold-forming, cold-straightening, or cambering, or from the uneven cooling of shapes following hot-rolling, welding, or cutting operations [3]. In cold-drawn bar production, the wire is uncoiled, rolled, shot-peened, and drawn through a drawing die [4]. These steps can cause uneven residual stress inside the bar. This process can be especially complex for rods with non-symmetric cross sections, such as rectangular cold-drawn steel bars, which can result in complex stress distributions.

Established destructive residual stress measurement methods are mostly relaxation methods, where deformation is measured after material removal. Common relaxation techniques are the hole-drilling method [5], the crack compliance method [6], and the contour method [8]. The contour method is based on Buckner's elastic superposition theory [7]. Here, the specimen is cut in half laterally, and the residual stress relief causes deformation. The stress necessary to force the deformed surface back to its original position is estimated as the residual stress. The deformed surface displacement is measured, and an opposite and equal deformation is applied in finite element modeling to predict residual stress using the contour method [8]. Even though the contour method assumes the error in stress prediction due to cutting is negligible, Sun et al. [9] presented that cutting-induced plasticity could affect the accuracy of the stress measurement significantly if the removed material had an asymmetrical stress distribution. It would cause uneven stress relaxation, and the error in stress estimation can be as high as 100 MPa. For drawn bars, Heyn and Bauer showed a high internal stress state by measuring the change in bar length after removing layer-wise material [10]. Midha et al. measured surface residual stress in bars by the slitting method, where a cut was made in the longitudinal direction and the opening of the slit was measured [11].

In this work, we performed surface and intermediate axial cuts on the bar, and the resulting warping was modeled by incorporating the finite element method to estimate residual stress. This method can be advantageous in optimizing residual stress, as the deflection can be calculated at different points along the length of the specimen. This allows us to determine a unique distribution of stress using FEM. It can also provide the flexibility of estimating both axial and transverse residual stress in three-dimensional models.

2. Methodology

2.1 Experiment method

To estimate the residual stress condition along the thickness of the sample under consideration, the sample was cut axially at various locations along the thickness, and the warp on the sample was measured. Two types of material removal were explored: (a) surface milling, which removes material to different depths, and (b) electric discharge machining (EDM), which only introduces a cut along the thickness. The surface-milled samples were machined at a slow speed by holding the sample in a vice. Therefore, due to machining, the sample did not deflect. When released from the vice, it deflected. 1018 cold-drawn steel samples of dimensions 6" (L) × 0.5" (W) × 0.375" (T) were used for the study. The surface-cut samples were prepared by removing the material only for a certain length (4" of the 6"-long sample) and leaving a small reference volume (2" in length of the sample that was not machined). This reference makes it easier to measure the deflection. A total of three surface-cut samples were prepared by changing the depth of surface machining: 0.04", 0.08", and 0.12". After machining, the end deflections were measured.

The EDM cuts were parallel to the top and bottom faces along half of their length with a thin EDM wire measuring ~390 μm in diameter. The depths of the cut lines were 0.047", 0.094", 0.141", and 0.1875" below the top surface. Figure 1 below shows a photograph of the EDM cut specimens and highlights the warp of the top and bottom segments after cutting. These segments were referred to as "leaves" here. All the thicker sections bent outward (red arrows), and only the two thinnest leaves bent inward (yellow arrows).

Figure 1. EDM cuts at different positions along the thickness of cold drawn samples showing warping resulting from residual stress.

2.2 Deflection measurement using dial gauge

The deflection profile of the top and bottom leaves in each specimen was measured along the cut length using a dial gauge. The dial gauge was attached to the moving arm of a 3-axis gantry, and the deflection of the cut leaves was measured at 0.125" intervals along the length and across the width of the specimens. Figure 2 shows the photograph of the setup used for the deflection measurements.

Figure 2. Photograph of the experimental setup used for the deflection measurements.

2.3 Deflection measurement using 3D camera

For one of the samples (surface cut of 0.12 inch), a 3D image of the deformed bar was taken using a Revopoint 3D scanner. Here, a point cloud of the specimen was generated while the sample rotated about the longitudinal axis. These meshed points were converted into a solid geometry in the FE software (Figure 3a). The warping of the *yz1234* plane with respect to the *yz5678* plane was measured to obtain the deflection. An origin point was set at *yz6* approximately, with a local coordinate axis. Based on the origin, the position of *yz1, yz2, yz3, yz4, yz5, yz6, yz7*, and *yz8* was calculated. Figure 3b shows the front view of the *yz1234* and *yz5678* planes, and it is visible that the *yz1234* plane moved in both *Y* and *Z* directions. The translation from *center2* (center of *yz5678*) to *center1* (center of *yz1234*) in the *Y* and *Z* directions was 2.39 mm and 2.11 mm, respectively. However, as the corners were not sharp, the origin and other point positions may have errors, and this did not produce good agreement with the dial gauge data. Therefore, this data or process was not further utilized.

(a) (b)

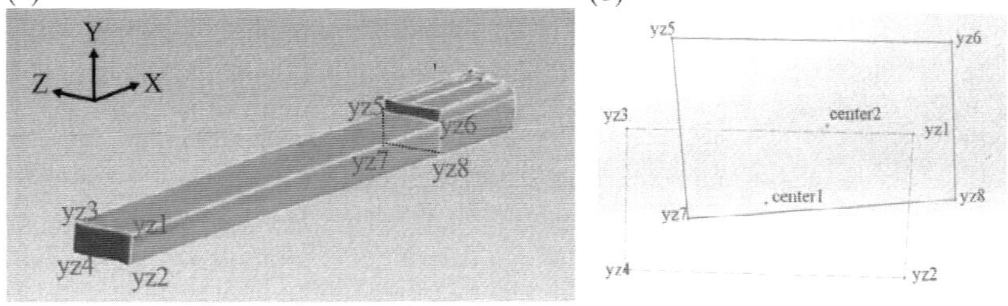

Figure 3: (a)3D scanned deformed bar with local co-ordinate (b) plane yz1234 and plane yz5678 with local position with respect to yz6

2.4 Finite Element Analysis

A 2D analysis was carried out using the FE software ANSYS. A rectangular area that simulates the *XY* plane of the sample was designed such as the length is 101.5 mm (which was the cut length), and the thickness was 9.5mm which was the thickness of the bar before cutting. The material properties for the modeling were density 7850 kg/m^3, Poisson's ratio 0.3, Young's modulus 200GPa. The element had linear mesh with element size of 0.5mm.

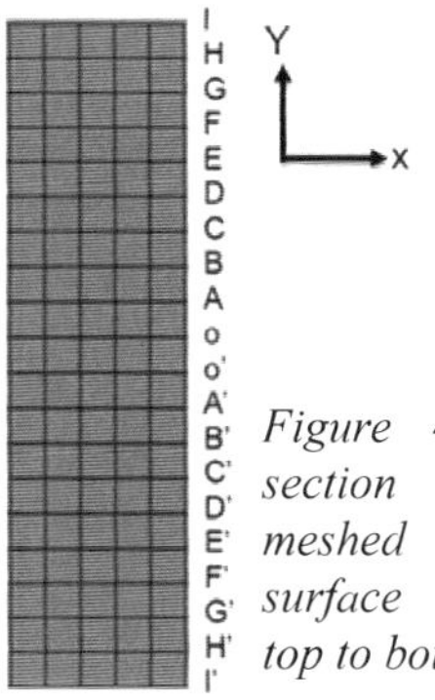

Figure 4: A section of meshed surface from top to bottom

A schematic showing the flow chart of the FE process to determine/model the stress is shown in Figure 5. A section of the meshed surface from top to bottom with specific labeling is shown in Figure 4. The labeling corresponds to different lines of nodes where displacements were applied in the 1st step of the analysis. The displacement magnitude on different nodal lines was changed to manipulate the final residual stress profile. A particular combination of displacement on different nodes can generate a specific stress profile through the thickness. In most models, the *o* and *o'* line nodes' *x* and *y* displacements were restricted to apply constraints. The generated *x* stress was transferred as the initial stress in the 2nd step. The 2nd step was performed to generate the residual stress profile. The deformed geometry of the 1st step was used at the input for the 2nd step to preserve the meshing. As the displacement was in micron in the 1st model, the geometry could be assumed to be undeformed. The 2nd step was performed without any load application to release stress and acquire stress equilibrium. Moreover, to apply constraints in this step, the left edge nodes were fixed similarly to the 3rd step. This equilibrated stress after stress release was defined as the residual stress inside the model for residual stress estimation. The generated residual stress after the stress release was exported to the 3rd step as the initial stress. In the 3rd step, the left edge nodes were assigned fixed boundary conditions to simulate the model as cantilever beam for the cutting. For the surface cut of 0.04-inch, 0.08-inch, 0.12-inch, element death was applied to top edge for surface depth of 1mm, 2mm and 3mm accordingly. For the intermediate EDM cut, 203 elements were selected for element death in a horizontal line of 500-micron thickness at different depths from the top surface. After the cut was applied, the deformation direction and *Y* directional displacements were calculated to compare with the warping of the experimental specimen.

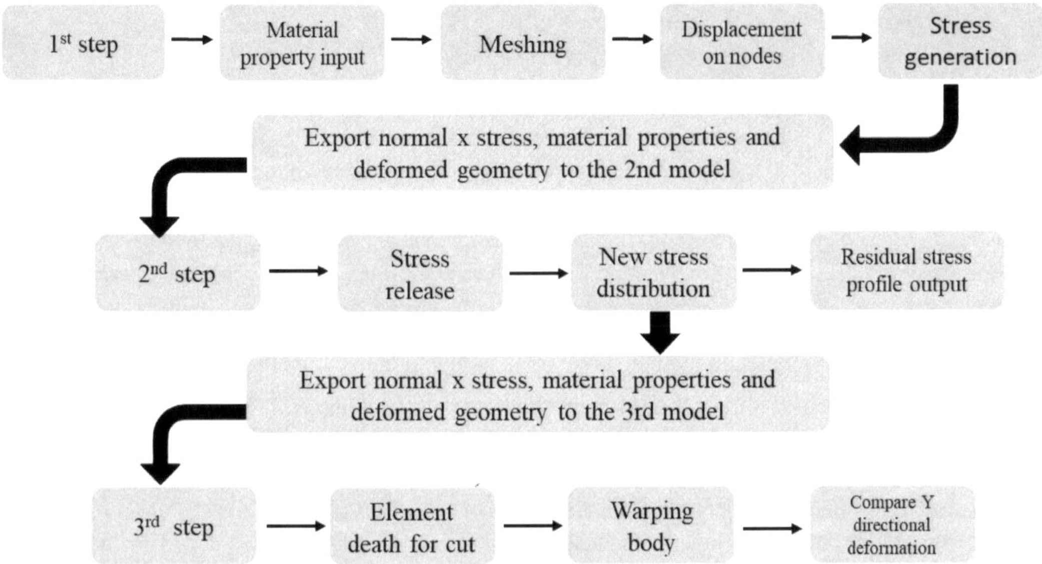

Figure 5: Flow chart showing the finite element analysis simulation steps

3. Result & Discussion

3.1 Tension Compression Profile Comparison

The first step was to determine if the specimen had tension or compression on the surface. Compression-tension-compression (CTC) and tension-compression-tension (TCT) profiles were compared for this reason. For the CTC profile, the displacement was applied in a gradual manner toward the core so that it would generate compression-to-tension stress toward the center after stress release. For the TCT profile, the loading was applied in the opposite direction. The generated residual stress profile is shown in Figure 6. After element death was applied on the top similar to the surface cut, the CTC model was warping in the anticlockwise direction, whereas the TCT model was warping in the clockwise direction, as shown in Figure 7. From the experimental results, the warping for the surface cut was in the clockwise direction, which matches the TCT model. Hence, only TCT models were considered for the residual stress estimation in the later part of this work.

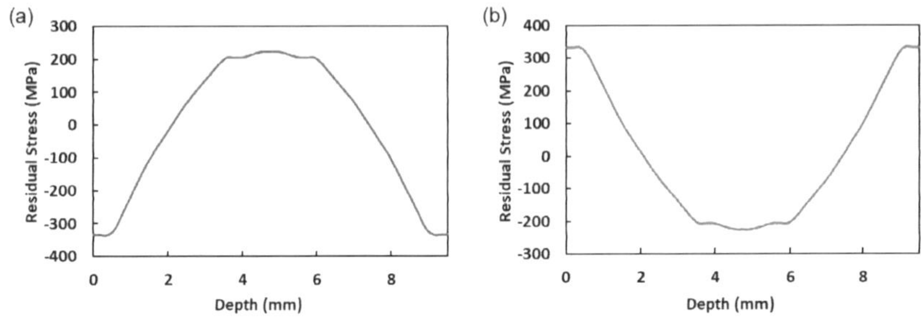

Figure 6: Residual x stress with depth (a) compression-tension-compression (b) tension-compression-tension

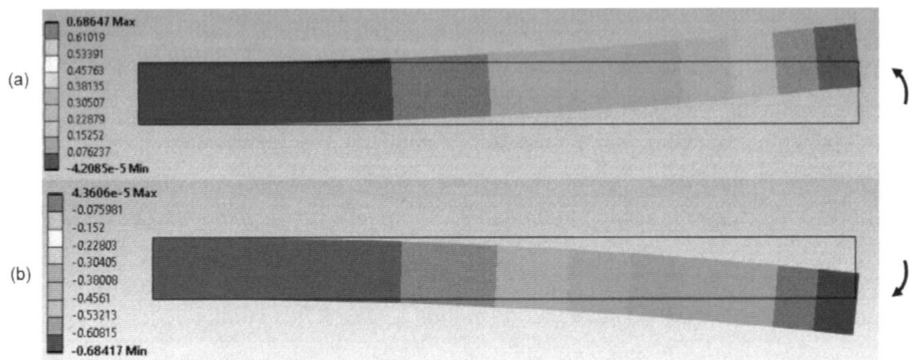

Figure 7: Y direction deformation on 2nd step (a) compression-tension-compression (b) tension-compression-tension

3.2 Residual stress estimation for surface cut

The end deflection of the surface cut samples was measured using a dial gauge and the deflections are listed in Table 1. Several FE models were developed to match the surface cut model deformation. The models were compared with all the experimentally measured deflections. The developed axial residual stress profile from the top edge to the bottom is shown in Figure 8. As the tension-compression-tension profile was already established, the 1st model was developed as a gradual tension to compression in the middle (Model 1). Even though the deformation direction is clockwise and the deflection matched the experiment for the 1mm surface cut, a significant difference in the results can be seen for other surface cuts (Table 1). Hence, the subsequent models were developed with different compression magnitude and region in the middle to reduce the *Y* directional deflection. In Model 2, the compression value is half of the maximum tension and the deflection for the 3mm cut showed lower error. However, the 2 mm cut result did not change much compared to Model 1. Hence, in Model 3, maximum tension and maximum compression were kept comparable. However, the deflection is high for the 2mm and 3mm cut. Therefore, Models 4 and 5 were developed such that these models had a large region of compression at the center. In Model 5, the deformation of all three cuts is close to the experimental results. In Figure 9, the deformation in the *Y* direction due to the 1 mm cut is presented for Model 5. The deflection direction and the deflection of 0.67 mm match the experiment.

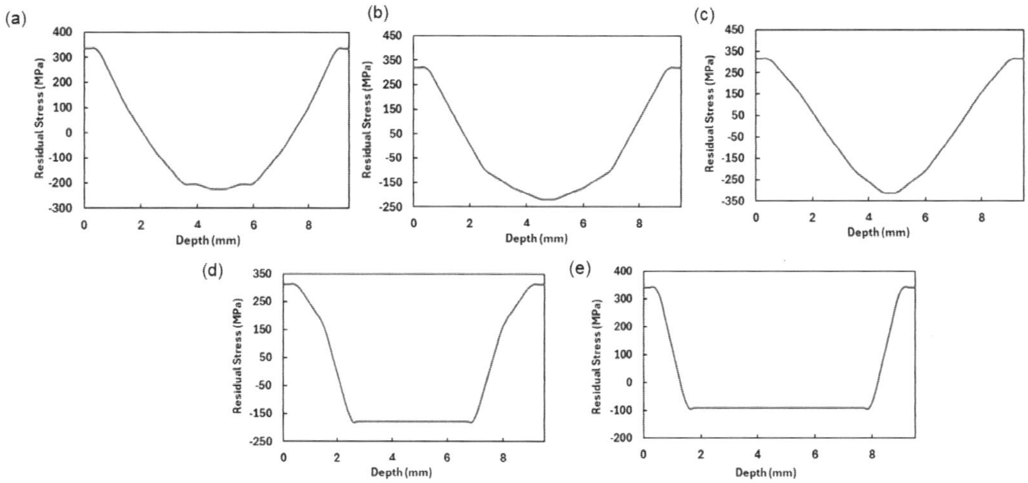

Figure 8: Residual stress profile (a) Model 1 (b) Model 2 (c) Model 3 (d) Model 4 (e) Model 5

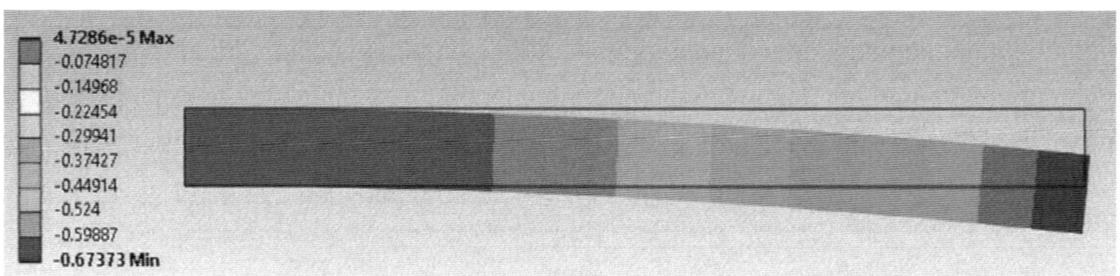

Figure 9: Y directional deformation for 1mm surface cut (Model 5)

Surface Cut	Experiment(mm)	Model 1	Model 2	Model 3	Model 4	Model 5
1mm (0.04in)	**0.68**	0.68	0.68	0.69	0.68	0.67
2mm(0.08in)	**0.99**	1.42	1.43	1.58	1.53	0.98
3mm(0.12in)	**1.42**	2.14	2.10	2.56	2.05	1.27

Table 1: Experimental and simulation deflection comparison for different surface cuts

3.3 Residual stress estimation for intermediate EDM cut

Although the deflections were measured for all four specimens, the specimen with the thinnest leaf was not considered for further analysis. This is because the top and bottom leaves were touching at the end of the specimen, and it is uncertain if the top leaf has its complete deflection or if part of it is obstructed by the bottom leaf. The measured deflections of both top and bottom leaves of the other three specimens are shown below in Figure 10. It is important to note that a reference flat surface of the specimen is required to measure the defection profile successfully. However, in the experimental setup used, it was challenging to position the uncut portion of the specimens perfectly flat. To circumvent this, the deflections of each specimen were measured from the tip of the leaf to 1" past the cut line in the length direction. The measurements of the 1" uncut region were used to establish a reference flat plane, and the rest of the measurements were transformed accordingly.

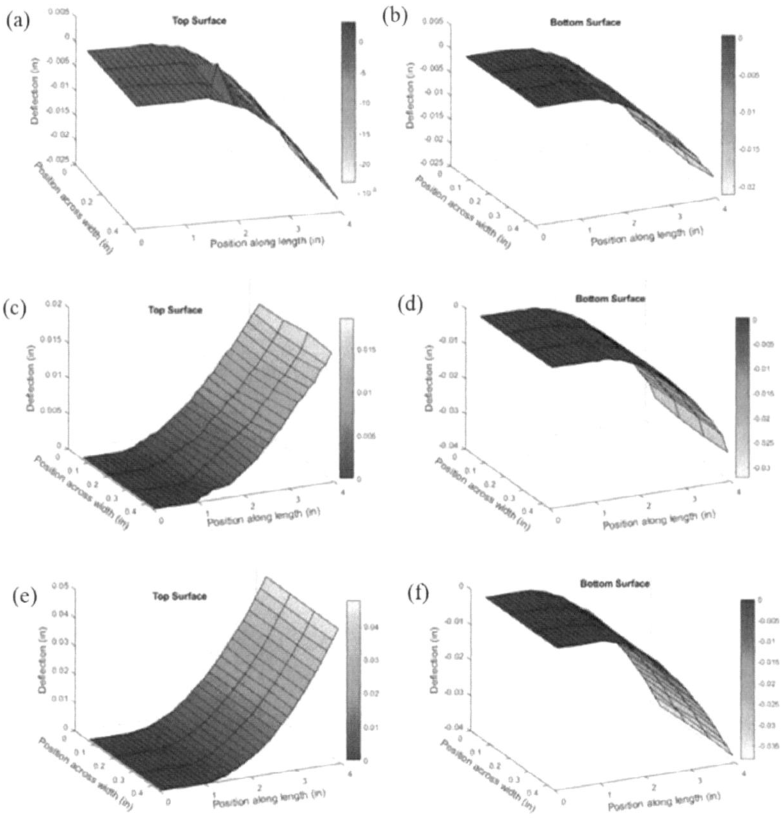

Figure 10: Measured deflection profile of the specimens with the depth of cut line at: (a-b) 0.047"; (c-d) 0.094" and (e-f) 0.1875". Positive and negative deflection values indicate outward and inward bending of the top surface, and vice versa for the bottom surface.

Residual stress profiles for the intermediate EDM cut were developed. As the Model 5 profile was the best match for the surface cut deflections, EDM cut results were compared to this model first. The deflection direction for the upper and lower leaf after element death matches the experiments. In every cut, the upper leaf and the lower leaf were bending outward (anticlockwise and clockwise, respectively). However, Model 5 showed high deflection in every case when compared with the experiment (Table 2). As the Y-directional deformation was higher than expected in this model, the next model was developed by considering lower compression to zero stress in the middle. Model 6 to Model 9 (Figure 11) have a tension-compression-tension profile with no stress in the center. Model 6 showed better results than Model 5 for the lower leaf. We gradually increased the zero-stress zone for Model 7 and Model 8. For Model 7, even if the lower leaf deflection showed less error, the error in the top leaf result was increasing. Hence, in Model 8, the compression was reduced to a smaller zone by applying the compressive displacement only in a line of nodes and fixing the adjacent lines in the model. This model showed less error for both the upper and lower leaf. However, the upper leaf deflection was still far from the experimental deflection. Thus, a new model (Model 9) was developed with higher compression for a smaller region of the specimen. The bottom leaf deflection for all the cuts showed the best match with this model. The deflection direction for the upper and lower leaf for Model 9 is presented in Figure 12. However, only the top leaf results for the middle or 4.5 mm thick top leaf EDM cut showed a close match, and the top leaf deflection increased for the other two EDM cuts at 2.5 mm and 3.5 mm depth from the surface.

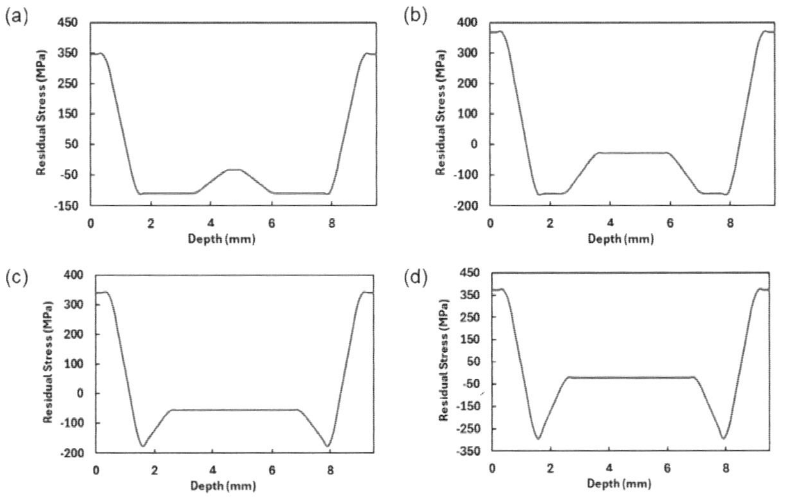

Figure 11: Residual stress profile (a) Model 6 (b) Model 7 (c) Model 8 (d) Model 9

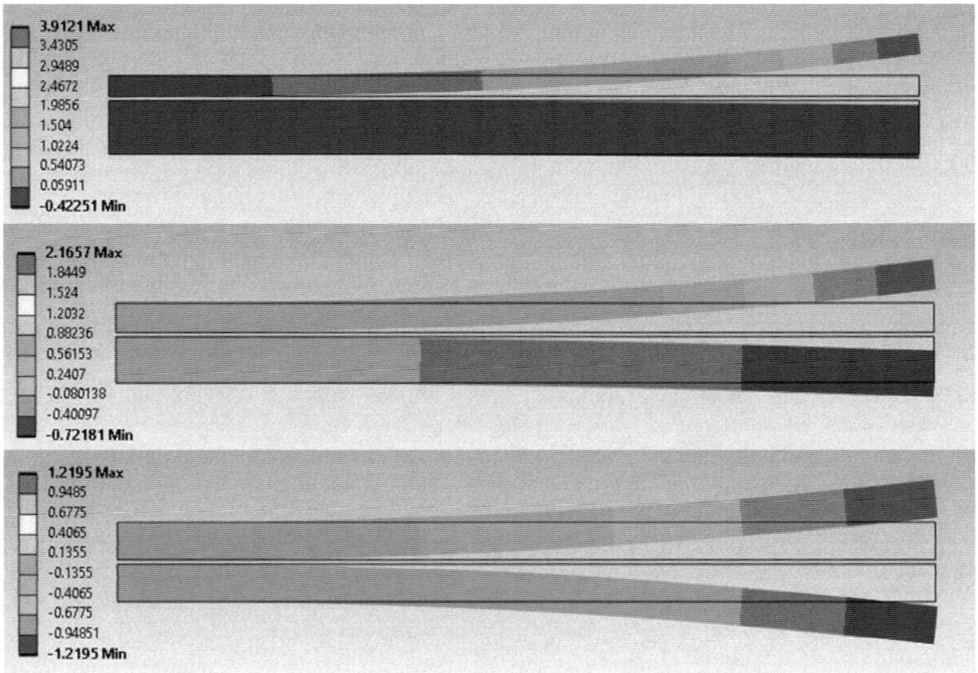

(a) Top leaf 2.5 mm

(b) Top leaf 3.5 mm

(c) Top leaf 4.5 mm

Figure 12: Y directional deformation for intermediate EDM cut (Model 9)

EDM Cut (top leaf thickness)		Experiment(mm)	Model 5	Model 6	Model 7	Model 8	Model 9
2.5mm (0.094")	top	**0.58**	3.38	3.65	4.2	3.47	3.91
	bottom	**0.508**	1.13	0.98	0.67	0.74	0.42
3.5mm (0.141")	top	**0.457**	2.81	2.89	2.91	2.33	2.17
	bottom	**0.813**	1.51	1.18	0.96	1.05	0.72
Middle (0.1875")	top	**1.19**	1.99	1.88	1.63	1.54	1.22
	bottom	**0.96**	1.99	1.88	1.63	1.54	1.22

Table 2: Experimental and simulation deflection comparison for different EDM cuts

In all the models, the deformation in the 2nd step due to the stress release was minimal. Figure 13 shows the deformation in the Y direction for Model 6. The endpoint Y deflection for model 6 in the 2nd step was 0.0020583 mm and in the 3rd step was 0.98346 mm. After subtraction, the Y directional deformation in the 3rd step changed to 0.98141mm or 0.98 mm. Since the change was minimal, this subtraction step may be neglected.

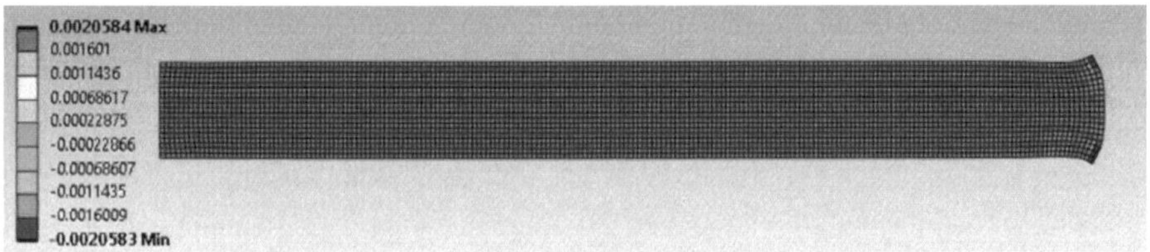

Figure 13: Y deformation in mm on 2nd step (model 6)

4. Conclusions

This work explores the use of surface and EDM cuts followed by finite element analysis to estimate residual stress in cold-drawn rectangular rods. For the surface cut, the deformation comparison provided good agreement with the experimentally measured deflections for the tension-compression-tension profile. Moreover, the residual stress profile for drawn bars is generally reported as a tension-compression-tension profile in the literature [12,13]. Model 5 is the best match with experimental deflections for the surface cut for a tension-compression-tension profile. However, in this approach, the deflection of the tip is only considered. The estimated stress profile can be optimized by automating the process using the line deformation. For the EDM cut samples, even if the lower leaf deflection showed a good match, the top leaf deflection had large errors. The proposed profile is a tension-compression-tension profile with no stress in the middle (Model 9). The machining for the surface cut can alter the profile, so the EDM cut profile can provide more accurate estimation. Hence, the surface milling and EDM cut were considered as two separate experiments with separate residual stress profiles. Plastic deformation can be inhomogeneous in the real drawing process, and it can cause a complicated residual stress state in the core. Thus, more closely spaced EDM cuts can provide more accurate stress distribution estimation. The stress profile can be modified further to reduce errors. Additionally, using a 3D geometry for the same modelling approach can provide more advantages in estimating residual stress in the longitudinal and transverse directions and a more accurate estimation as all directional deflections can be considered. We have already seen that the tip of the bar is deflected in both x and y directions (Figure 3c), and in this work, the Y directional deformation is considered. A 3D model can provide more scope to generate longitudinal and transverse stress, where stress relief can cause similar 2D deformation, leading to more accuracy in stress estimation.

Acknowledgments

The authors would like to acknowledge Subal Sharma from Michigan State University for their help with scanning the sample using the 3D scanner. The authors would also like to acknowledge the support provided by Iowa State University, Center for Nondestructive Evaluation through the Industry/University Cooperative grant.

References

[1] Yang, Zheng-Yan, et al. "In-situ imaging approach for investigating residual stress formation in rolling process." *International Journal of Mechanical Sciences* 248 (2023): 108220. https://doi.org/10.1016/j.ijmecsci.2023.108220

[2] European Convention for Constructional Steelwork. Committee on Stability. "Manual on stability of steel structures." European Convention for Constructional Steelwork, 1976.

[3] Huber, A. W., and L. S. Beedle. "Residual stress and the compressive strength of steel. Welding Journal, 33 (12), p. 589-s,(December 1954), Reprint No. 96 (54-3)." (1953). http://preserve.lehigh.edu/engr-civil-environmental-fritz-lab-reports/1510

[4] Dias, Vinicius Waechter. "Influência dos procedimentos de corte sobre as distribuições de tensões residuais em barras redondas trefiladas de aço ABNT 1045." (2013). Available from:http://hdl.handle.net/10183/109170.

[5] Niku-Lari, A., J. and Lu, and J. F. Flavenot. "Measurement of residual-stress distribution by the incremental hole-drilling method." *Journal of Mechanical Working Technology* 11.2 (1985): 167-188. https://doi.org/10.1016/0378-3804(85)90023-3

[6] Prime, Michael B. "Residual stress measurement by successive extension of a slot: the crack compliance method." Applied Mechanics Reviews (1999): Vol. 52, No. 2 pp. 75-96

[7] Bueckner, H. F., ''The Propagation of Cracks and the Energy of Elastic Deformation,'' Trans. ASME, 1958, **80**, pp. 1225–1230.
[8] Prime, Michael B. "Cross-sectional mapping of residual stresses by measuring the surface contour after a cut." J. Eng. Mater. Technol. 123.2 (2001): 162-168.
[9] Sun, Y. L., et al. "Evaluation of errors associated with cutting-induced plasticity in residual stress measurements using the contour method." Experimental Mechanics 57 (2017): 719-734. DOI 10.1007/s11340-017-0255-5
[10] E. Heyn and O. Bauer: *Int. Z. Metallographie 1* (1911) 16 50.
[11] Midha, P. S., and G. F. Madlen. "Residual stress relief in cold-extruded rod." *Metals Technology* 3.1 (1976): 529-533. https://doi.org/10.1179/030716976803392231
[12] Rocha, Alexandre da Silva, Rafael Menezes Nunes, and Thomas Hirsch. "Analysis by design of experiments of distortion potentials in drawn and induction hardened wire." *Materials Research* 15 (2012): 266-276. DOI: 10.1590/S1516-14392012005000020
[13] He, Shen, et al. "Residual stress determination in cold drawn steel wire by FEM simulation and X-ray diffraction." *Materials Science and Engineering: A* 346.1-2 (2003): 101-107. https://doi.org/10.1016/S0921-5093(02)00509-9

ICRS 12: Proceedings from the 12th International Conference on Residual Stresses, 2025
October 20–23, 2025; Detroit, Michigan, USA
https://doi.org/10.31399/asm.cp.icrs2025p0103

Effects of Multi-Axis Oscillations – A Novel Approach to Residual Stress Measurements via X-Ray Diffraction

James Pineault, Jeff Taptich
Proto Manufacturing Inc., 12350 Universal Drive, Taylor, Michigan, USA, 48180
xrdlab@protoxrd.com, jtaptich@protoxrd.com
Jeff Nantais, Mohammed Belassel
Proto Manufacturing Ltd., 6150 Morton Industrial Parkway, Lasalle, Ontario, Canada, N9J 3W3
jnantais@protoxrd.com, mbelassel@protoxrd.com

Abstract

X-Ray Diffraction (XRD) has been used for the measurement of residual stress (RS) since the 1920s, with numerous advancements in accuracy, precision, and speed since. XRD methods rely on the queried material to be quasi-isotropic, homogeneous, and to have a near-random grain orientation distribution; however, often this is not the case. When the coherent domain size (or grain size) is large relative to the irradiated volume, a non-random grain orientation distribution may be sampled, and the diffraction Debye ring may become "spotty." The presence of a spotty Debye ring introduces errors in the apparent diffraction peak position and, by extension, results in errors in the determination of RS. Traditionally, oscillations have been conducted in the same plane as the tilting plane in either Omega (Iso-inclination) or Chi (Side-inclination) modes to minimize experimental errors. However, there is no existing literature on Omega oscillations while measuring RS in Chi mode, or vice versa, or the use of simultaneous oscillations in both planes during measurements. In this novel study, experimental data were collected in both modes, with oscillations both individually and simultaneously, in both axes over a range of oscillation angles. Empirical predictive modeling was employed to compare with experimental data with the goal of optimizing RS measurement accuracy.

Keywords

X-Ray Diffraction, Residual Stress, Errors, Oscillation, Coarse Grains

Introduction

The collection of "spotty" x-ray diffraction (XRD) patterns has been observed since the inception of XRD as applied to the initial attempts by Max von Laue to characterize single crystal materials in 1912 [1]. More precise analyses of XRD patterns were soon more rigorously explained by Bragg [2,3,4]. With the research that followed in polycrystalline materials, the rapid development of purpose-built film cameras for use in powder diffraction analyses, such as the Debye-Scherrer camera in 1916 [5], were developed and refined for use on fine grain polycrystalline materials rather than single crystals. However, in many instances where polycrystalline materials were being analyzed, the characteristic Debye rings collected were found not to be continuous, unbroken rings, but rather Debye rings with "spotty" characteristics [6]. This observation will occur when the volume of material sampled by the incident x-ray beam does not contain a sufficient number of near-randomly oriented coherent domains or grains (see Figure 1).

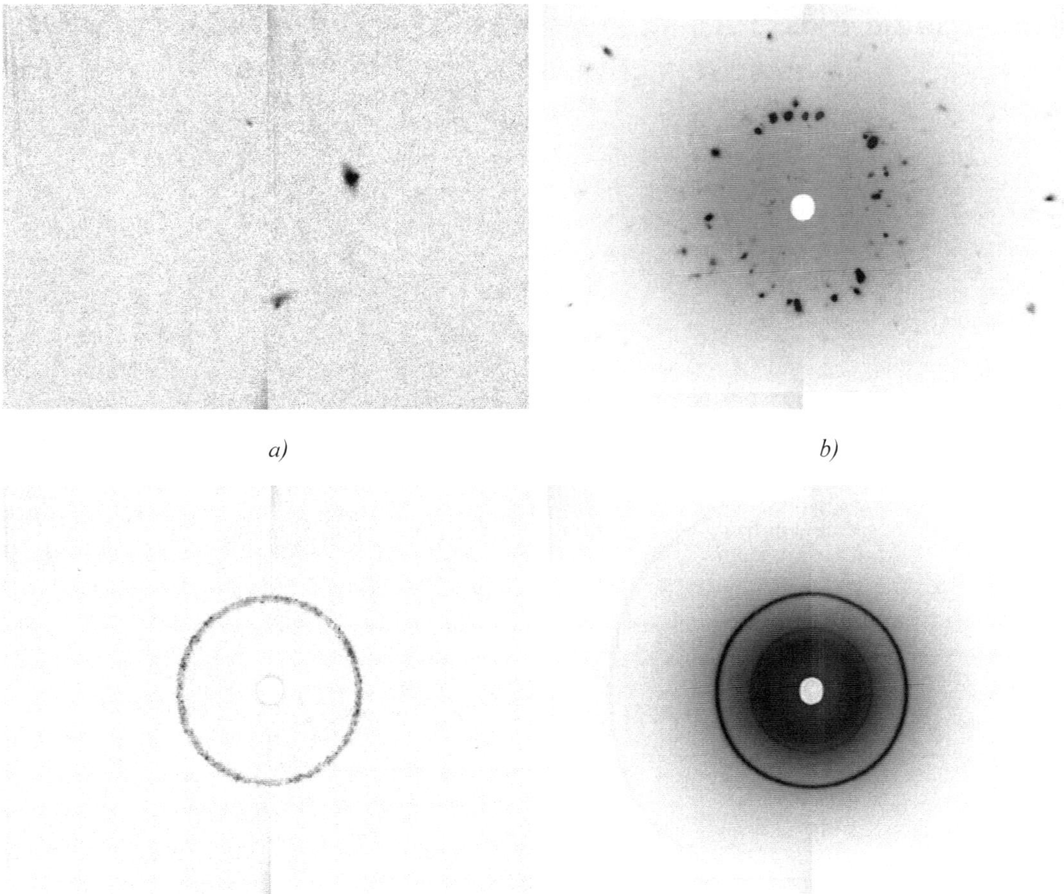

Figure 1 – Back-reflection XRD patterns from a Proto Laue-COS instrument using Mn x-radiation on nickel alloy specimens with grain size decreasing in the order a), b), c), d).

This scenario makes an objective analysis of the diffraction peak position more difficult and leads to inaccuracies in its determination. This effect can be caused by either preferred orientation or where the coherent domain size is large relative to the sampling volume. To mitigate this problem, and to better randomize the grain orientation distribution sampled, test samples were spun, effectively randomizing their orientation and making "spotty" patterns smoother, simplifying the analysis and improving the accuracy of diffraction peak position determination. This effect (and associated corrective measures) was persistent through the development of the modern diffractometers in the 1970's [7].

Although commonly used for phase analysis, XRD has also been used for the measurement of residual stress (RS) since the 1920s [8] with recent technological developments leading to its widespread use for the measurement of RS. As with powder diffraction analyses on polycrystalline materials, the accuracy of XRD based RS measurements may also be affected when diffraction patterns are "spotty".

As such, oscillation techniques have been integrated into the measurement of RS. Several investigations have been carried out to assess the effects and efficacy of various oscillation techniques in different geometries and experimental conditions [9,10,11,12] with the measurement of residual stress as the desired output.

Traditionally, oscillations have been conducted in the same plane as the tilting plane in either Omega (Iso-inclination) or Chi (Side-inclination) modes to minimize experimental errors [13]. However, there is no existing literature on

Omega oscillations while measuring RS in Chi mode, or vice versa, or the use of simultaneous oscillations in both planes during measurements.

The following sections explain how experimental data were collected in both modes, with oscillations in both planes over a range of angles. Comparisons to empirical predictive modeling are also presented.

Experimental Methods

The x-ray diffraction method uses the distance between crystallographic planes, i.e. d-spacing, as a strain gage. When the material is in tension, the d-spacing increases and when the material is under compression the d-spacing decreases. This method can only be applied to crystalline, polycrystalline and semi-crystalline materials. The diffraction angle 2θ is measured experimentally and the d-spacing is then calculated using Bragg's law $n\lambda=2d\sin\theta$ [2,3,4]. The presence of residual stresses in the material produces a shift in the x-ray diffraction peak angular position which is directly measured by the detector [14]. The determination of the shift in the Bragg angle is thus interpreted in terms of shifts in the lattice spacing. Once the d-spacing is measured for unstressed (d_0) and stressed (d) conditions, the strain $\varepsilon_{\phi\pi}$ is calculated. Using the $\sin^2\psi$ technique where a series of d-spacings are calculated, stresses can be determined using an equation derived from Hooke's law for quasi-isotropic, homogeneous materials [15]:

$$\varepsilon_{\Phi\psi} = \frac{1}{2}S_{2(hkl)}.[\{\sigma_\Phi - \sigma_{33}\}.sin^2\psi + t_\Phi.sin2\psi] + \varepsilon_0 \quad \textbf{(1)}$$

Where ½ $S_{2(hkl)}$ is the x-ray elastic parameter of the material for the selected crystallographic plane [16], σ_Φ is the stress in the direction of the measurements, σ_{33} is the stress in the direction of the normal to the surface of the specimen in the reference frame, ψ is the angle subtended by the bisector of the incident and diffracted x-ray beams and the specimen surface normal, ε_0 is the normal crystallographic strain, and $\varepsilon_{\Phi\psi}$ is the crystallographic strain at a given ψ tilt (see Figure 2).

The detailed RS measurement parameters used in this study are shown in Table 1. A spot size of 1-2 mm diameter (or in some cases a 2 mm x 5 mm rectangular spot) was typically used with ~ 22Ψ tilts distributed from 0 to ~±45°.

Table 1: Stress measurement parameters of different material analyzed.

Material	X-Ray Tube Anode	Diffraction Plane (hkl)	2θ Bragg Angle (°)	X-Ray Elastic Constant ½$S_{2\ hkl}$ (MPa^{-1})
Martensitic Steel	Cr	211	156	5.92 x 10^{-6}
316L Stainless Steel	Mn	311	153	7.14 x 10^{-6}
IN718	Mn	311	152	6.33 x 10^{-6}
Ti-6Al-4V	Cu	213	142	11.89 x 10^{-6}

Prior to applying XRD to the characterization of RS, a comprehensive approach to the measurement protocols is employed to address instrument calibration and the validation of results. The five steps required to calibrate instrumentation and validate stress measurement results were performed as per ASM Handbook Volume 11 to obtain results with a high level of confidence [17]. In this study, all residual stress measurements were performed using the Proto LXRD Micro-Area instrument which can measure RS in both Omega and Chi geometries (modified Chi may also be used but was not employed in this study), thus making it possible to oscillate the Chi, the Omega, or both axes during a RS measurement in either Omega or Chi modes (see Figure 2).

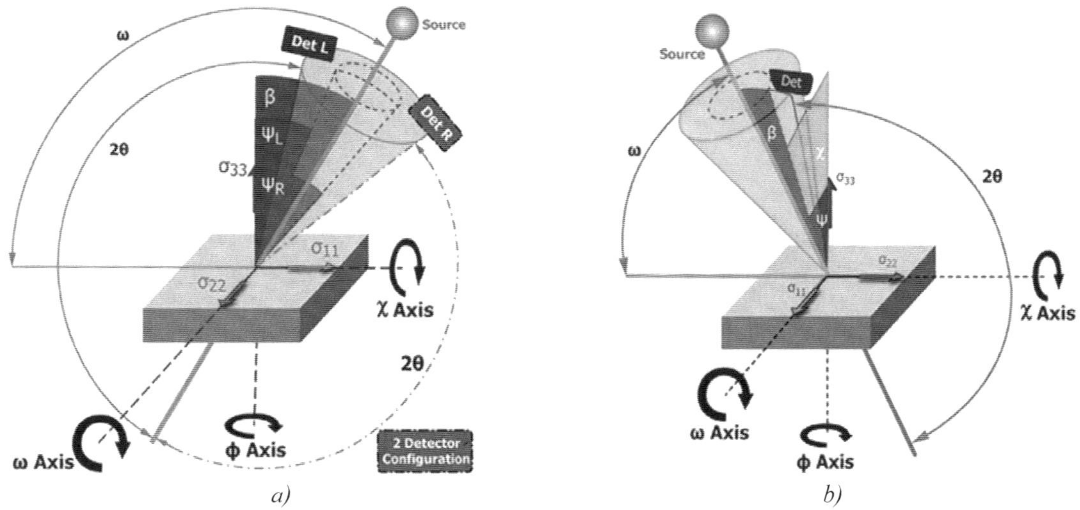

Figure 2 – Schematic representation of RS measurements in a) Omega and b) Chi modes [15].

Oscillations up to +/-13° for each Omega and Chi axes, each individually, and simultaneously were used in the collection of the RS data in both modes. Additionally, proficiency standards as well as samples with material conditions typically encountered in engineering applications were also characterized using select, optimized oscillation conditions.

Results and Discussion

Although predictions have been made regarding the relative error in RS results via analytical models [13], these were limited to either Chi oscillations in Chi mode, or Beta (i.e. the x-ray beam angle β shown in Figure 2, $\psi_{1,2}=\beta\pm(\pi-2\theta)/2)$) oscillations in Omega mode. To better understand the systematic errors introduced when Chi oscillations are employed in Omega mode (or Beta oscillations in Chi mode) a series of RS measurements were collected using various "Beta only" oscillations, "Chi only" oscillations, and "Beta + Chi" oscillations in both Omega and Chi modes on martensitic steel high stress proficiency standards using Cr x-radiation. The results obtained can be seen in Figure 3.

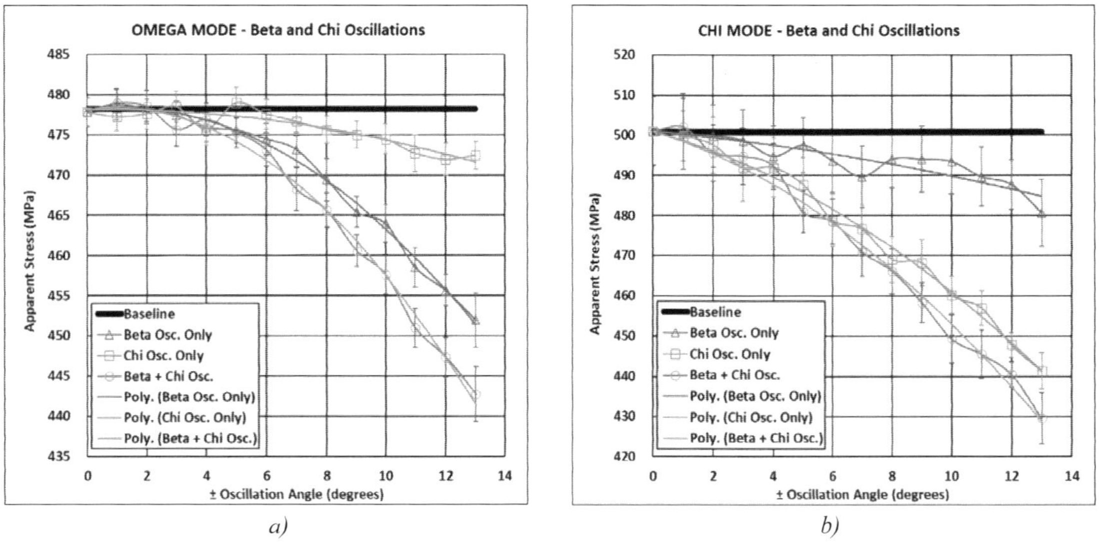

Figure 3 – Plot of apparent RS for various axis oscillations (±) in a) Omega mode and b) Chi mode on martensitic steel high stress proficiency standards.

The RS measurements in Omega mode were performed on a high stress proficiency standard with an established RS value of -478 ± 35 MPa. The data in Figure 3 indicate that Chi oscillations in Omega mode have a smaller effect on the apparent systematic error than Beta oscillations in Omega mode for a given oscillation angle (±), and that their effect is approximately additive when performed simultaneously. The RS measurements in Chi mode were performed on another proficiency standard with an established RS value of -501 ± 35 MPa. These data indicate that Beta oscillations in Chi mode have a smaller effect on the apparent systematic error than Chi oscillations in Chi mode for a given oscillation angle (±), and that their effect is also approximately additive when performed simultaneously.

When comparing the results obtained in Chi mode vs. Omega mode, the use of an Omega mode configuration results in smaller relative systematic errors than when similar oscillations are performed in Chi mode. For example, when both Beta and Chi oscillations of ±10° were used simultaneously, the relative errors are approximately 3.9% and 10.2% for Omega and Chi modes respectively. A "Beta only" oscillation of ±10° translates into relative errors of approximately 3.12% and 2.18% for Omega and Chi modes respectively. Thus, in Omega mode, keeping Beta oscillations low minimizes systematic errors whereas Chi mode is more tolerant of higher magnitude Beta oscillations as compared to Omega mode. Similarly, a "Chi only" oscillation of ±10° translates into relative errors of approximately 0.81% and 7.99% for Omega and Chi modes respectively. Thus, in Chi mode, keeping Chi oscillations low minimizes systematic errors whereas Omega mode is significantly more tolerant to higher magnitude Chi oscillations as compared to Chi mode. These comparisons indicate that Chi mode is more susceptible to systematic errors than Omega mode overall for similar oscillation ranges. This should be accounted for when defining a data collection and oscillation strategy to be used when collecting RS data in either Omega or Chi geometry. Careful selection of oscillation ranges can maximize their effect insofar as limiting the systematic errors introduced, while concomitantly minimizing the random experimental standard errors due to coarse grain effects.

While it is known that random experimental standard errors due to coarse grain size may be minimized by increasing the x-ray beam aperture size (and thus the number of grains sampled), in instances where a steep surface RS gradient exists, it can often be useful to keep the x-ray spot size as small as possible while applying oscillations during data collection. By carefully combining Chi and Beta oscillations, systematic errors can be limited while significantly improving random experimental standard errors in RS measurements. Based on the data obtained in Figure 3, an empirical predictive model was established. Table 2 provides an example of what the empirical predictive model predicts the relative systematic errors to be for a RS measurement given certain the selected axis oscillation schemes.

Table 2 – Predicted systematic relative error for various oscillation schemes in Omega mode.

± Oscillations (Degrees)		**Relative Error (%) in RS**
Beta (Ψ)	**Chi**	
0	0	0
5	0	0.56
0	10	0.81
5	10	1.37

Using these oscillation schemes, their effect on the actual experimental standard errors was evaluated for RS measurement on various samples. The four oscillation schemes defined in Table 2 were applied to the measurement of RS on both Low and High stress proficiency standards of a few example materials. The resultant experimental errors in the RS results are plotted in Figure 4.

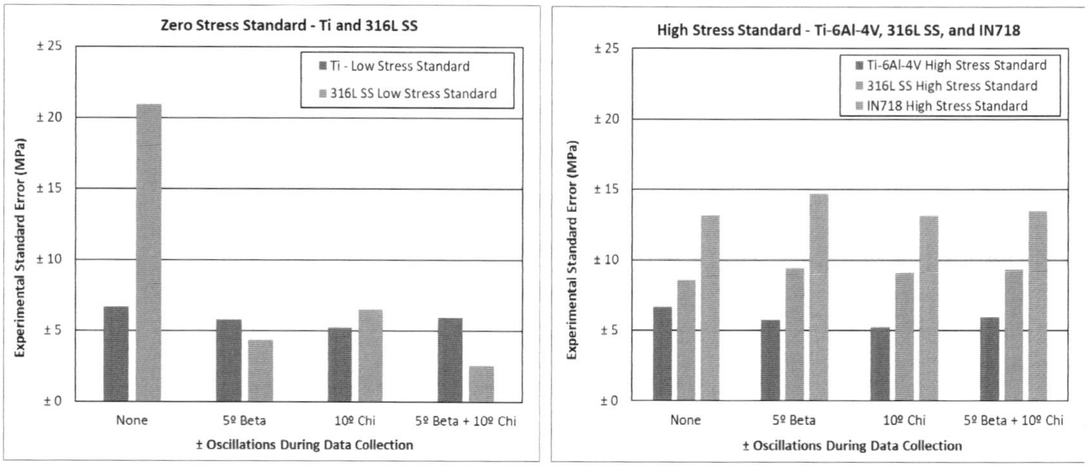

a) *b)*

Figure 4 – Experimental standard errors from RS measurements using various oscillation schemes on a) Low and b) High stress proficiency standards.

In the case of the Low stress proficiency standards, the data in Figure 4 indicate that the RS measurement results may benefit from a small amount of oscillation since such "zero stress powders" sometimes exhibit grain size effects. In the case of the High stress proficiency standards, their method of preparation and associated cold working typically results in a relatively fine grain size distribution, thus oscillations seem to have little or no discernable effect on the experimental standard errors found in the measurement of RS on them. Moreover, in the case of the Ti-6Al-4V, 316L SS, and IN718 High stress proficiency standards, the absolute RS results were within 3.7 MPa, 14.3 MPa, and 17.2 MPa of their nominal values respectively, while within 6.3 MPa, 4.7 MPa, and 4.5 MPa respectively of the empirical predictive model for the maximum oscillation ranges employed of ±5° Beta and ±10° Chi simultaneously. This result indicates that the predictive trends found in Figure 3a) work within a reasonable tolerance of less than ~±7 MPa.

Given the confidence that the oscillation schemes defined in Table 2 maintain systematic errors as determined by the data plotted in Figure 3, they were applied to the collection of data on samples with an unknown RS in Omega mode. The experimental standard errors in these RS measurements are plotted in Figure 5 for each of the various oscillation schemes defined in Table 2.

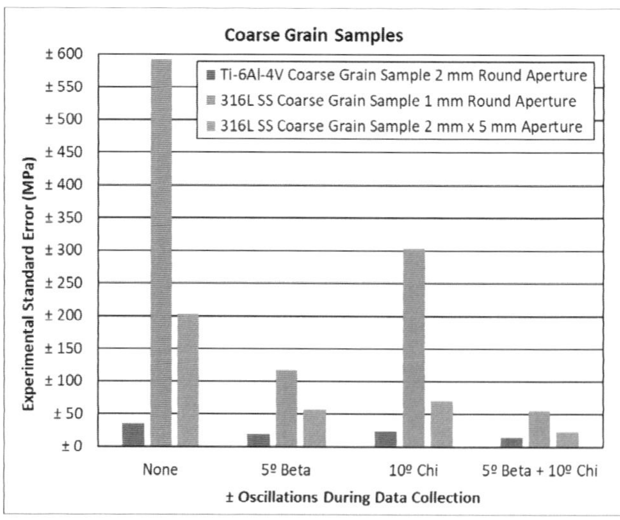

Figure 5 – Experimental standard errors from RS measurements using various oscillation schemes on samples with relatively coarse grains.

The results in Figure 5 indicate that experimental errors were reduced significantly with ±5° Beta oscillations only, and more so using ±10° Chi oscillations only, with the maximum effect in reducing the experimental standard error found using both ±5° Beta and ±10° Chi oscillations simultaneously. It is expected that doing so incurs a systematic error of 1.37% but reduces the experimental standard errors by 57%, 91%, and 89% (or ±20 MPa, ±537 MPa, ±180 MPa) for the Ti-6Al-4V coarse grain sample using a 2 mm round aperture, the 316L SS coarse grain sample using a 1 mm diameter round aperture, and the 316L SS coarse grain sample using a 2 mm x 5 mm rectangular aperture respectively. It is important to note that two sets of experiments performed on the 316L SS sample, one using a 1 mm round aperture and one using a 2 mm x 5 mm rectangular aperture, show that even where a smaller aperture is used, experimental errors can be significantly reduced. In such instances, increasing the amount of oscillation, thus increasing systematic errors, may be balanced and even improved upon by the decrease in the experimental standard errors.

For example, in the case of the 316L SS coarse grain sample using a 1 mm diameter round aperture, the standard experimental error was found to be ±56 MPa using both ±5° Beta and ±10° Chi oscillations simultaneously. Because the systematic error is predicted to be 1.37% of the RS in this case (with the RS determined to be -669 MPa) it would translate into a systematic error of ~±10 MPa (or a potential ~±17 MPa if the ~±7 MPa maximum expected deviation from the empirical predictive model is accounted for). This means that the oscillation ranges used while collecting these data could be further optimized until the experimental standard error found in the RS measurement approaches the expected systematic error. However, in the case of the 316L SS coarse grain sample using a 2 mm x 5 mm rectangular aperture, the experimental standard error was found to be ±23 MPa using both ±5° Beta and ±10° Chi oscillations simultaneously. Therefore, there is likely not much to gain by increasing the oscillation angles in terms of an improved experimental error because the ±23 MPa experimental standard error is very close to the potential ~±17 MPa variance if all empirical predictive model tolerances are accounted for.

In the case of the Ti-6Al-4V coarse grain sample using a 2 mm diameter round aperture, the standard experimental error was found to be ±15 MPa using both ± 5° Beta and ± 10° Chi oscillations simultaneously. Because the systematic error is predicted to be 1.37% of the RS in this case (with the RS determined to be -206 MPa) it would translate into a systematic error of ~±3 MPa (or a potential ~±10 MPa if the ~±7 MPa maximum expected deviation from the empirical predictive model is accounted for). Therefore, there is likely not much to gain by increasing the oscillation angles in terms of an improved experimental error because the ±15 MPa experimental standard error is very close to the potential ~±10 MPa variance if all empirical predictive model tolerances are accounted for.

Conclusions

Since XRD methods rely on the queried material being quasi-isotropic, homogeneous, and having a near-random grain orientation distribution, deviations from these conditions typically lead to increased experimental standard errors. This is because when the coherent domain (or grain) size is large relative to the irradiated volume, a non-random grain orientation distribution may be sampled, and the diffraction Debye ring may become "spotty". In such cases, errors in the apparent diffraction peak position, and thus errors in RS measurement results, may ensue. The observations herein confirm the advantageous effects of oscillations conducted in the same plane as the tilting plane in either Omega (Iso-inclination) or Chi (Side-inclination) modes to minimize experimental errors, as well as Beta oscillations while measuring RS in Chi mode, Chi oscillations while measuring RS in Omega Mode, and the use of simultaneous Beta and Chi oscillations during RS measurements in either mode.

Improvements in observed standard experimental errors in RS measurements on coarse grain materials were found when both Beta and Chi Oscillations were employed, with the most significant improvements found when applied simultaneously in an optimized fashion. In Omega mode, it was found that Chi Oscillations had the greatest effect in reducing experimental errors while incurring the smallest systematic errors, whereas in Chi mode, Beta oscillations had the greatest effect in reducing experimental errors while incurring the smallest systematic errors in that mode. However, the smallest errors overall were found to be achievable in Omega mode because the systematic errors were lower for similar oscillation ranges as compared to those in Chi mode. Thus, Omega mode enables the collection of data over wider oscillation ranges resulting in smaller overall errors given the inherent systematic errors for the specific oscillation ranges employed. To enable the selection and use of the optimal Beta and Chi oscillations to be employed during the measurement of RS in coarse grain materials, an XRD instrument capable of tilting in both Omega and Chi modes is therefore required.

Although none were included in this study, similar experiments could be performed on coarse grain samples with unknown RS using Beta and Chi oscillations while operating in Chi mode. However, based on the increase in systematic errors expected for oscillations in Chi mode, it is likely that the benefits will not be as marked as those found using oscillations in Omega mode.

Acknowledgments

The authors would like to acknowledgments Brian Simpson for collecting the images in Figure 1.

References

[1] Friedrich W., Knipping P., von Laue M., "Interferenz-Erscheinungen bei Röntgenstrahlen," Sitzungsberichte der Mathematisch-Physikalischen Classe der Königlich-Bayerischen Akademie der Wissenschaften zu München, 1912: 303.

[2] W.L. Bragg, "The diffraction of short electromagnetic waves by a crystal," Proc. Cambridge Philos. Soc. 17 (1913) 43-57.

[3] W.L. Bragg, "The structure of some crystals as indicated by their diffraction of X-rays," Proc. Roy. Soc. Lond. A 89 (1913) 248-277.

[4] W.H. Bragg, W.L. Bragg, "The reflection of X-rays by crystals," Proc. Roy. Soc. Lond. A vol. 88, No. 605 (1913) 428 438.

[5] P. Debye, P. Scherrer, "Interferenzen an regellos orientierten Teilchen im Röntgenlicht," Phys. Z. 17 (1916) 277-282.

[6] B.D. Cullity, "The Elements of X-Ray Diffraction", 2nd Edition, Addison-Wesley, Reading., 1978

[7] C.F. Jatczak, J. A. Larson, S.W. Shin, "Retained Austenite and Its Measurements by X-Ray Diffraction", SAE/SP-80/453, Section 4, Warrendale PA, 1980.

[8] H.H. Lester, R.H Aborn, Army Ordnance, Vol. 6, 1925-26.

[9] M.R. James, "The Use of Oscillation on PSD-Based Instruments for X-Ray Measurements of Residual Stress", Experimental Mechanics, Vol. 27, No, 2, 164–167, 1987.

[10] Mitchell, C. M., "Stress Measurements by X-Ray Diffractometry", United States Patent Number 4,561,062, 1985.

[11] J.A. Pineault, M.E. Brauss, "Measuring Residual and Applied Stress Using X-ray Diffraction on Materials with Preferred Orientation and Large Grain Size", Advances in X-Ray Analysis, Vol. 36, New York 1993.

[12] D. Backman, L. Li, J.H. Sa, J. Pineault, S. Carlson, M. Stanfield, "A Novel X-ray Diffraction Procedure for Determining Residual Stresses Around Cold Expanded Holes", Journal of Materials Engineering and Performance, Volume 33, pages 4074–4087, ASM, (2024)

[13] B. Dionnet, M. Belassel, "Influence de la Taille des Grains, de la Géometrie, et du Biphasage sur la Détermination des Contraintes par DRX", Laboratoire de Microstructure et Méchanique des Matériaux LM-URA CNRS 1219, Ensam, Paris, Février 1995.

[14] Noyan, I. C., Cohen, J. B., "Residual Stress Measurement by Diffraction and Interpretation", Springer-Verlag New York Inc., 1987.

[15] ASTM E2860, "Standard Test Method for Residual Stress Measurement by X-Ray Diffraction for Bearing Steels".

[16] ASTM E1426, "Standard Test Method Determining the X-Ray Elastic Constants for Use in the Measurement of Residual Stress Using X-Ray Diffraction Techniques".

[17] J. Pineault et al, "X-Ray Diffraction Residual Stress Measurement in Failure Analysis", ASM Handbook, Vol. 11, (2021), pp.221-239.

ICRS 12: Proceedings from the 12th International Conference on Residual Stresses, 2025
October 20–23, 2025; Detroit, Michigan, USA
https://doi.org/10.31399/asm.cp.icrs2025p0111

Geometric effects of linear friction welding on residual stresses

Jiaxi Zhang[1], Anastasia Vasileiou[1], Xenofon Gogouvitis[2], Michael Preuss[3,4], Mike Smith[1]

[1] The University of Manchester, Department of Mechanical and Aerospace engineering, Manchester, United Kingdom

[2] Safran landing system, Gloucester, United Kingdom

[3] Monash University, Department of Materials Science and Engineering, Clayton, Victoria, Australia

[4] The University of Manchester, Department of Materials, Manchester, United Kingdom

Abstract

Linear friction welding is a promising solid-state joining technology, which has been widely used for the manufacturing of blisks and preforms of complex components in aerospace engineering. As other welding processes, strong local heat input leads to the formation of high tensile residual stresses during linear friction welding. Numerical prediction of welding residual stresses is critical for the assessment of structure integrity. Compared to traditional welding, LFW includes unique processing variables that have potential influence on residual stresses distribution. For instance, presented research shows that applied pressure has significant effect on residual stresses. However, the geometric effects of components on residual stresses has not been investigated yet. One key attraction of LFW is to provide the possibility to manufacture preforms of complex parts with less raw material, thus, the welding interface with various cross-sections are inevitable. Based on validated finite element modeling, we try to further investigate the geometric effects on residual stresses and build the relationship between geometric effects and residual stress distribution. It deepens the understanding of residual stress evolution patterns of linear friction welding and potential optimizing directions.

Introduction

Linear friction welding (LFW) is a solid-state joining technology where components are joined by using linear reciprocating motion to generate high enough frictional heat. It extends the utilization of friction welding from circular parts in traditional rotary friction welding to non-circular parts. A classic application case of LFW is the manufacturing of bladed disks (blisks), which replaces traditional mechanical joining of dovetail joint between disks and blades. Thanks to LFW, both performance and weight of blisks are improved [1]. LFW also shows great potential for other applications, such as the repair of damaged components and manufacturing of preforms of complex components [2].

Due to local heat input followed by rapid cooling, LFW also forms strong tensile residual stresses at the weld line (WRS). It is generally recognized that tensile WRS are detrimental to welding structure integrity [3]. More specifically, WRS have been coded into internationally-recognized structural integrity assessment procedures, such as R5 and R6. Thus, characterization and modelling of WRS are crucial to the application of LFW.

Most LFW modelling research currently focuses on processing simulations to understand influences of various processing parameters, such as applied pressure and rubbing velocity. These works are usually validated by axial shortening [4] [5] or flash morphologies [1] [6]. However, this type of work does not consider the cooling processes during which WRS are generated. To date, there has been relatively little research on developing WRS models for

LFW apart from Bühr et al. [7] [8] who undertook WRS experimental and modelling analysis of Titanium LFW.

As mentioned, LFW provides potential to reduce buy-to-fly ratio by producing preforms of complex components. Thus, various welding interface are inevitable. It is meaningful to explore geometric effect on WRS. In this paper, a simplified LFW modelling method is presented, which has been validated with experiment results [7]. The model has been used here for four parts with reprehensive and relevant cross-sections to show the impact of geometric effects at the welding interface on WRS in LFW.

Methodology

A generic pattern reported previously for LFW is that WRS are strongly related to geometric effects [2], where the component of WRS is largest in the direction of the longest contacting surface, followed by the direction of the shortest contacting surface, while the direction normal to the weld interface has the lowest value. Such pattern is related to a single component with a simple rectangle cross-section. In this paper, four different cross-sections modelling objectives were designed, as shown in Figure 1, including two rectangles with different aspect ratio, one square and one diamond shaped cross-sections. To control variables, four components have same area and input heat flux with same time and power.

A 3D fully-coupled thermal-mechanical modeling approach has been developed to predict WRS in LFW based on commercial FEA software ABAQUS. In this model, two complex physical factors during LFW, severe deformation and frictional behavior, are neglected. They are secondary factors in terms of WRS prediction, because 1) most interface material experienced severe deformation would be finally expelled to form flash and has no significant effect on residual stress. 2) frictional behavior can be replaced by a simple surface heat flux to present the relevant thermal effect. Therefore, they would only have to be considered if the friction welding processing behavior was to be predicted. Compared to similar thermal-mechanical modelling [7] [9], one of the main differences of the present model is that it neglects the axial shortening [7] and flash formation [9], which helps to further reduce meshing workload and allows using a relatively coarse interface element to reduce computational cost.

Linear brick elements with reduced integration, coupled temperature and displacement (Abaqus designation C3D8RT) is used to adapt fully-coupled way. Due to the symmetry of all proposed parts, only one eighth of the two parts joined together is considered in the finite element model. To model effects of heat generation and pressure during LFW, a series of boundary conditions referred as [7], such as symmetry constraints, applied pressure, clamping pressure and surface heat flux are applied to the part, as shown in Figure 1. Note, the applied pressure and clamping forces are maintained for extra 15 s after the heating process and then released, which follows recorded scenario in practical manufacturing [7]. Finally, the part is cooled to room temperature to predict WRS. The model was validated with "weld 4" case by residual stresses prediction [7], as well as predicted temperature fields also validated with measured thermocouple data in [8]. More detailed modelling implementations and validation results will be presented in a subsequent publication.

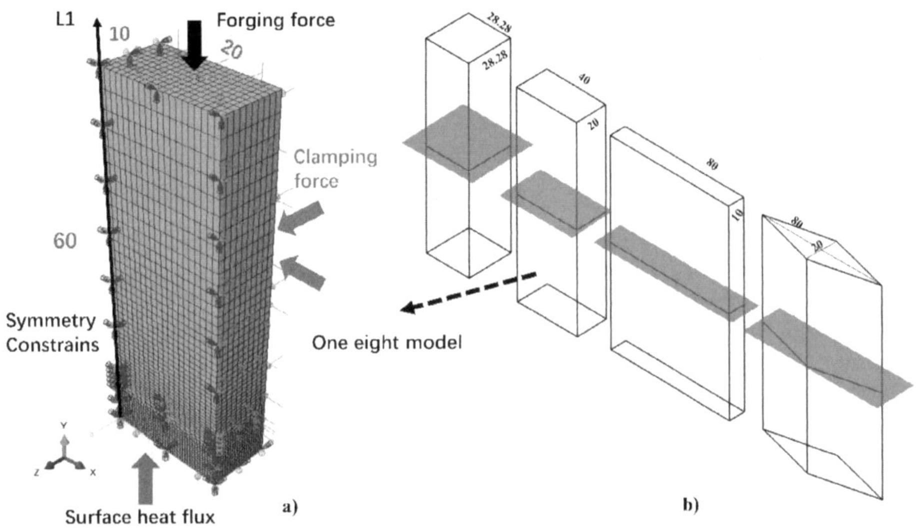

Figure 1: a) The schematic of one eight FEA model arrangements of meshing and boundary conditions, b) The geometric sizes of different modelling parts

Results and Discussion

For the sake of fast and convenient comparison of four different parts, the macro WRS distribution of two representative cross-sections, transverse-normal plane and longitudinal-normal plane are shown in Figure 2 and Figure 3 respectively. Firstly, based on comparison of single component itself, all modelling results confirm the mentioned distribution patterns of residual stress components again. WRS in the direction of the longest contacting surface has the largest peak value. The WRS in the direction normal to the weld interface has the lowest value and apparently lower than WRS along welding interface by an order of magnitude, which can be seen in both Figure 2 and Figure 3. Transverse-normal planes in Figure 2 present WRS tend to concentrate at the edge of components and attenuate to the center direction in all cases of cube component. It should be noted that concentration areas of WRS are vertical to the corresponding direction of residual stress components. On the contrary, area along residual stress components direction has lower corresponding WRS component values and central areas have larger WRS values than edge area.

Based on above pattern, the aspect ratio of rectangle obviously plays a crucial role in WRS distribution. The cross-sections with the large aspect ratio, such as 80× 10 case in Figure 2 would make two WRS concentration areas overlap or further separated, which causes that most cross-section areas are at high WRS state or low WRS area. Furthermore, owing to the same length and width, S11(X-direction WRS) and S33 (Z-direction WRS) in square part has same distribution and magnitude. As the increase of aspect ratio of interface, from 40×20 to 80× 10, the ratio of area of high residual stress components also grows. However, diamond cross-section part exhibits a different distribution patterns as cube parts, where most transverse-normal planes show the high WRS, which is contrast to cube parts, especially for secondary residual stress component S33, namely WRS along Z direction.

In the view of longitudinal-normal planes as shown in Figure 3, these plots help to further detect the WRS distribution along depth direction (Y-axis). It is not surprising that S11 has strong concentration effect at the edges, since it can be deduced from above pattern in Figure 2. In general, the width of stress band of high residual stress components, such as S11 and S33, are very narrow and no more than 5 mm, which is significantly lower than fusion welding.

Theoretically, this feature challenges the residual stress measurements since it asks higher resolution to characterize dramatic stress gradient. It is noteworthy that S33 keeps much larger values throughout the most area in diamond part, which is in sharp contrast to other cube parts.

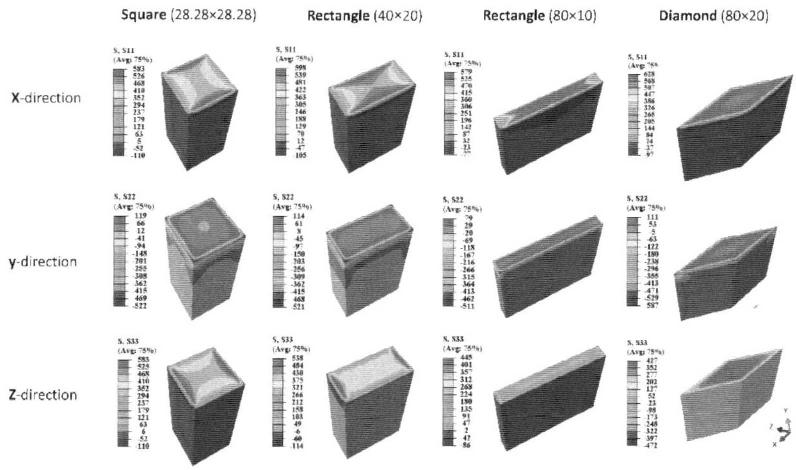

Figure 2: Residual stress components at four different geometric cross-sections, shown in the mid-length transverse-normal planes

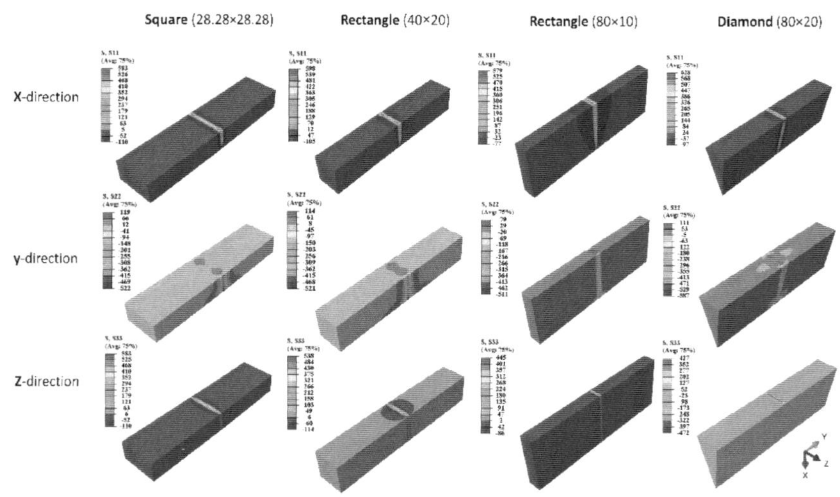

Figure 3: Residual stress components at four different geometric cross-sections, shown in the through thickness longitudinal-normal planes

Figure 4 and Figure 5 are drawn to further quantitatively analyze the WRS pattern along thickness direction, namely along L1 line direction in Figure 1. Based on WRS in single component in Figure 4, it can be seen that the width of stress band of S11 and S33 are almost same at all cases. The stress band of S22 is a little bit wider but the value is nearly negligible. The relationship of stress values between S11 and S33 are dramatically influenced by geometric effects.

In Figure 5, it generally presents that geometric effects have very limited influence on distributions and magnitudes of S11 (largest residual stress component) along L1 line, where peak S11 WRS are from 424MPa to 547MPa and widths of stress band are nearly same. There are some obvious variations in S22 (Y-direction WRS), however, the value is negligible comparing to other two WRS components, shown in Figure 5 b). In terms of S33, shown in Figure 5 c), three components have very similar distributions and peak values 345MPa (80×20), 398MPa (Diamond) and 424MPa (28.28×28.28)), except the case of rectangle (80×10), which has the apparently lower peak value of 224 MPa.

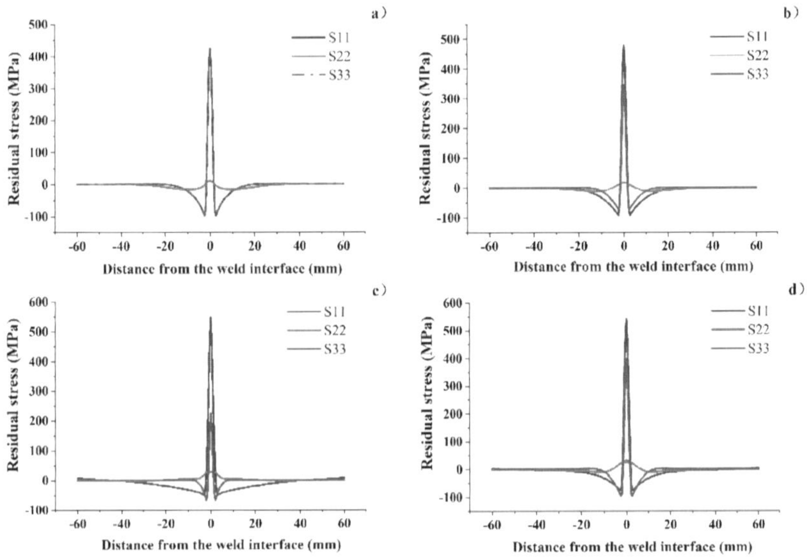

Figure 4: Predicted WRS profiles of four components along L1 line, a) Square , b) Rectangle with 40×20, C) Rectangle with 80×10, d) Diamond . Note, model is one eight and data is mirrored.

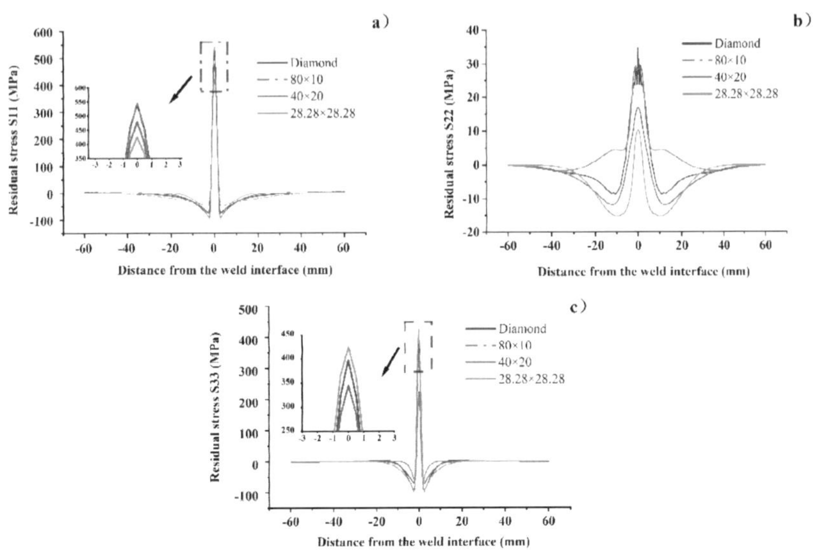

Figure 5: The influence of geometric effects on WRS profiles, a) WRS at X direction, b) WRS at Y direction, c) WRS at Z direction

Conclusions

In this work, a validated FEM model is extended to explore the geometric effects on WRS in LFW. Four components with same area but different cross-sections (rectangles with different aspect ratios and diamond) are designed and corresponding WRS are predicted. The following conclusion can be drawn:

i) Weld geometry has strong influence on the WRS distributions and magnitudes on weld interfaces. Predicted patterns can help to fast qualitative analysis. In all cases, WRS is largest in the direction of the longest contacting surface, followed by the direction of the shortest contacting surface and the direction normal to the weld interface has the lowest value.

ii) The peak WRS at parts with rectangle or square cross-sections tend to concentrate at edge areas of components. Furthermore, the distribution of high-stress area is dramatically influenced by aspect ratio of rectangle cross-sections. These characterizes should be taken into account when conducting structure integrity assessment or selection of means of residual stress measurement to avoid neglect of peak WRS areas.

iii) In comparison to the WRS distribution on weld interface, geometric impact on WRS distribution along thickness direction are relatively weak. The width of stress band is very narrow along depth direction (lower than about 5mm), which challenges the WRS measurements.

iv) Diamond components (blade-like) shows a significant discrepancy of WRS distribution on weld interfaces compared with cube components, where the most areas present a high WRS value sate, rather than the high WRS concentrating at edge area in cases of cube part.

Acknowledgements

The authors are grateful and wish to acknowledge the PGR studentship funded by the University of Manchester.

References

[1] R. Turner, J. C. Gebelin, R. M. Ward, and R. C. Reed, "Linear friction welding of Ti-6Al-4V: Modelling and validation," *Acta Mater*, vol. 59, no. 10, pp. 3792–3803, Jun. 2011, doi: 10.1016/j.actamat.2011.02.028.

[2] A. R. McAndrew, P. A. Colegrove, C. Bühr, B. C. D. Flipo, and A. Vairis, "A literature review of Ti-6Al-4V linear friction welding," Mar. 01, 2018, *Elsevier Ltd*. doi: 10.1016/j.pmatsci.2017.10.003.

[3] P. J. Withers, P. J. Withers, and H. K. D. H. Bhadeshia, "Residual stress. I-Measurement techniques Residual stress Part 1-Measurement techniques," 2014. [Online]. Available: https://www.researchgate.net/publication/30414745

[4] X. Song *et al.*, "Residual stresses in Linear Friction Welding of aluminium alloys," *Mater Des*, vol. 50, pp. 360–369, Sep. 2013, doi: 10.1016/J.MATDES.2013.03.051.

[5] P. Geng, G. Qin, and J. Zhou, "A computational modeling of fully friction contact-interaction in linear friction welding of Ni-based superalloys," *Mater Des*, vol. 185, p. 108244, Jan. 2020, doi: 10.1016/J.MATDES.2019.108244.

[6] A. R. McAndrew, P. A. Colegrove, A. C. Addison, B. C. D. Flipo, M. J. Russell, and L. A. Lee, "Modelling of the workpiece geometry effects on Ti-6Al-4V linear friction welds," *Mater Des*, vol. 87, pp. 1087–1099, Dec. 2015, doi: 10.1016/J.MATDES.2015.09.080.

[7] C. Bühr, B. Ahmad, P. A. Colegrove, A. R. McAndrew, H. Guo, and X. Zhang, "Prediction of residual stress within linear friction welds using a computationally efficient modelling approach," *Mater Des*, vol. 139, pp. 222–233, Feb. 2018, doi: 10.1016/J.MATDES.2017.11.013.

[8] C. Bühr, P. A. Colegrove, and A. R. McAndrew, "A computationally efficient thermal modelling approach of the linear friction welding process," *J Mater Process Technol*, vol. 252, pp. 849–858, Feb. 2018, doi: 10.1016/J.JMATPROTEC.2017.09.013.

[9] Y. Fu, W.-Y. Li, X.-W. Yang, T.-J. Ma, and A. Vairis, "The effects of forging pressure and temperature field on residual stresses in linear friction welded Ti6Al4V joints," *Adv Manuf*, vol. 4, no. 4, pp. 314–321, 2016, doi: 10.1007/s40436-016-0161-6.

https://doi.org/10.31399/asm.cp.icrs2025p0118

Effect of residual stresses on the fatigue behavior of surface-hardened steel wheels under rolling contact

Hans-Jakob Schindler
Mat-Tec AG, Rueti, Switzerland
schindler@mat-tec.ch,

Abstract

The present paper reports on the failure analysis of the steel wheel of a crane, where numerous cracks have been found. Since the residual stresses were likely to be one of the most important influencing factors, the distribution of the tangential residual stresses, which were introduced by inductive case-hardening, were measured to a depth of 30 mm by means of the Cut-Compliance (CC-) method. Unexpectedly, significant tensile residual stresses were found even in the hardened surface layer, where usually compressive stresses prevail. In fact, in an unused wheel that was investigated for comparison, the expected compressive stresses were present. This means that the original residual stresses were significantly altered by the service load of the damaged wheel. The phenomenon known as "ratchetting" is likely to be the cause for the unusual redistribution of the residual stresses. The region near the rolling surface is prone to ratchetting because of very high tensile residual stresses and the superimposed cyclic stresses due to rolling contact. By a rough engineering analysis based on fundamental relations of fracture mechanics, it could be shown that the present residual stresses are responsible for crack initiation as well as crack propagation, including the observed bifurcation in a depth of about 15 mm. It also could be shown that the cracks are unable to penetrate into the wheel deep enough to become unstable and to cause a catastrophic rupture of the wheel. In order to prevent such cracks, first of all the thickness of the hardened layer has to be increased.

Introduction

After a few years in service, numerous cracks were found on the surface of a wheel of a large crane (Fig. 1). The wheel is a circular disc of about 80 mm in thickness and 450 mm in diameter, made of a heat-treated steel, fixed to the axle by warm shrinking (Fig. 1). The area of contact between the rail and the wheel is visible as the smooth, reddish-shiny part of the cylindrical surface of the disc. Two types of cracks can be clearly distinguished: Type I is a group of numerous cracks, axially aligned and occurring predominantly within the contact area. They are mainly closed, thus hardly visible by eye. Type II are gaping cracks that extend axially across the entire width of the cylindrical surface. They are clearly visible even on the lateral surfaces of the disc. In the example shown in Fig. 1, the crack bifurcates at a depth of about 17 mm and propagates subsequently in opposite circumferential direction. Five cracks of this type, about evenly distributed along the circumference, are present in the considered wheel.

The author was asked by the manufacturer of the crane to investigate the cause of these cracks. The main questions of interest were whether or not these cracks could lead to a catastrophic failure of the wheel and what measures can be taken to prevent them from occurring. These questions can only be answered if the damage and cracking mechanisms are understood sufficiently. This means that the observed cracks need to be explained in terms of fracture mechanics at least semi-quantitatively. This is the topic as well as the scope of the present paper.

A key role in the cracking process is surely played by the residual stresses and the hardness of the material. Accordingly, the main focus was placed on measuring the corresponding profiles. With regard to a fracture mechanics analysis, the residual stresses need to be known at least up to the depth of the cracks, which seems to be about 25 mm (see Fig. 1). For comparison, the same measurements were carried out on a second wheel (denoted in the following as "Wheel 2"), which was a new wheel taken directly from the manufacturing process. The hardening process of Wheel 2 was not exactly the same as the one of Wheel 1, since it was changed such that a larger hardening depth was achieved. However, it still can serve as a reference in the present investigation.

As common in a "real-world" case, the failure analysis had to be conducted under tight constraints in terms of available time and money. Only a few residual stress measurements could be performed, and for the theoretical analysis simplifying physical models had to be applied. However, as shown in the following, the latter turned out to be efficient and sufficient for the present purpose, which was first of all understanding the mechanisms that led to the observed cracks.[1]

Figure 1: Cracks on the cylindrical surface of the wheel (left) and their appearance on the lateral surface (right).

Figure 2: Cuboid 80 x 30 x 35 mm removed from the wheel and cut in three pieces A, B and C (left). Due to the inner network of cracks, piece B fell apart in three main parts and several smaller fragments (right).

Macro-fractography

In order to expose the cracks in the interior of the wheel, a cuboid measuring 80 mm (i.e. full width) x 30 mm (circumferentially) x 35 mm (radially) was removed from the wheel by electrical discharge machining (EDM). Subsequently, it was divided by two additional radial EDM-cuts in the three pieces A, B and C shown in Fig. 2 (a). Since the tangential branches of the type-II crack were longer than the width of the cuboid in the interior of the wheel, the central piece (B) decayed in three big parts and several smaller fragments (Fig. 2 (b)) after the radial cuts. The holes that appear on the cut surface shown in Fig. 2 (b) are caused by small pieces of material that fell off due to small

[1] The investigation is subjected to a secrecy agreement, so there are some restrictions concerning the disclosure of technical details in this report.

branches of type-I cracks meeting or crossing each other. This unusual behavior of cracks is possible only under mainly compressive stresses, where the cracks are predominantly closed during the damaging process, thus unable to shield each other.

The topography of the surfaces of the type-II crack differs fundamentally between the hardened zone next to the rolling surface the rest of the crack (see Fig, 1, right). indicating different mechanisms of crack extension. Whereas the crack surface is nearly plane and appears as brittle in the outermost layer of about 5 mm, it is very rough at larger depths (see Figs. 3 (a) and (b)). Nevertheless, the two crack surfaces perfectly fit to each other when hold together (see Fig. 3), indicating that essentially no macro-plasticity was involved in the fracture process.

Figure 3: Piece C with two parts of piece B (a) and the third part of B (b). Edge between the contact area and the EDM-cut B-C (c). "R" indicates the rolling surface.

From Fig. 2 and Fig. 3, it can be seen that the cracks of type I appearing on the rolling surface are the visual part of an inner network of numerous cracks in a layer of about 5 mm thickness. As sketched in Fig. 4, They seem to originate from a circumferential main crack in a depth of about 5 mm, (in the following called the "primary type-I crack"). It bifurcates at distances of a few dozens of millimeters, some branches turning towards the surface, eventually penetrating it (Fig. 3 (c)). Crack types I and II appear as nearly independent from each other. It seems that the type II crack was formed only when the primary type-I crack already existed, since it is very unlikely that the latter would have crossed the gaping type-II-crack without any sign on the crack line.

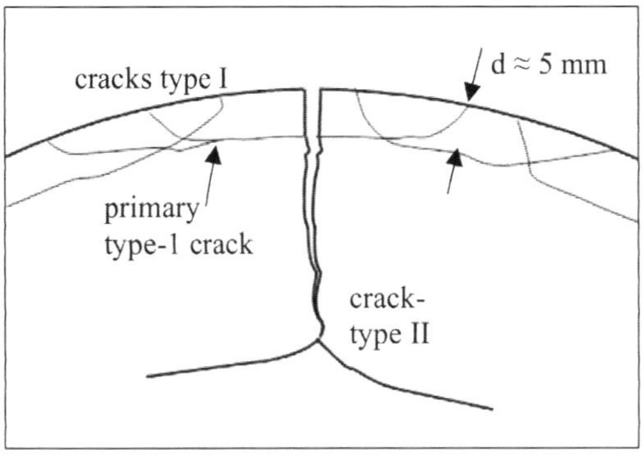

Figure 4: Crack pattern beneath the rolling surface in the central part of the wheel (schematic).

Material

The wheels are made of a quenched and tempered steel. The rolling surface is hardened by induction hardening. The hardness-profiles are shown in Fig. 5. Wheel 1 has a hardening depth of about 4 mm, whereas Wheel 2 – a new, unused wheel that was investigated for the purpose of comparison – has one of about 7.5 mm. The narrow peak in hardness at the surface (x=0) of Wheel 1 is expected to be due to the local strain-hardening caused by the highly localized contact stresses ("Hertzian stresses") that moves around the circumference, causing incremental micro-plastic deformation of the contact surface.

The hardness of the parent material of Wheel 1 is about 320 HV (Fig. 5). Therefore, based on the generally known and well proven correlation $R_m \approx 3.1 \cdot HV$, the ultimate strength R_m is estimated to be about 990 MPa. For the present quenched and tempered steel, the yield stress R_p is known to be about 80% of R_m, thus about 800 MPa.

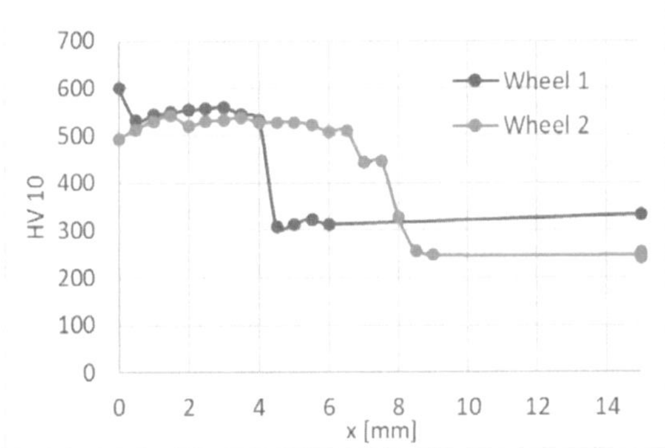

Figure 5: Hardness HV 10 in the central part of the wheel as a function of depth x

Residual Stresses

Measurement

The residual stress profile was measured by the cut-compliance method [1 - 3]. The measurement principle is shown schematically in Fig. 6: A narrow cut of current depth a is introduced incrementally along the plane where the residual stresses are to be measured. The strain ε_M that is caused by the release of the residual stresses on the cut surfaces is measured at a suitable distance s from the cut edge by a strain gage (SG). From the measured strain as a function of a, $\varepsilon_M(a)$, the stress intensity factor (SIF) $K_{rs}(a)$ due to the residual stresses is obtained by the general relation

$$K_{rs}(a) = \frac{E}{Z(a/s)} \cdot \frac{d\varepsilon_M}{da} \tag{1}$$

Z(a/s) is the so-called influence function that depends on the geometry of the considered piece of material, on cut depth a and the location of the strain-gage. E denotes Young's modulus.

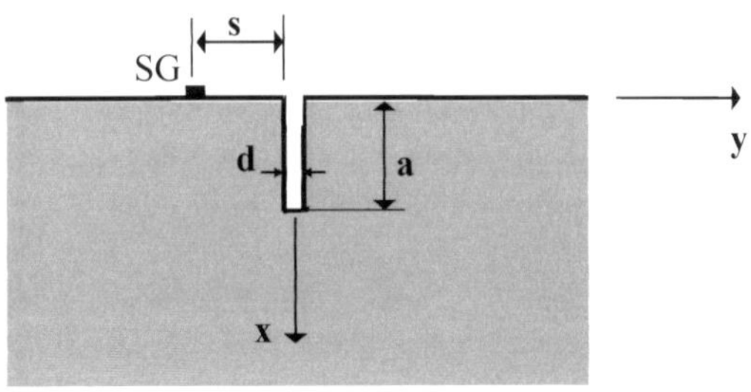

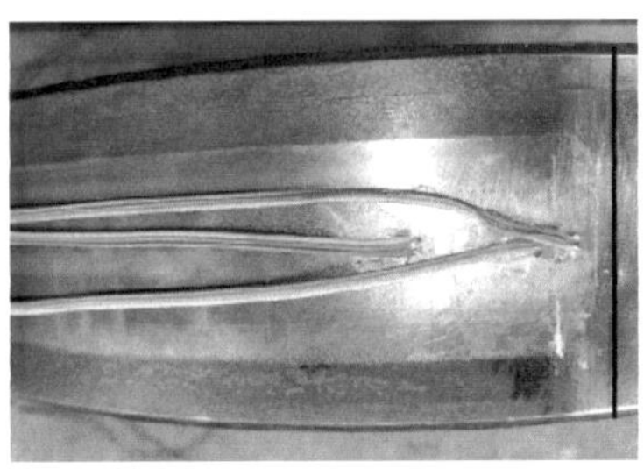

Figure 6: Principle of the CC-Method (left) and arrangement of the three strain-gages used in the present measurement (right). The black line indicates the EDM-cut.

From $K_{rs}(a)$ one obtains the residual stress $\sigma_{rs}(x)$ in y-direction (i.e tangential) as a function of x by inversion of the fundamental relation [3]

$$K_{rs}(a) = \int_0^a \sigma_{rs}(x) \cdot h(x,a) \cdot dx \tag{2}$$

In (2), h(x,a) denotes the weight function for a surface crack of depth a [4]. The sensitivity of the measurement depends mainly on a/s. Therefore, in order to measure the residual stress profile to a depth of about 30 mm with sufficient accuracy, it is advisable to use more than just one SG. In the presented cases, strain measurements at s = 5 mm, 14 mm and 40 mm were made (Fig. 6, right).

Representative residual stresses are only obtained if no cracks are present between the cut and the SG and if no type-II crack is present in a distance of at least twice the measurement depth (i.e. 60 mm) from the cut. $K_{rs}(a)$ as obtained from $\varepsilon_M(a)$ by (1) is shown in Fig. 7. The residual stress profiles obtained by numerical inversion of (2) are shown in Fig. 8.

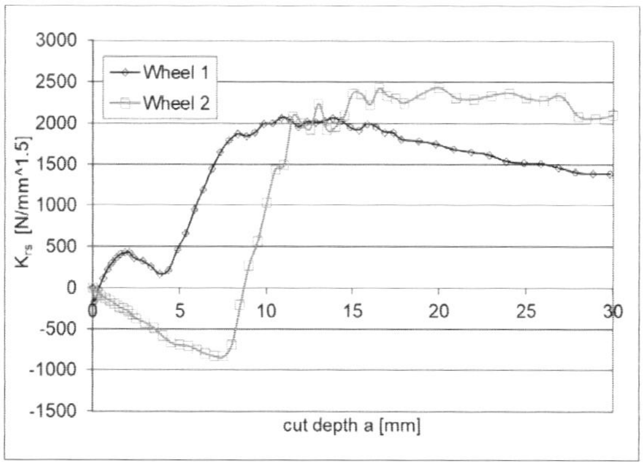

Figure 7: Stress intensity factor $K_{rs}(a)$ obtained from (1).

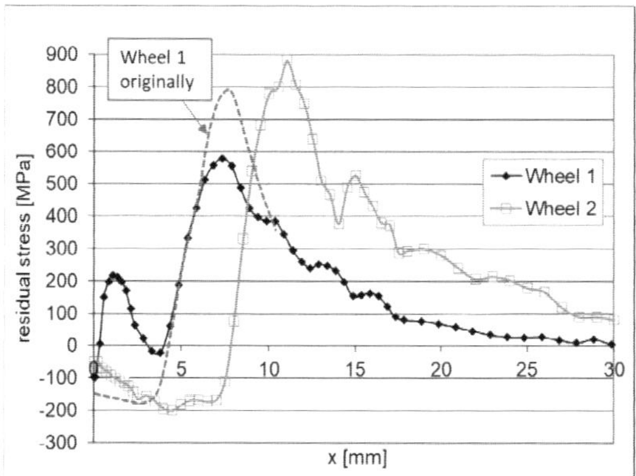

Figure 8: Residual stress σ_{rs} as a function of depth x obtained from (2).

Discussion

Consider first the residual stress profile of Wheel 2 (Fig. 8). As expected, up to the hardening depth of about 7.5 mm (see Fig. 5), the residual stresses are compressive. Subsequently, they change to tension by a high gradient. A peak about as high as the yield stress is reached at a depth between 10 mm and 12 mm.[2] For x > 12 mm, the residual stresses decrease rapidly with increasing x, falling below 100 MPa for x > 30 mm. This shape of the residual stress profile can be explained as follows: The very high tensile peak is due to the locally high temperature induced by induction heating, while the main part of the wheel remained at ambient temperature. Only in the outermost layer, where water quenching caused a martensitic transformation, compressive stresses are obtained due to the corresponding increase in volume. In the adjacent zone, where the temperature was equally high, but the temperature gradient not sufficient for martensitic transformation, the high tensile residual stresses remained.

From the above considerations, a residual stress profile similar to the dashed red line was expected for Wheel 1 (Fig. 8). Surprisingly, the measurement does not correspond to the expectation: First, significant tensile stresses prevail in the hardened zone, and second, the peak tensile stress stays far below the yield stress. This unexpected shape of the residual stress profile of Wheel 1 suggests that it was altered during service. However, residual stresses introduced by hardening are known to be very stable, so such a change is unusual as well. The reasons for this effect are evaluated and discussed in the following.

Stresses due to service load

The loading-system of the wheel in service is shown schematically in Fig. 9. The nominal load F = 235 kN acts on the axle. At the moving contact area with the rail, it causes Hertzian contact stresses that act on the cylindrical surface of the wheel. According to [6], the maximum pressure σ_{yH} and shear stress τ_{xyH} are

$$\sigma_{yH} = -976 \text{ MPa}; \qquad \tau_{xyH} = \pm 325 \text{ MPa} \tag{3}$$

[2] In the range of about 11 mm < x < 12 mm for Wheel 2 (Fig. 8), the residual stresses exceed the yield stress, which is unlikely for physical reasons. The corresponding peak of nearly 900 MPa is rather an artefact that can be explained with local plasticity in the vicinity of the cut front [5]. The actual peak is expected to be only about 800 MPa.

The maximum shear stress occurs at $x = x_H = 3.0$ mm. In the range $0 < x < x_H$ the stresses are highly nonlinear with respect to F, and not much dependent on x. Outside this range, the stress field has to match Boussinesqu's solution for a single force acting on a half-plane [7] asymptotically. Fig. 10 shows the corresponding stresses for the exemplary depth x=5 mm. Here, due to rolling, the stresses change cyclically between the maxima and minima of these curves. Fig. 11 shows the corresponding stress ranges as a function of x. In order to cope with the elastic stress singularity for x -> 0, it is assumed that the elastic stresses apply up to the limits given by the Hertzian stresses (3), which is a conservative assumption. In the range $0 < x < 20$ mm, the effective width B_{eff} of the disc is assumed to increase linearly from B_R to B (i.e. from 40 to 80 mm, see Fig. 9).

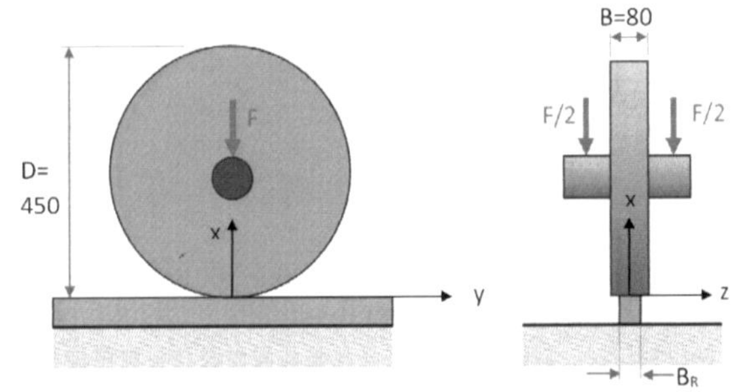

Figure 9: Disc-shaped wheel loaded by a bearing force F, rolling on a steel rail of an effective width B_R = 40 mm.

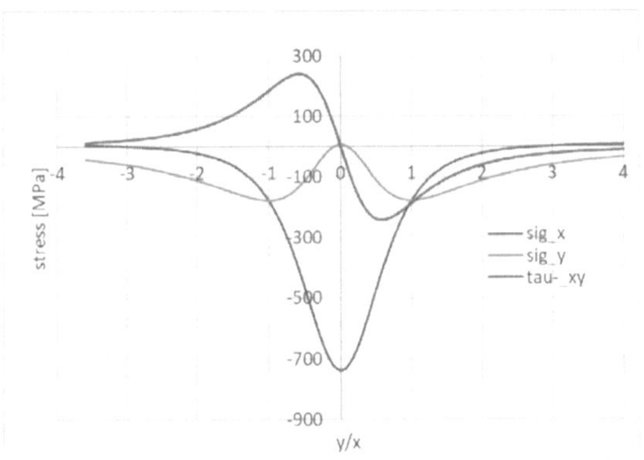

Figure 10: σ_x, σ_y and τ_{xy} as a function of y/x at x = 5 mm for a force F=335 kN concentrated at x=y=0

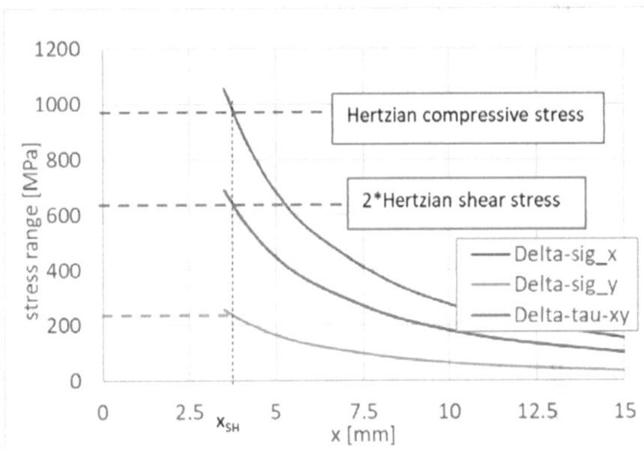

Figure 11: Cyclic stress ranges $\Delta\sigma_x$, $\Delta\sigma_y$ and $\Delta\tau_y$ as a function of x, truncated by the Hertzian stresses.

Analysis of Crack Initiation and Growth

Crack initiation

Fatigue damage is caused by reversed dislocation movements, thus by cyclic shear stresses beyond the elastic limit. In the presence of tensile stresses, it can lead to crack formation and growth. As a well-known rule of thumb, endurance (i.e. infinite fatigue life) under reversed bending is guaranteed if the amplitude of the bending stress is less than $0.4 \cdot R_m$. Using once again the well-proven correlation $R_m \approx 3.1 \cdot HV$, $\Delta\sigma_y = 2 \cdot \Delta\tau_{xy}$ (for the case of bending) and assuming that endurance is controlled by crack initiation, one finds to the following approximate condition for crack initiation:

$$\Delta\tau_{xy} > 1.24 \cdot HV \tag{4}$$

As shown in Fig. 12, this condition is fulfilled for Wheel 1 in the range of about 4 mm < x < 5 mm. This means that in this range of x, fatigue damage that sooner or later leads to crack initiation, is likely to occur. In Wheel 2, due to the different hardness profile (see Fig. 5), this limit is not reached, so no crack initiation is expected.

Formation and Propagation of Cracks Type I

Due to rotational symmetry of the wheel, the fatigue damage that leads eventually to crack initiation, is the same all along the circumference. If the condition for crack initiation (4) is fulfilled in a certain range of x), a circumferential crack can be formed just by coalescence of neighbouring crack-seeds. This means that a primary type I crack can grow circumferentially in the corresponding depth without the condition for fatigue crack growth (FCG) (see below, eq. (5)) being fulfilled. However, the branches of the type I cracks that propagate towards the outer surface (Fig. 5) have to obey the rules of FCG. Correspondingly, the residual stresses in the range $0 < x < 4$ mm play an important role with regard to the pattern of type I cracks: Tensile residual stresses in circumferential direction promote bifurcations of the primary crack and enable the branches to propagate towards the surface and to exit from it. In contrast, compressive residual stresses in the hardened layer (like those in Wheel 2, or Wheel 1 in the original state) would prevent the primary type I crack from bifurcating and turning towards the surface.

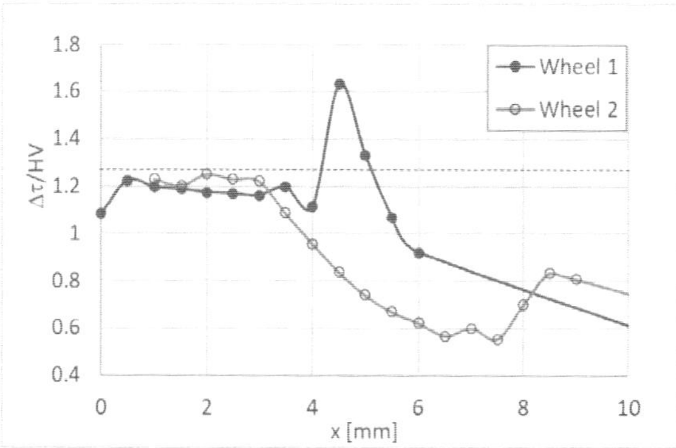

Figure 12: $\Delta\tau_{xy}/HV$ as a function of x. The red line indicates the limit of endurance according to condition (4).

Formation and Propagation of Crack Type II

Unlike the primary type I, type II cracks must follow the rules of FCG. This means that cracks only grow if the condition

$$\Delta K_{eff} = K_{max} - K_{min,\, eff} > \Delta K_{th,int} \tag{5}$$

is fulfilled. ΔK_{eff} denotes the effective range of K_I, which means that K_{min} in the phase of crack closure is assumed to be zero, or in mathematical terms: $K_{min,eff} = \max(K_{min},0)$. $\Delta K_{th,int}$ denotes the "intrinsic" (i.e. closure-free) threshold of FCG. The latter can be estimated by $\Delta K_{th,int} \approx 0.89 \cdot E \cdot \sqrt{\beta}$ (with β denoting Burgers vector), which represents essentially the empirical relation proposed in [8], adjusted to additional experimental data by the factor 0.89. For the present steel, $\Delta K_{th,int} = 2.93$ MPa·m$^{0.5}$ (= 93 N/mm$^{1.5}$) is obtained.

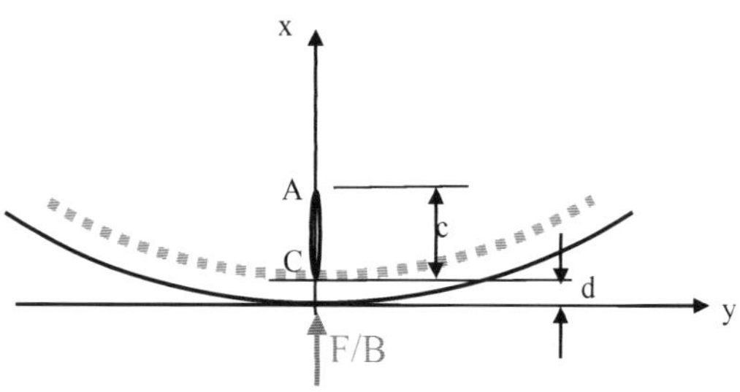

Figure 13: Radial crack of length c, emanating from an initial crack-seed at x = d, loaded by the service load F/B.

Consider the radial crack shown in Fig. 13. Since the stresses σ_y due to the service load always stay in the negative range (see Fig. 10), the maximum and minimum K_I, respectively, at the crack-tip A are obtained as

$$K_{max}(c) = \int_{d}^{d+c} \sigma_{rs}(x) \cdot h(x,c) \cdot dx \tag{6}$$

$$K_{min,}(c) = \int_{d}^{d+c} \left(\sigma_{rs}(x) - \Delta\sigma_y(x)\right) \cdot h(x,c) \cdot dx \tag{7}$$

The adaption of the weight function for an internal crack as given in [3] to the crack-parameters and the coordinate system used in Fig. 13 leads to

$$h(x,c)=\sqrt{\left(\frac{2}{\pi\cdot c}\right)\cdot\left(\frac{2x-2d}{2c-2x+2d}\right)} \tag{8}$$

With (6) – (8), $\sigma_{rs}(x)$ from Fig. 7 and $\Delta\sigma_y(x)$ from Fig. 11, the effective fatigue load ΔK_{eff} of crack-tip A is obtained from (5) and shown in Fig. 14 as a function of crack size c. There it can be seen that an initial defect size c_i of less than 0.5 mm is sufficient to initiate FCG in radial direction. A defect size as small as 0.5 mm can surely not be excluded in the region where multiple cracks of type I were initiated, so crack growth in the radial direction is likely to occur. However, K_{max} (i.e. the maximum crack load K_I) is limited to about 1300 $N/mm^{1.5}$ (Fig. 15). This is far below the fracture toughness of the parent material, so an unstable crack extension at tip A is not possible for any c.

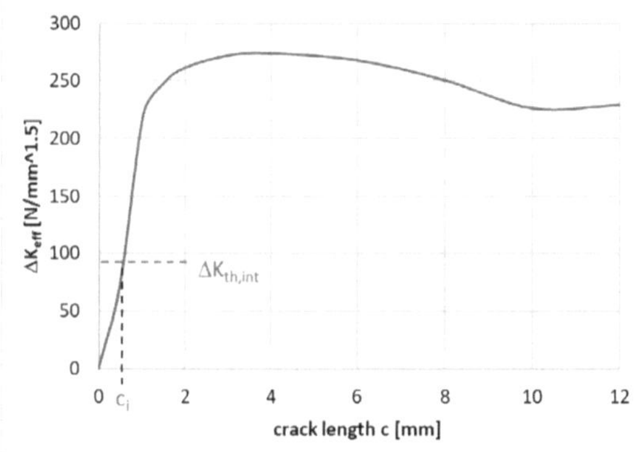

Figure 14: ΔK_{eff} as a function of crack length c for the crack configuration shown in Fig. 13.

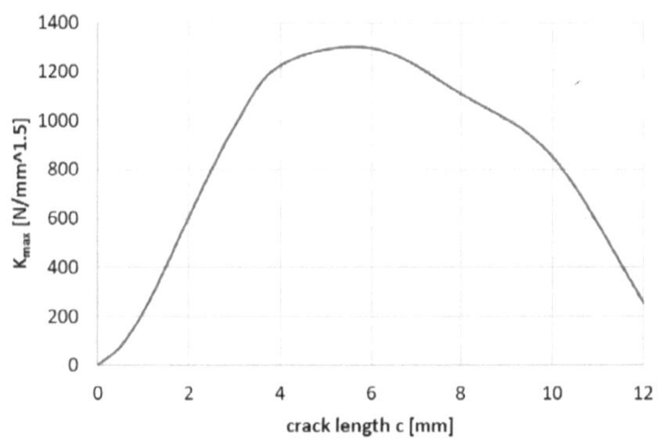

Figure 15: K_{max} as a function of crack length c for the crack configuration shown in Fig. 13.

Near tip C of the radial crack, a circumferential type-I-crack or at least a layer of heavily damaged material is present, indicated by the dotted line in Fig. 13. This represents a kind of an inner surface, so the singular stress-field at C is disturbed by reduced crack-tip constraints. Therefore, at tip C, crack initiation and growth does not follow the usual rules of FCG. Anyway, as c is increasing due to FCG at crack tip A, the stresses in the ligament $0 < x < d$ rise with increasing c due to the stress re-distribution. Their resultant F_t can be estimated from the following equation:

$$K_{\max}(c)=K_{rs}(a=c+d)-F_t\cdot h(x={}^{d}\!/_{2},a=c+d) \tag{9}$$

where $K_{max}(c)$ is the maximum SIF as given in Fig. 15, $K_{rs}(a)$ the SIF due to the residual stresses (Fig. 7) and h(x,a) the weight function for a surface crack of depth a [4]. $F_t(c)$ obtained from (9) is shown in Fig. 16. With regard to the stability of crack-tip C located on the inner edge of the hardened layer (Fig. 13), the system can be considered as a deep crack with a ligament width d loaded by the lateral force F_t (9) acting on the ligament without bending the latter. The corresponding SIF can be found in [9], page 9.2, to be

$$K_I^{\,C}(c)=\frac{\pi}{\sqrt{\pi^2-4}}\cdot\frac{2\cdot F_t(c)}{B\cdot\sqrt{d}} \tag{10}$$

In Fig. 17, $K_I^C(c)$ is shown for d = 4 mm. Fracture toughness of the hardened layer is known to be less than 100 $MPa\cdot m^{0.5}$ (or 3162 $N/mm^{1.5}$, respectively). This value is reached by K_I^C at about c = 9 mm, and rapidly increasing for increasing c (or a decreasing d in case of FCG at C). Anyway, an unstable fracture of the hardened ligament is nearly inevitable in the range of about 8 mm < c < 15 mm.

By the unstable fracture of the hardened surface layer, the internal crack shown in Fig. 13 is immediately transformed to the gaping surface crack of length a = c + d shown in Fig. 18. This sudden change of the crack configuration is accompanied by a fundamental change of the loading state of crack-tip A: It switches from Mode I to predominantly Mode II. The corresponding SIF K_{II} can be obtained by decomposition of the analytical solution for an anti-symmetrical pair of forces acting on the crack mouth as given in [9], page 8.3. One finds

$$K_{II} = 1.3 \cdot \frac{F}{B \cdot \sqrt{\pi \cdot a}} \tag{9}$$

For the nominal force F = 235 kN, B = 80 mm and, for example, a = 16 mm, (9) results in K_{II} = 536 N/mm$^{1.5}$. When F rolls over the crack mouth, K_{II} switches the sign, so $\Delta K_{II} = 2 \cdot 536$ N/mm$^{1.5}$ = 1072 N/mm$^{1.5}$, which is much higher than ΔK_{Ieff} in the previous state of the crack (Fig. 14). Therefore, the transition from an internal crack (Fig. 13) to an edge crack (Fig. 18) is accompanied by a huge acceleration of FCG.

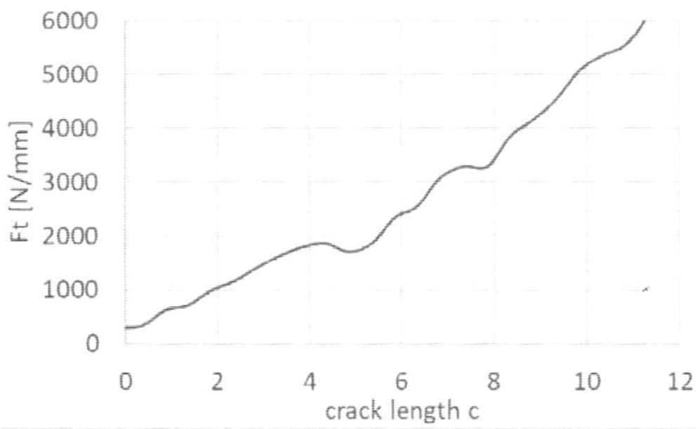

Figure 16: Resultant F_t of the tangential stresses in the layer 0<x<4mm as a function of c

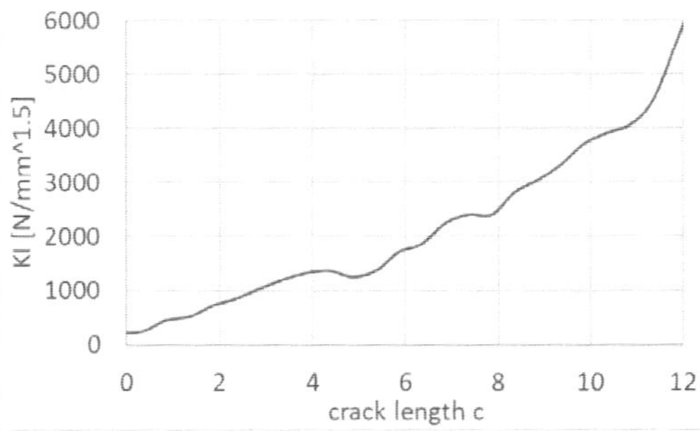

Figure 17: K_I^C due to $F_t(c)$ for the crack shown in Fig. 13 at tip C for a ligament width b=4 mm as a function of c.

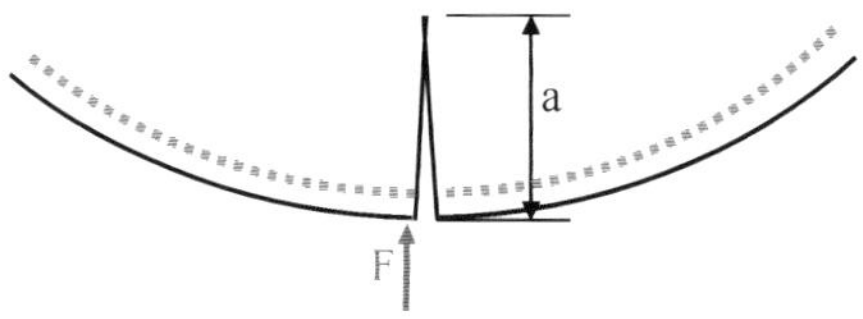

Figure 18: Surface crack resulting from the internal crack shown in Fig. 13 by rupture of the ligament.

Under cyclic Mode-II, a crack tends to deviate from its original plane, because crack extension follows principally the direction of maximum energy release rate, which in case of Mode II is not the original direction. A positive K_{II} causes the crack to grow to the right, whereas a negative K_{II} to the left. Therefore, a bifurcation of the growing crack as shown in Fig. 19 is likely to occur in the following way: When F acts on the left side of the crack, the left branch of the crack is closed due the compressive stresses in radial direction, thus is inactive, whereas the right branch is loaded by ΔK_I, which causes it to grow by an increment. When F rolls over the crack-mouth to its the right side, the situation is the same, but reversed: The left branch grows by an increment, whereas the right one is inactive.

After bifurcation and some growth in tangential direction, both of the crack branches are loaded mainly in Mode I. The corresponding ΔK_I is highly affected by the radial stresses in the vicinity of the crack tips. The cyclic stresses due to F are compressive, thus causing crack closure. The predominantly tensile residual stresses in circumferential direction in the range 0 < x < a cause additional radial compression. Therefore, the ΔK_I is decreasing with increasing branch length, so tangential crack growth is expected to decelerate and come to rest at a certain branch length. The bifurcated crack shown in Fig. 1 seems to have reached its final length.

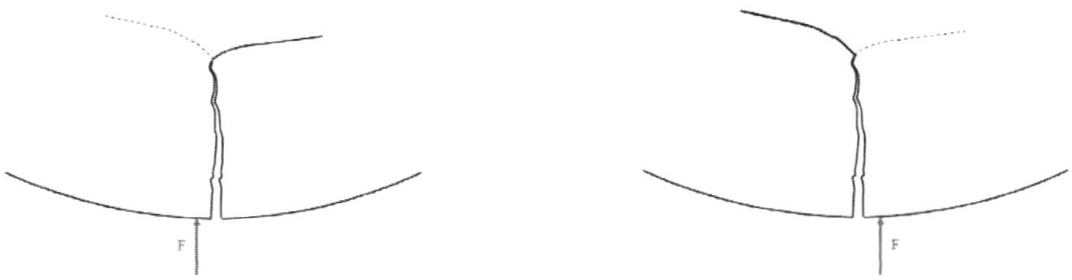

Figure 19: Surface crack after bifurcation: Depending on the position of F relative to the crack, one of the two branches is alternately growing, while the other is inactive due to crack closure.

Discussion

From the above theoretical considerations and physical explanations, it is evident that the residual stresses played an important role in the fatigue and cracking behavior of the wheel. In particular, the atypical tensile stresses in the hardened surface layer are crucial for the occurrence of the observed cracks. Actually, with compressive residual stresses in the hardened layer (like in the case of Wheel 2), the type-I cracks would not have occurred, because they could not grow through a compressive stress-field. Compressive residual stresses in the hardened layer would also have prevented the initial radial crack from breaking through the surface layer, so the transition of the radial internal crack to the type-II-crack would not have occurred either. The same is true for Wheel 1, if the residual stresses were still the original ones as indicated in Fig. 8. So, with regard to the root cause of the failure, the main question is, why the beneficial original stresses were changed to those measured in Wheel 1.

The original tangential residual stresses exhibit a tensile peak of nearly the yield stress at about x = 7 mm. (Fig. 8). At this depth, compressive stresses in x-direction of about σ_x = -450 MPa and shear stresses of about τ_{yx} = ±150 MPa are acting cyclically due to the service load (Fig. 11). Superimposed to the high tangential residual stress in the range 5 mm $< x <$ 9 mm, they cause local microplastic deformations in the circumferential and radial directions, which lead incrementally to irreversible plastic straining. This effect is known as "ratchetting" and results in a relaxation of the tensile residual stresses. The corresponding extension in tangential direction forced the adjacent hardened layer to extend as well, which gave rise to tensile tangential stresses such that the residual stresses changed from compression to tension. Alternatively, the shift from compressive to tensile residual stresses in the hardened layer can also be explained by the condition of equilibrium: Relaxation of the tensile stresses in the range 5 mm $< x <$ 10 mm has to be accompanied by an increase of the stresses in the range $0 < x < 4$ mm, otherwise the stresses $\sigma_y(x)$ on the section y=0 would no longer be in equilibrium.

Conclusions

The following general conclusions can be drawn from this investigation:

- The observed cracks could be explained. It is shown that the residual stresses introduced by surface hardening played a crucial role. They need to be known not only near the surface, but at least to the depth of the cracks.
- The Cut-Compliance-Method has proven once again to be suitable and efficient to measure the residual stresses to the required depth.
- The residual stresses tuned out to be changed by the contact stresses in service even at depths far beyond the zone affected by the Hertzian stress field.
- Because of their simplicity and physical transparency, analytical models are efficient and of high value in the analysis of a complex case of failure like this.

References

[1] Schindler, H.J, Cheng, W., Finnie, I., Measurement of Residual Stress in a Roller of High Strength Steel, Proc. of the 4th Int. Conf. on Residual Stress, Baltimore, 1994, p. 1266 – 1274.

[2] Schindler, H.J., Cheng, W., Finnie, I., "Experimental Determination of Stress Intensity Factors due to Residual Stresses", J. Experimental Mechanics, Vol. 37, No. 3, 1997, pp. 272- 279.

[3] Schindler, H.J., Bertschinger, P., Some Steps Towards Automation of the Crack Compliance Method to Measure Residual Stress Distributions, Proc. of 5th Int. Conf. on Residual Stresses (ICRS 5), Stockholm, 1997, p. 682-687.

[4] Wu, X.-R., Carlsson, A.J., "Weight functions and stress intensity factor solutions", Pergamon Press, Oxford, 1991.

[5] Schindler, H.J., "Residual Stress Measurement in Cracked Components: Capabilities and Limitations of the Cut Compliance Method", Proc. of 5th European Conf. on Residual Stresses, Materials Science Forum Vols. 347-349, 2000, 150-155.

[6] Warren C. Young, Roark's Formulas for stress and strains, McGraw Hill, 6th Edition, 1989.

[7] Timoshenko, Sp., Goodier, J.N., Theory of Elasticity, McGraw-Hill, 3rd Edition, 1970, pp. 97 – 100.

[8] Hertzberg, RW. "On the calculation of closure-free fatigue crack propagation data in monolithic metal alloys", Mater. Sci. Engng., A190:25–32., 1995.

[9] H. Tada, P.C. Paris, G.R. Irwin, "The stress analysis of cracks handbook", Del Research Corp., Hellertown, PA, 1974.

ICRS 12: Proceedings from the 12th International Conference on Residual Stresses, 2025
October 20–23, 2025; Detroit, Michigan, USA
https://doi.org/10.31399/asm.cp.icrs2025p0128

An Overview of the Engineered Residual Stress Implementation (ERSI) Working Group

Dallen L. Andrew, Ph.D.
Hill Engineering, Marriott-Slaterville, UT, USA
dlandrew@hill-engineering.com

Jacob Warner
United States Air Force, A-10 Aircraft Structural Integrity Program, Hill Air Force Base, UT, USA
jacob.warner@us.af.mil

Abstract

The Engineered Residual Stress Implementation (ERSI) working group was formed in 2016 with a mission to "develop a holistic paradigm for the implementation of engineered residual stresses into lifing of fatigue and fracture critical components". ERSI emerged from within the United States Air Force (USAF) aircraft structural integrity community as a forum for individuals and organizations to collaborate constructively, transition technology and data to the public sphere, and consult on policy/best practices concerning the incorporation of residual stresses with other entities such as the FAA, DoD, ASTM, SAE, etc. ERSI members represent a broad diversity of interests and backgrounds, both domestic and international, from military, academia, and industry.

The primary focus of ERSI so far has been the transition of a classic engineered residual stress technology, cold expansion of holes, into life extension for USAF weapon systems. Although hole cold expansion is known to provide significant structural fatigue life extension, the full potential improvement has not been included in certified airworthiness limits. With extensive support from ERSI, the USAF recently issued a Structures Bulletin which allows aircraft structural integrity managers to utilize cold expansion benefits for initial and recurring inspection intervals, a significant achievement for both platform availability and fleet-wide cost savings.

This achievement is a holistic product from the three primary focus areas, or committees, within ERSI that represent different technical disciplines of aircraft structural integrity: 1) analysis and test, 2) residual stress characterization, and 3) nondestructive inspection/evaluation, quality assurance, data management.

These three committees work together to identify and address technical gaps, define the requirements and guidelines for implementation, and collaboratively develop and accomplish new round robin activities that advance the state-of-the-art. This paper describes recent work performed by the ERSI working group across these three committees.

Introduction

The overarching vision of ERSI is to help to develop a holistic paradigm for the implementation of engineered residual stresses into lifing of fatigue and fracture critical components through the following key objectives:

- Define a common vision for the accounting of engineered residual stress for the cold expansion (Cx) of holes
- Provide forum to collaborate on new developments, best practices, lessons learned
- Develop an implementation roadmap
- Identify, define, and enable the resolution of gaps in the state-of-the-art

Since its inception in 2016, ERSI has grown to over 120 participants from different countries, Department of Defense organizations, national labs, universities, original equipment manufacturers, industry partners, and USAF Aircraft Structural Integrity Program (ASIP) managers.

ERSI Analysis and Test Committee

The mission of the analysis and test committee is to establish analytical & testing guidelines to support implementation of engineered residual stress, with key objectives of:

- Develop and document best practices for integration of engineered residual stress in crack growth prediction methodologies
- Establish testing requirements considering the impacts of residual stress on fatigue crack growth
- Develop datasets and case studies to support analysis methods validation
- Identify, define, and enable the resolution of gaps in the analytical methods state-of-the-art
- Support the development of an implementation roadmap

The following sections discuss recent achievements of the analysis and test committee.

Interference Fit Fastener Round Robin

In 2021, the ERSI analysis and test committee performed a round robin activity for analysis of interference fit fasteners. The loading and geometry information are listed below.

- Constant amplitude, R = 0.1, max stress = 192.3 MPa (27.9 ksi)
- 7075-T651 aluminum alloy, 6.35 mm thick (0.25 inch)
- Two conditions tested: Open hole, 0.4% interference
- Three conditions predicted: Open hole, 0.4% interference, 0.6% interference

The results from the various contributors are given in Figure 1. Overall, these results showed tight grouping of the open hole predictions, although all analysts under-predicted the test data. The surface correction approach shows promise for the open hole condition. The stress approach used by Raider closely matched life and crack growth curve shape. More evaluations and testing are planned by ERSI to further look at these cases and other interference fit cases in the future.

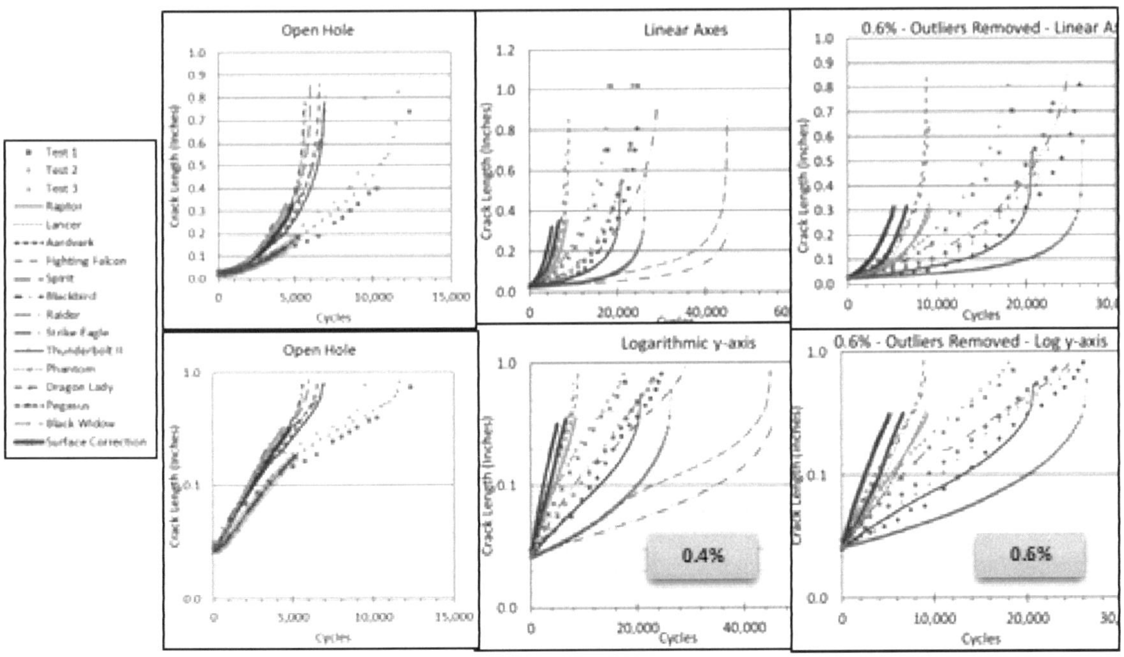

Figure 1: Interference fit round robin results [1]

Stress Intensity Factor Round Robin

In 2021, the ERSI analysis and test committee completed a round robin activity for the comparison of stress intensity factors. The primary objective of the Stress Intensity Factor (SIF) round robin was to evaluate differences between available SIF solutions for a single corner crack at a fastener hole with remote uniform tension loading. The evaluations included not only the root SIF solution but any corrections to account for single vs multiple cracks, finite width, and hole offset. These solutions were compared to explicit Finite Element Analysis (FEA) results of each case. Any findings were intended to drive improvements to solutions available to the fracture mechanics community.

Seven different cases of corner cracks at a hole were developed and SIF solutions along the crack front were requested from participants (see configuration details in Figure 2). A building block approach was utilized (see Table 1) with Case 1 representing the root SIF solution, and each case added an additional level of complexity with corrections to the root solution.

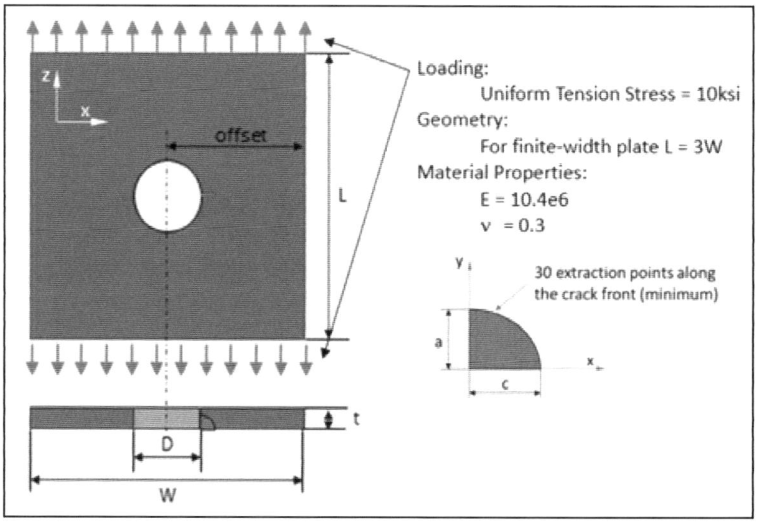

Figure 2: Round robin corner crack at a hole configuration [2]

Table 1: SIF round robin matrix of cases [2]

Case	Surface Crack Length (c) (inch)	Bore Crack Length (a) (inch)	Corner Crack Configuration	Width (inch)	Thickness (inch)	Hole Diameter (inch)	Hole Offset (inch)
1	0.050	0.050	Double Symmetric	100.0	0.25	0.50	50.0
2	0.050	0.050	Single	100.0	0.25	0.50	50.0
3	0.050	0.050	Single	4.0	0.25	0.50	2.0
4	0.050	0.050	Single	4.0	0.25	0.50	0.6
5	0.050	0.050	Single	1.20	0.25	0.50	0.6
6	0.050	0.075	Single	100.0	0.25	0.50	50.0
7	0.100	0.050	Single	100.0	0.25	0.50	50.0

Nine submissions were received from eight participants which included solutions utilized by AFGROW and NASGRO, solutions from Newman/Raju and Fawaz/Andersson, and explicit FEA of each case. FEA approaches utilized various tools and methods which provides an additional opportunity to evaluate the different approaches and their impact on the accuracy of the SIF. Seven reference solutions (correlating with the seven cases) with relative errors in KI on the order of 0.03% or less were provided by Andersson (Submission 6) under USAF contract and were utilized as the reference solutions for all cases.

Case 6 are the only results included in this paper. The Mode I SIF is plotted along the crack front as a function of normalized parametric angle in Figure 3, while the percent difference relative to Submission 6 from Andersson is presented in Figure 4.

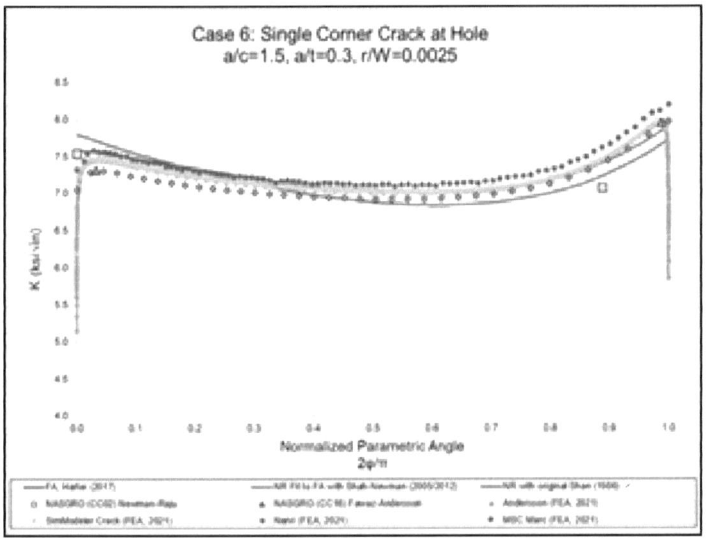

Figure 1: Stress intensity vs. normalized parametric angle (Case 6) [2]

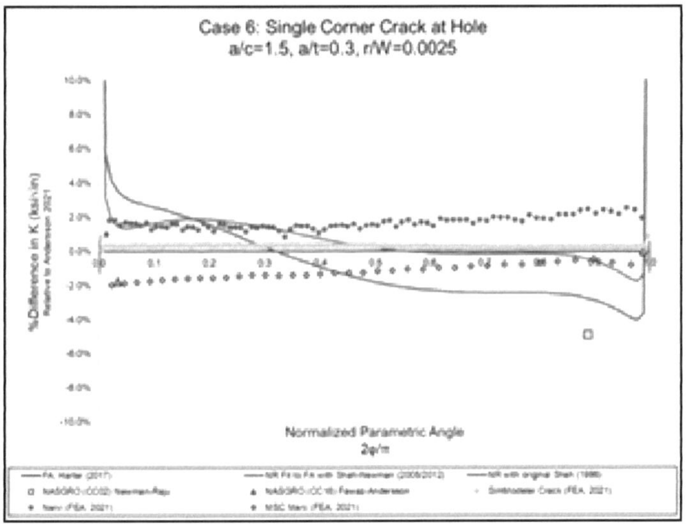

Figure 2: Percent difference in stress intensity vs. normalized parametric angle (Case 6) [2]

The overall summary and conclusions include:

- Successful SIF comparisons completed utilizing a wide array of available solutions and toolsets, with submissions provided by eight different participants
- Overall, results were within 2% of the reference case, however, deviations were observed for narrow width and varying aspect ratio cases exceeding 10% in some cases
- Significant discrepancies were discovered with commonly utilized finite width corrections, with differences/error up to 10%
- Follow-on actions are in work to resolve discrepancies uncovered by the round robin effort
- A robust dataset was developed that can be utilized as a reference set for follow-on studies
- Comparisons between varying FEA approaches have highlighted the opportunity to identify modelling best practices and provide guidance to the community

A full report of the results of this round robin is available and can be shared upon request to the author.

ERSI Residual Stress Characterization Committee

The mission of the residual stress characterization committee is to help ERSI stakeholders (e.g., end users and aircraft programs) design and implement fit-to-purpose residual stress measurement and simulation efforts. The committee has an established group of residual stress measurement and simulation professionals available to review, define, engage, and/or document:

- Repeatability of residual stress measurement data (in lab variability)
- Reproducibility of residual stress measurement data (lab-to-lab variability)
- Inter-method residual stress comparisons (e.g. neutron diffraction to x-ray to contour)
- Measurement model comparisons (e.g. for Cx holes)
- Uncertainty quantification & statistical methods relative to residual stress data (connect to inter-method as well as model-measurement)

The following sections discuss recent achievements of the residual stress characterization committee.

2 inch Cx Residual Stress Determination for Process Simulation Validation

In 2017 a research program was started that would allow for cross-validation of residual stress measurement methods and for the validation of finite element simulations of the Cx process [3]. Two aluminium alloys were selected, 2024-T351 and 7075-T651. The level of applied expansion was varied from approximately 3.16% to 4.16%. Multiple strain measurement methods were used during the Cx process, including Digital Image Correlation (DIC), a fiber optic system, and strain gages. In addition to the strain measurements that were performed during the Cx process, many additional measurements have been performed over the last several years. A high-level summary of what measurements have been performed is listed below:

- 2017 Performed Cx
- 2017 Argonne National Lab performed Energy-dispersive X-ray diffraction (EDXRD)
- 2018 Through Transmission Neutron Diffraction performed at Coventry in UK
- 2018 7075 Cx Coupon Processed at the CHESS EDXRD Facility
- 2019 Proto and NRC Performed Surface XRD
- 2020 Neutron Diffraction was Performed at Joint Physics Analysis Center (JPAC)
- 2021 2024-Low Cx Coupon Contour Cut at Stress-Space in UK
- 2021 7075 Cx Coupons Provided to Oakridge NL for Neutron Diffraction

In addition to the listed work that has already been accomplished, additional work remains to be done with these coupons in the future, including the writing of multiple journal papers to discuss various aspects of the results. It is noted that as part of being an ERSI round robin, there has been no central funding source, and all work provided was done at cost to the process/data owning organization. Additional information and data can be provided upon request to the author.

Contour Method Reproducibility Experiment

One area of research that has received some focused attention is the uncertainty quantification of the methods by which residual stresses within a body are determined. One method that has demonstrated excellent spatial resolution and that can be easily integrated into a fatigue crack growth analysis is the contour method. An area that has not been published on is the interlaboratory variability that is associated with the contour method. In an effort to begin this quantification effort, a group of international contour method practitioners was formed.

The scope of the effort was regarding bulk stress fields, neglecting machining or other near-surface stresses. Several coupon blanks cut from a single residual stress bearing bar of 7050-T74 high-strength aluminium alloy, where the residual stress was generated during the quench/age process. 14 samples were fabricated, numbered A00 to A13, of identical geometry of 50x75x24 mm.

Several coupons were used for planning measurements to help guide the remainder of the program and what would be given to the round robin participants. The initial contour results were performed by UC Davis on samples A01, A07, A13. The A01 and A07 results were nearly identical, with a higher magnitude for A13, likely due to proximity to end of bar (and was distant from participant samples). It was also observed that the spatial distribution of stress was similar along length of bar. Neutron diffraction results were also performed at the Oak Ridge National Laboratory facility on sample A08, where similar spatial form was observed with an offset of ~25MPa (within expectation). Lastly, hole-drilling results were performed by UC Davis using sample A00, where it was observed that the near surface stresses were very symmetric. Results are shown in Figure 3.

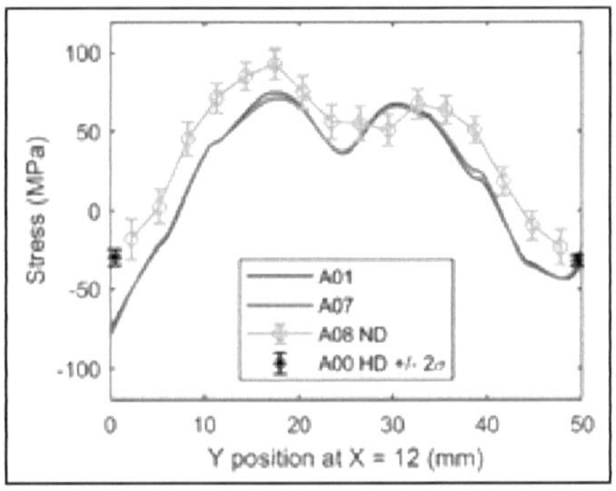

Figure 3: Planning measurement results [4]

An international group of eight participants from industry and academia were then asked to provide contour measurement results from the eight different but replicate samples. The observed interlaboratory reproducibility was reported as 8.1 MPa average for all locations, 6.1 MPa on interior, and 17.6 MPa near boundary (within 1 mm). Results are shown in Figure 4. The differences from the group mean varied among the participants (see Figure 5) with RMS differences range 7.8-14.1 MPa and maximum differences range 35.5-107 MPa.

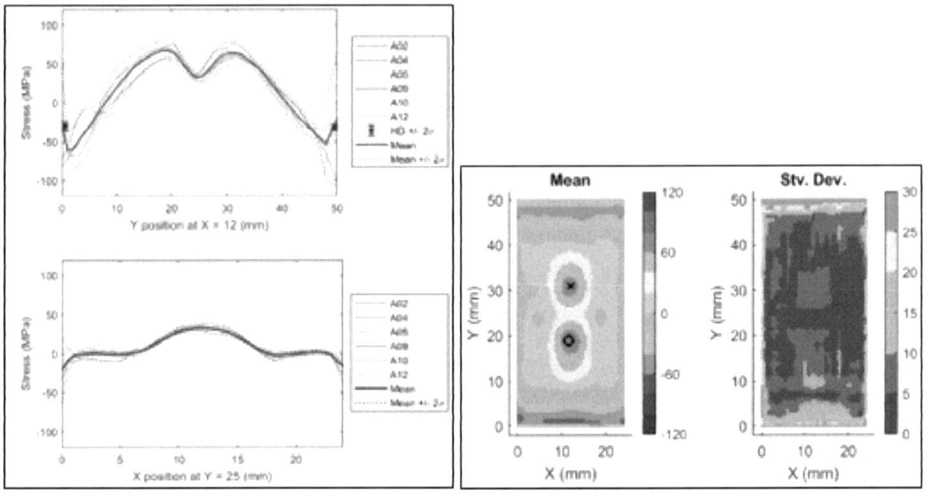

Figure 4: Participant results [4]

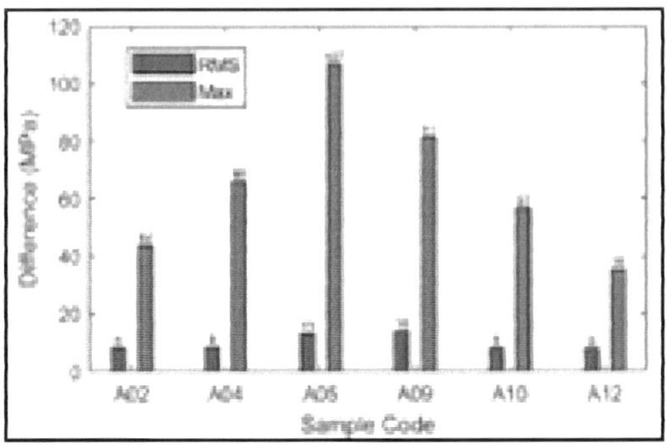

Figure 5: Participant result differences [4]

For further details on this research, please see [5].

ERSI NDI, NDE, Quality Assurance and Data Management Committee

Validated quality assurance (QA) and nondestructive evaluation (NDE) methods continue to be a key focus on the path to "full credit" for engineered residual stresses. A valid QA/NDE method is necessary to verify the engineered residual stresses are attained as intended, both when the residual stress is first imparted into the structure and during sustainment of that structure. There are several programs under way which will help to establish a valid NDE method. One program that is under way is described in the following section. The last section discusses the development and update of a residual stress database and the associated data management related challenges in dealing with all the data associated with Cx holes and residual stresses.

Nondestructive Evaluation for Quality Assurance and Surveillance of Cx Fastener Holes

The program objectives for this effort included:

- Develop NDE techniques for quantifying the residual stress state at Cx holes
- Evaluate and rank NDE techniques for quantifying residual stress state at Cx holes
- Investigate key confounding factors and their influence on NDE response
- Applied expansion, diameter, thickness, material, edge margin, coatings, etc.
- Optimize, demonstrate, and verify NDE techniques for Cx hole evaluation

The key objective was to be able to verify residual stress is present at the hole post-Cx (i.e., go/no-go), which is necessary to be able to obtain "full-credit" for residual stress benefit from Cx. The NDE technology used in the program included (see
Figure 6):

- Eddy current surface probe
- Measures gradient of conductivity at the surface
- Clear distinction between Cx and non-Cx holes in all cases
- Eddy current low frequency in-hole probe
- Measures gradient of conductivity caused by the split-sleeve ridge
- Clear distinction between Cx and non-Cx holes in most cases
- Ultrasonic probe
- Ultrasonic critically refracted longitudinal (LCR) wave probe in pitch-catch configuration
- Clear distinction between Cx and non-Cx holes in most cases

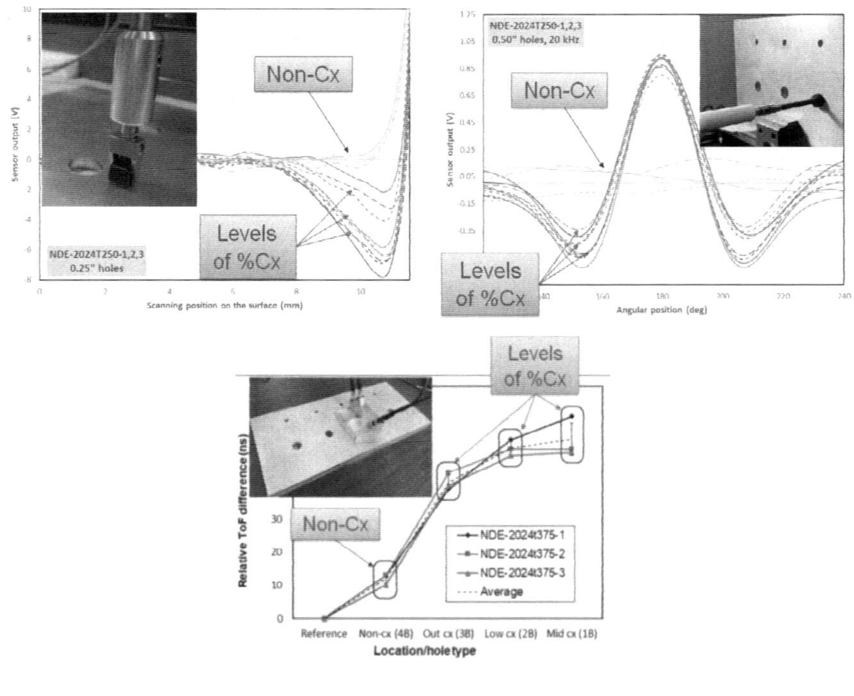

Figure 6: NDE technologies [6]

Extensive development and testing of NDE technology prototypes for detection of compressive residual stress at Cx holes was performed throughout the program. Testing was performed under a variety of conditions and confounding factors. The probes were able to successfully differentiate between non-Cx and Cx holes. Several were capable of accurately detecting Cx in the presence of confounding factors including, shot peening, paint, and low edge margin. These NDE technologies appear to be a promising technique for NDE of residual stress near Cx holes. Future work is already being planned to further pursue this topic.

Residual Stress Database

The residual stress database is a publicly available dataset of residual stresses from the contour method, which consists of a folder with a collection of files each representing a residual stress database entry. The database includes visualization, library, searching, and interpolations of multiple files, with the capability of allowing the user to add profiles (see screenshot of the user interface in
Figure 7).

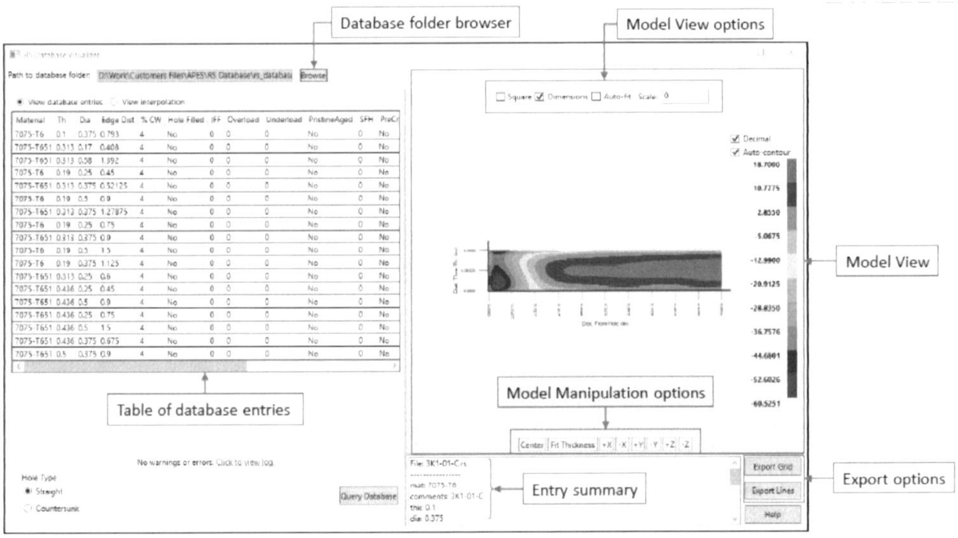

Figure 7: Screenshot of residual stress database interface [7]

The original residual stress database was organized in the ~2015 timeframe and included filtering of only five parameters: material, applied expansion, hole diameter, thickness, and edge margin. The newly updated database includes new capabilities, including:

- Filtering of over 15 new parameters
- Over/underload, pre-cycles, filled holes, pristine/aged, Cx countersink process, etc.
- Custom residual stress data output
- Export lines, along angles or at offset distances
- Handling of data replicates

The residual stress database is available to anyone and is completely free. While the original database had 47 total residual stress profiles, the updated version includes 323 total residual stress profiles. For access, contact the author.

Conclusion

To summarize the value of the ERSI working group and its members, Dr. James Castle (The Boeing Company) wrote the following:

"Engineered residual stresses provide a significant opportunity to extend the life of existing DoD platforms. With the increased number of assets grounded for maintenance, the ability to develop engineered residual stress techniques to extend airframes and lengthen intervals between inspections is essential technology. However, it has been demonstrated repeatedly that the ability to properly analyze, apply, and measure engineered residual stresses requires advanced knowledge to ensure appropriate application. Typically this has been accomplished through an extensive test and analysis program on each individual case with significant cost. This working group provides the opportunity to share the best practices the community has experienced in individual case by case insertions enabling tools and processes to be developed for the general cases that benefits all stakeholders especially the DoD which will benefit in improved platform availability at less investment per insertion."

Acknowledgments

Acknowledgements are given to the entire ERSI working group for their contributions to the technical body of work and research related to engineered residual stresses.

References

[1] Pilarczyk, R., Walker, K. (2022) "Fatigue Crack Growth Analysis Methods & Validation Testing, 2021 Achievements – Interference Fit Round Robin," *Proceedings of the 2022 Engineered Residual Stress Implementation Workshop*. https://residualstress.org/images/6/65/2022_FCG_Analysis_Methods_and_Testing.pdf.

[2] Pilarczyk, R., Walker, K. (2022) "Fatigue Crack Growth Analysis Methods & Validation Testing, 2021 Achievements – Stress Intensity Factor Round Robin," *Proceedings of the 2022 Engineered Residual Stress Implementation Workshop*. https://residualstress.org/images/6/65/2022_FCG_Analysis_Methods_and_Testing.pdf.

[3] Carlson, S. (2022) "Residual Stress Measurement, 2 inch Cx Residual Stress Determination for Process Simulation Validation," *Proceedings of the 2022 Engineered Residual Stress Implementation Workshop*. https://residualstress.org/images/1/10/2022_RS_Measurement.pdf.

[4] D'Elia, C. (2022) "Residual Stress Measurement, Contour Method Reproducibility," *Proceedings of the 2022 Engineered Residual Stress Implementation Workshop*. https://residualstress.org/images/1/10/2022_RS_Measurement.pdf.

[5] D'Elia, C.R., Carlson, S.S., Stanfield, M.L. *et al.* (2020) Interlaboratory Reproducibility of Contour Method Data Analysis and Residual Stress Calculation. *Exp Mech* **60**, 833–845. https://doi.org/10.1007/s11340-020-00599-0.

[6] Andrew, D. (2022) "NDI, NDE, Data Management, and Quality Assurance, Tools and Methods – Nondestructive Evaluation for Quality Assurance and Surveillance of Cx Fastener Holes," *Engineered Residual Stress Implementation Screamer*, Volume 4, Issue 1 (2022), p. 19-20. https://residualstress.org/images/9/9a/ERSI_Screamer_Volume_4_Issue_1_%282022%29.pdf.

[7] Hunt, L. (2022) "Risk Analysis and Uncertainty Quantification, New Residual Stress Database," *Proceedings of the 2022 Engineered Residual Stress Implementation Workshop*. https://residualstress.org/images/1/1b/2022_Risk_%26_UQ.pdf.

ICRS 12: Proceedings from the 12th International Conference on Residual Stresses, 2025
October 20–23, 2025; Detroit, Michigan, USA
https://doi.org/10.31399/asm.cp.icrs2025p0137

VIRTUAL STRENGTH TESTING OF CRANKSHAFTS TAKING INTO ACCOUNT RESIDUAL STRESSES FROM THE SURFACE HARDENING PROCESS

Dipl.-Ing. H. Dannbauer, Dr. O. Grieshofer, Dr. W. Meindl *Engineering Center Steyr GmbH & Co KG, St. Valentin, Austria*

Abstract

Crank shafts are subjected to a multi axial stress history during real engine operation, but because of increased efficiency, testing is performed for bending and torsion separately. The sustainable endurance moments are an important characteristic used for designing strength and optimizing the production process. Determining the bending and torsion endurance moments accurately with numerical means is only feasible if the surface hardening is considered. For this both the residual stress field as well as the altered material properties of the hardening layer need to be considered. For strength assessment the "Critical plane - Critical component" hypothesis was used which distinguish normal stress and shear stress type loading in an adequate way related to the material behavior.

Introduction

The development of crankshafts is characterized by a complex interaction of geometry, material, and manufacturing process, with costs naturally playing a significant role in the materials and methods used. Often, there are conflicting requirements, such as with the fillet radii between the journals and the webs: from a fatigue strength perspective, a larger radius would be optimal, while for the durability and wear of the bearing shells, a smaller radius is desirable to achieve the largest possible bearing surface. Similar contradictions need to be resolved, for example, with journal diameters concerning friction and stiffness, or with the web widths concerning the strength of the crankshaft and the design of the engine block.

At the end of the development cycle at the supplier, the endurance strength test is conducted on the test bench. If the requirements are not met, optimizations must be implemented to achieve the required strength. Each development cycle is very costly and time-consuming. In contrast, virtual testing could reduce both costs and time that possibly uncover untapped potential [4].

CAE - State of the Art

Crankshafts are subjected to complex loads during engine operation. On one hand, the combustion process results in broadband excitation, and on the other hand, the crank mechanism and the engine housing are oscillatory systems, typically leading to resonances within the operating range.

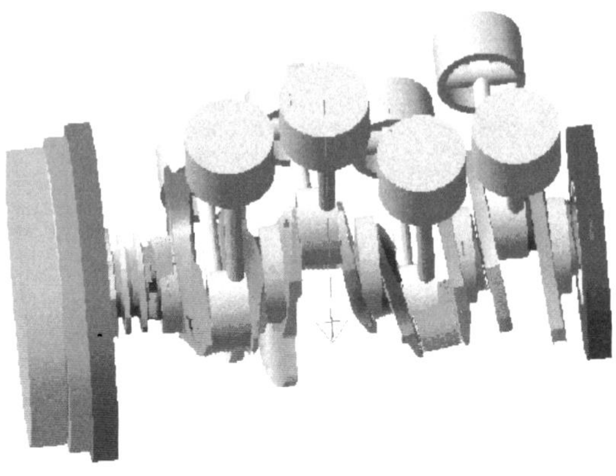

Figure 1: Multi-Body Simulation Model - Crank Mechanism

The Figure 1, shows a deformed snapshot of the multi-body simulation model of an entire crank mechanism, including a nonlinear vibration damper and a dual-mass flywheel. The crankshaft is modeled as a flexible body, and the main bearings of the crankshaft are mounted in the crankcase using an elastohydrodynamic (EHD) bearing model. The calculation is performed in a time-domain modal reduction, significantly reducing computation time and the amount of result data.

The following images exemplify the results of such a simulation: Figure 2 shows displacements in mm over crank angle. The curves are relative displacement between crank shaft and bearing. For this simulation an elastohydrodynamic bearing model has been used as described in the paragraph above and the Figure 1.

The bearing eccentricity of a main bearing journal is shown as a polar diagram over 720 degrees of crank angle. The complex pattern arises from the strong interaction between the crankshaft and the bearing under gas pressure excitation; the marked positions are used in the strength calculation of the housing.

The Figure 3 shows (dimensionless) Endurance Safety factors over rpm. The endurance safety factor analyses are carried out for stationary operating conditions in terms of rpm and the result of the operational strength analysis is graphically displayed for all main bearings over the crankshaft speed. One can observe a significant drop in endurance safety in the lower speed range due to an increase in combustion pressure, while in the middle and higher speed ranges, there are dips in safety mainly caused by bending and torsion resonances. Gyroscopic effects also play a role, as can be seen at high speeds at main bearing 1 (flywheel side).

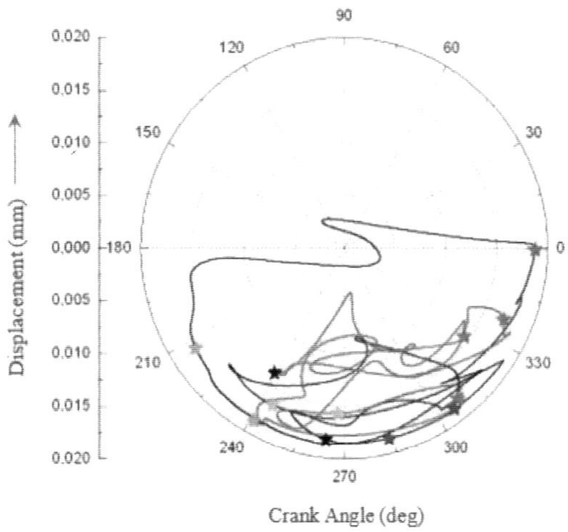

Figure 2: Bearing Orbit

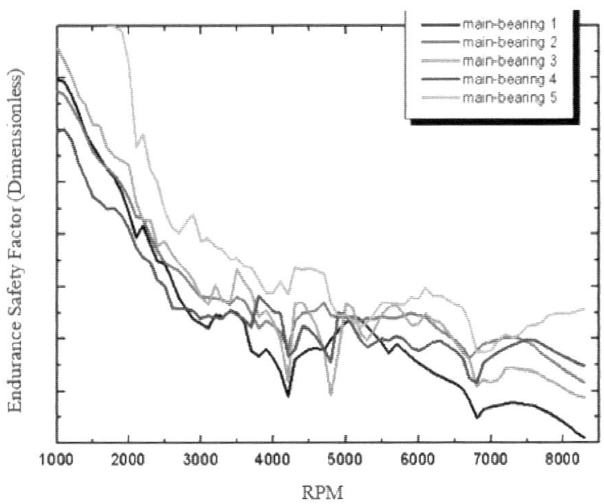

Figure 3: Fatigue Safety Factors Vs Rpm

Concept Component Test

As explained in point 1, a practical testing concept is of great importance for an efficient development process. Since test runs in a fired engine are not possible in the initial development phase of the crankshaft, and component testing with realistic loads fails due to the very high effort required, substitute loads must be defined for component testing. This involves using analytically or numerically calculated section forces in the crank pins and exaggerating the maximum occurring bending and torsion moments with scaling factors to account for both the dynamic effects and the lack of moment superposition due to two separate component tests.

Another surcharge factor (ranging from 1.5 to 2, depending on the material and process) is necessary due to the required high survival probability of the crankshaft during engine operation (> 99.999%) compared to the well-determined 50% values on the test bench. This also incorporates empirical values regarding the variability in geometry, material, process, and load.

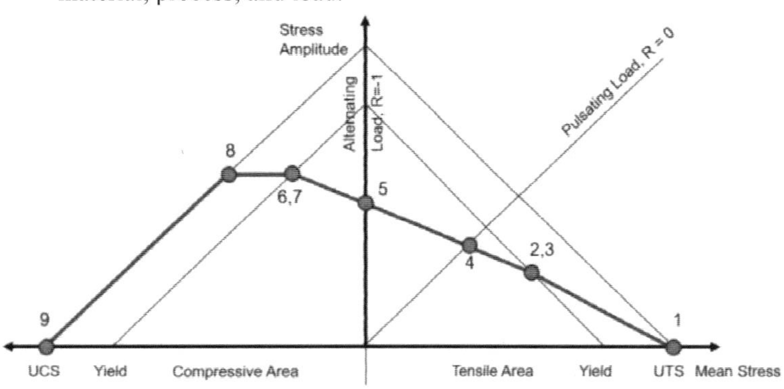

Figure 4: Haigh diagram

The test setup is typically realized through a resonance test, which results in another difference from real engine operation. Resonance tests of crankshafts can only be conducted without preload, expressed in the Haigh diagram (Figure 4), at point 5. However, real engine loads lead to stresses between point 4 (quasi-static bending at low speeds) and point 5 (torsion or bending resonances), depending on the operating condition. This influence would result in a reduction factor for the required bending and torsion endurance moments. However, as explained in last chapter, the critical states during engine operation are mostly resonance-dominated, making this approach unjustified due to conservative considerations. Additional advantages of resonance tests include the high load frequency, low test bench wear, simple control, and low energy consumption.

Virtual test bench

The goal of the virtual test bench is to replicate the component load as accurately as possible. For this purpose, a finite element model was created that includes all relevant components with the necessary level of detail, as shown in Figure 5. The outer beams represent the oscillating masses of the test bench, through which the load is ultimately applied to the crank pin being tested. The adapter serves to adjust the test bench to different crankshafts and has a different shape depending on the load case (bending or torsion), as shown in the Figure 6 and 7.

The calculation could theoretically be performed in two ways: either through an eigenvalue analysis or a quasi-static analysis. The difference in computational effort is not significant (real eigenvalues are somewhat more complex than linear statics), but an eigenvalue analysis assumes linear system behavior. Since there are bolted connections between the beams, adapter, and flange, and the crank pin is held in the flange by an interference fit, there could be nonlinear effects (lifting and sliding). Therefore, a quasi-static, nonlinear analysis was preferred and contact with friction was defined between all components. In the first two calculation steps, the interference fit of the pin and the bolt preload were applied, and in the subsequent steps, the moment loads were applied in both positive and negative directions.

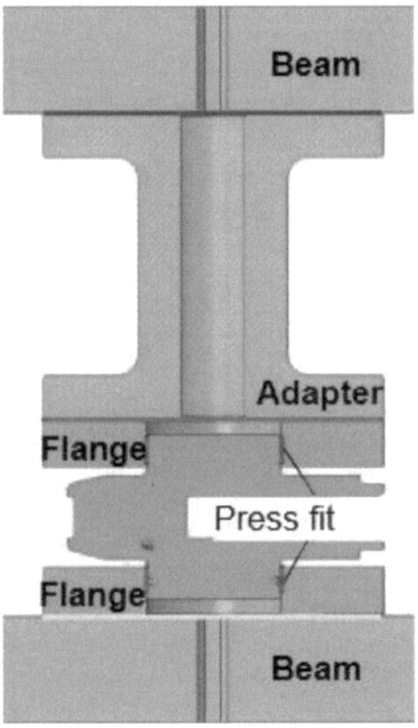

Figure 5: Virtual Test Bench

Approximating the natural vibration mode through a quasi-static deformation is permissible because the mass and vibration participation of the adapter, flange, and crank pin components are insignificant compared to the overall system. The Figure 6 and 7 shows the adapter for bending load and for torsion load.

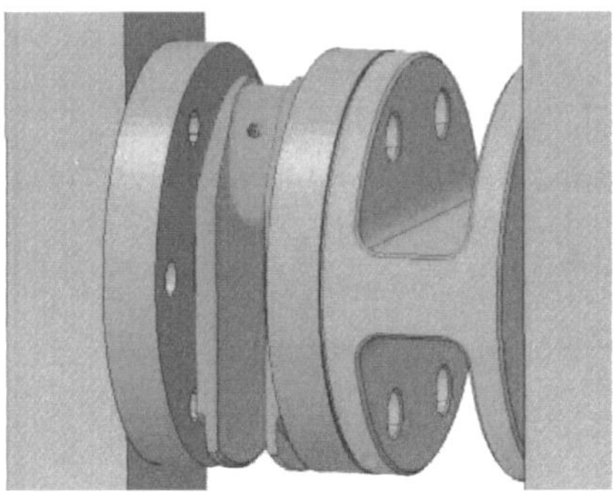

Figure 6: Bending Test

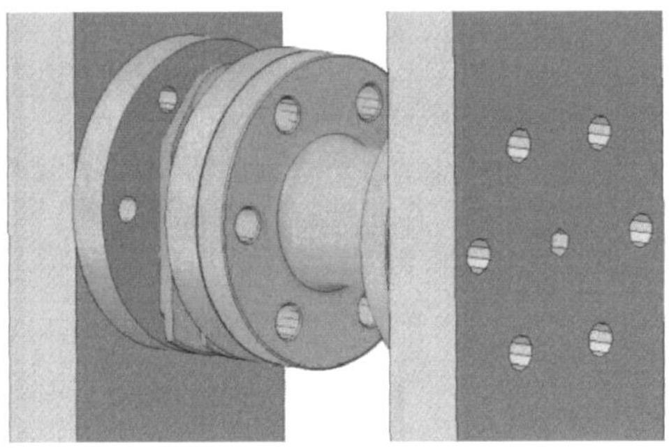

Figure 7: Torsional Test

Finite Element Analysis and Results

For a subsequent absolute fatigue strength assessment, a high-quality stress calculation is necessary; otherwise, a loss of accuracy must be expected. The meshing was done finely in the critical areas and adjusted to the occurring gradients using 10-node volume elements, as shown in the Figure 8, the entire crank pin along with a detailed enlargement.

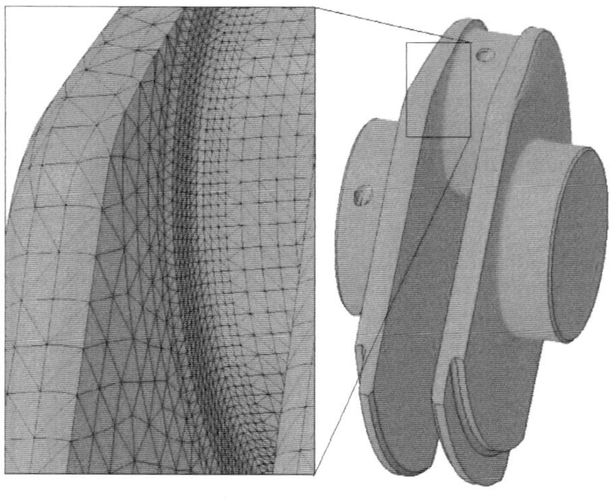

Figure 8: Mess Model – Critical area - fillet

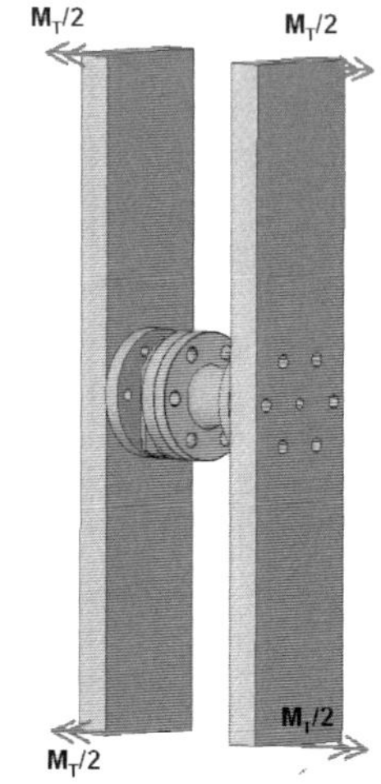

Figure 9: Torsional Test

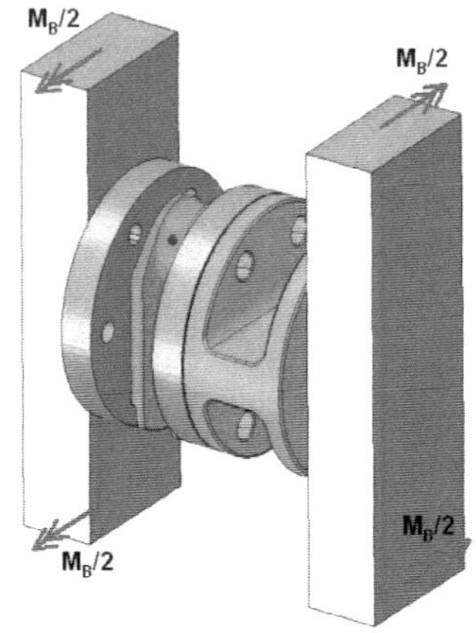

Figure 10: Bending Test

In the Figure 9 and Figure 10, the application of the loads is visible; moreover, the entire model is held in space without constraints using couplings. In total, 2 operating load cases each for bending and torsion were calculated to capture the nonlinear effects in the contact zones. The calculation was performed using ABAQUS Standard with linear geometry (small displacements) and linear material behavior.

In the Figure 11, the von Mises equivalent stresses for the assembly condition are shown. The maximum values exceed 1000 N/m^2 in the flange and in the oil holes of the crank pin. In the fillet radii of the main bearing journal, a slight influence of the press-fit can still be observed, while in the fillet radii of the connecting rod bearing journal, the influence is even smaller.

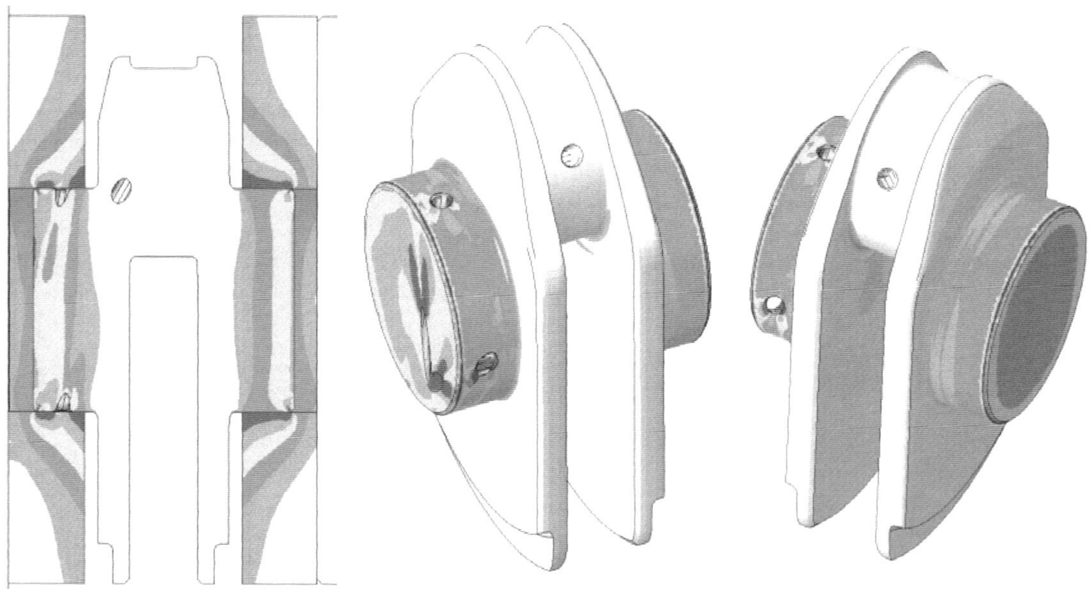

Figure 11: Von Mises equivalent stresses for the assembly condition

In the Figure 12, the von Mises equivalent stresses for the bending load case are shown, with the pre-stress condition subtracted. The stress values in the crank pin fillet radius are slightly higher than those in the main bearing journal fillet radius. Furthermore, the stress is highly concentrated in the immediate vicinity of the fillet radius, indicating a high notch effect.

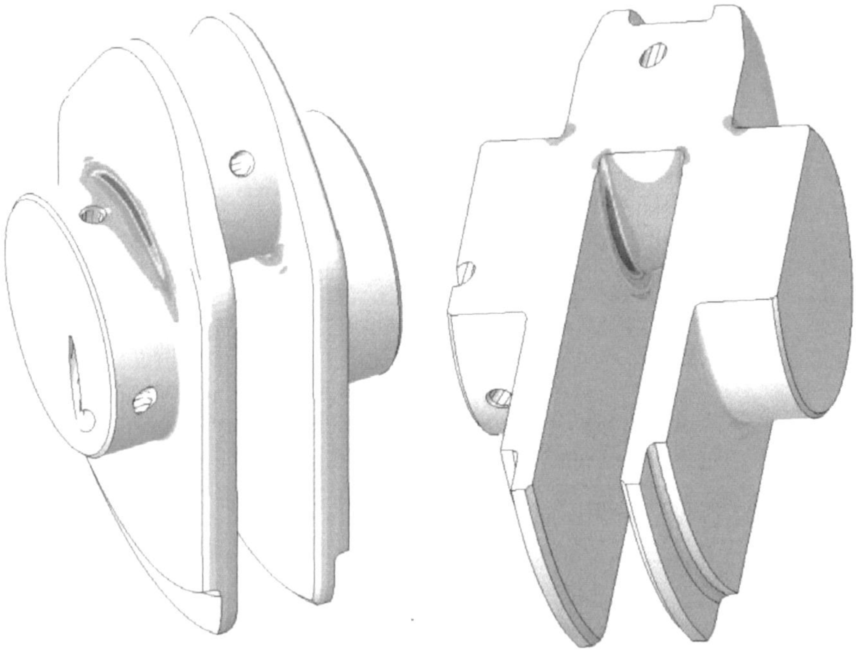

Figure 12: Von Mises equivalent stresses for the bending load

In the Figure 13, the von Mises equivalent stresses for the torsion load case (without assembly) are shown. The peak values are concentrated in the fillet radius of the connecting rod pin, and the distribution is significantly more uniform. A detailed analysis of the stress maxima shows that the locations for the assembly and operating load cases are clearly separated, ensuring a minimal influence of the clamping on the test results.

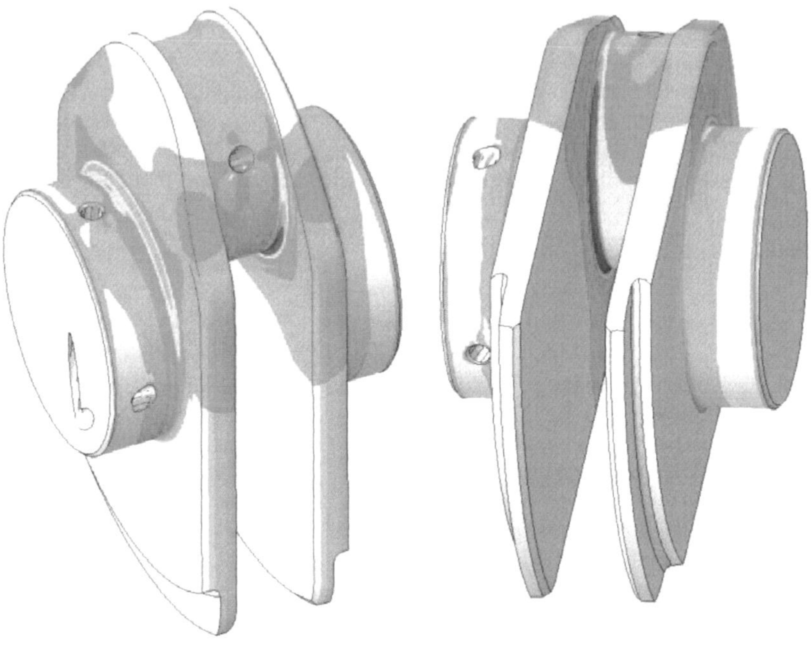

Figure 13: Von Mises equivalent stresses for the torsion load case

Fatigue strength analysis

The material properties, in addition to the stresses from the finite element analysis, play a crucial role in the accuracy of the fatigue strength prediction. The material parameters of 42CrMo4 for the non-hardened material were determined based on existing test results from static tests and bending fatigue tests. The values not directly available (alternating strength, mean strength, shear alternating strength) were determined using the FEMFAT Material Generator based on the FKM guideline. For the hardened surface layer, the material parameters were estimated using a study by Hertter [1] from measured hardness values (HV), with the alternating strength increasing by a factor of 1.7.

However, the Haigh diagrams show a significantly different shape, indicating the higher sensitivity to mean stress of the hardened material vs non-hardened material in Figure 14.

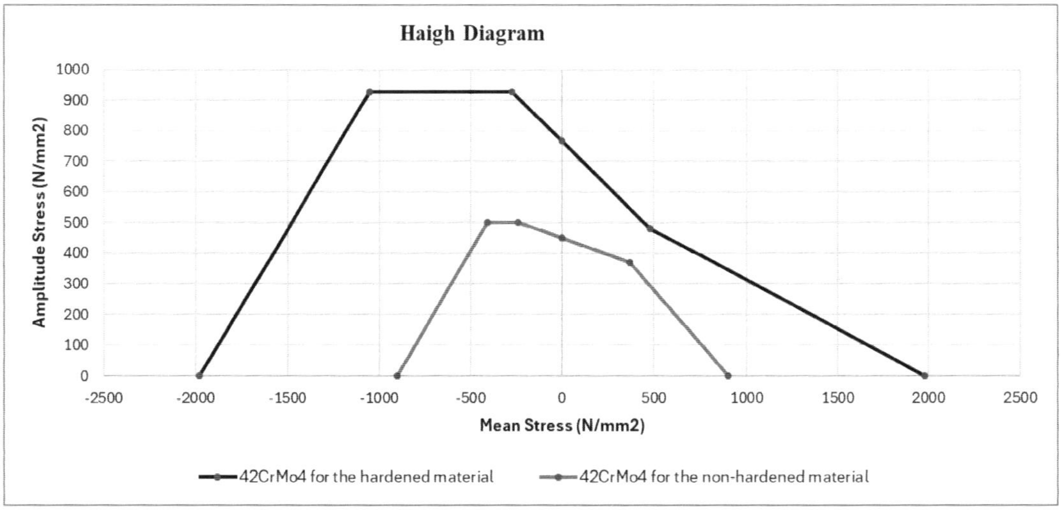

Figure 14: Haigh Diagram: 42CrMo4 for the non-hardened Vs hardened Steel

For the actual fatigue strength analysis, the "Critical Plane - Critical Component" method according to Gaier [2] was applied, which offers significant advantages, especially under shear stresses combined with mean stresses.

This patented method can be considered an extension of the usual cutting plane methods. Here (Figure 15), in the respective cutting plane (n), the stress vector $S_n(t)$ is projected onto a variable directional vector (d). The length $S_{n,d}(t)$ represents the signed equivalent stress, and the angle **θ** represents the polar angle. The polar angle characterizes the type of stress and allows for the distinction between normal and shear mean stresses in a 3-dimensional Haigh diagram in Figure 16.
As additional important effects, the micro-support effect in notches according to Eichlseder [3] and the statistical influence (Gaussian logN) were considered.

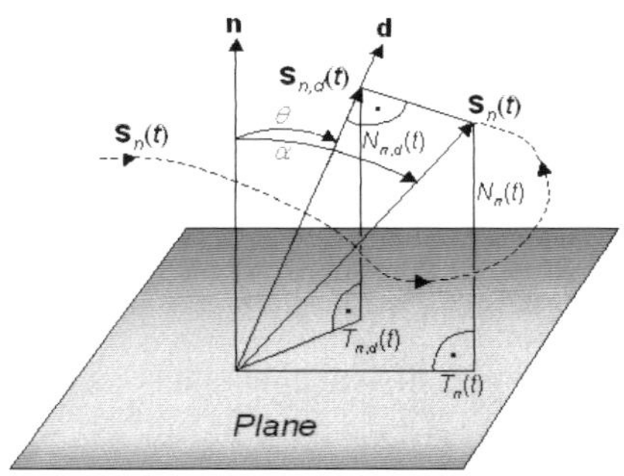

Figure 15: Critical Cutting Plane

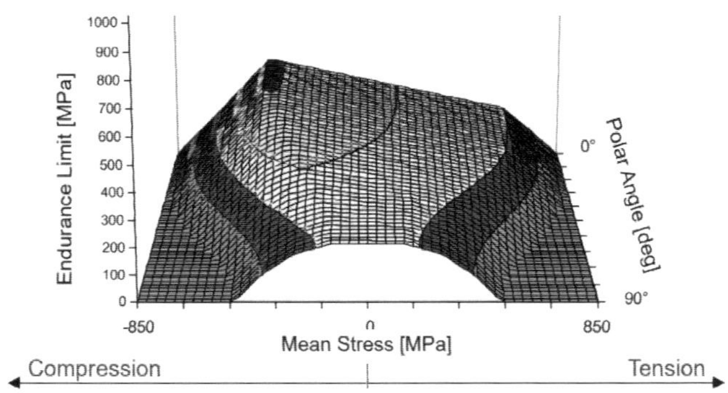

Figure 16: Haigh Diagram: Steel

Validation
To compare the results of the real and virtual test benches, the endurance safety factors at the actual endurance moments were calculated. For both the bending tests and the torsion tests, endurance moments were available for component samples with and without induction hardening.

The calculated endurance safety factors **without hardening** show very good agreement with the experimental results. The prediction is only 6% too progressive for bending in Figure 17 and 4% too conservative for torsion in Figure 18.

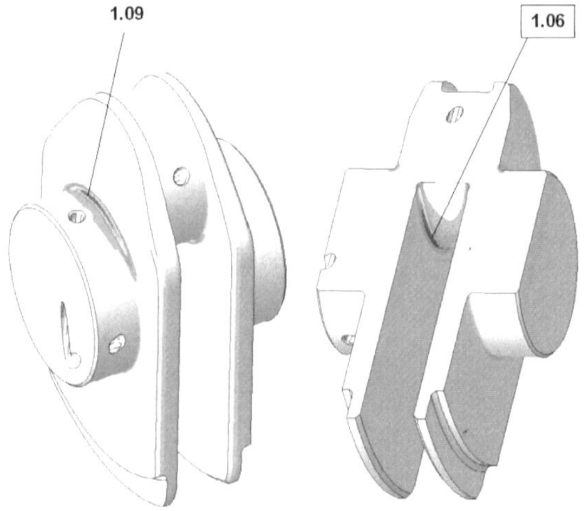

Figure 17: Test Vs Virtual Test Bench – Bending (W/O Induction Hardening)

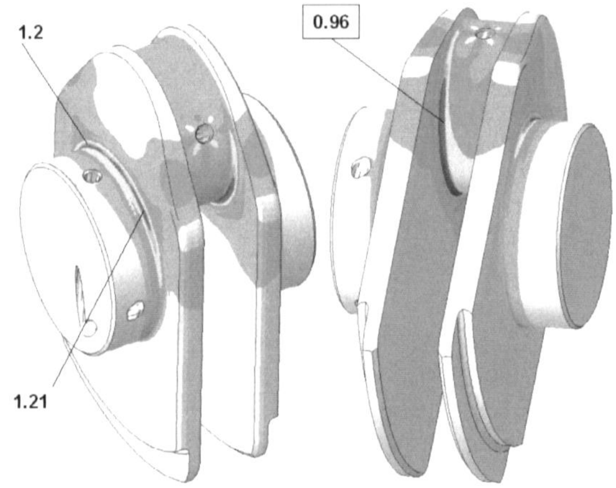

Figure 18: Test Vs Virtual Test Bench – Torsional (W/O Induction Hardening)

The calculated endurance safety factors **with hardening** show unsatisfactory agreement with the experimental results. The deviation is -41% for bending in Figure 19 and +5% for torsion in Figure 20.

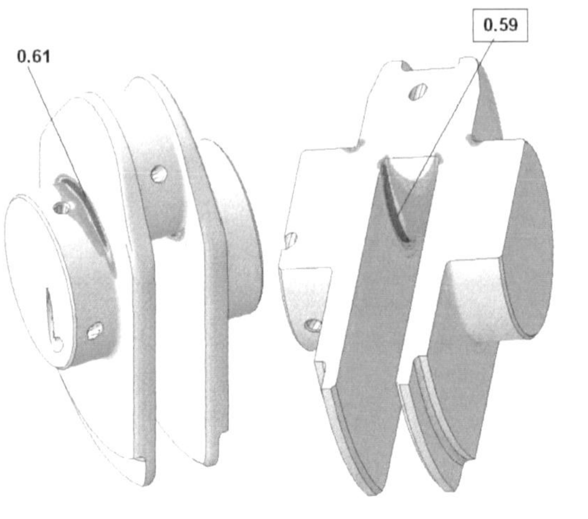

Figure 19: Test Vs Virtual Test Bench – Bending (with Induction Hardening)

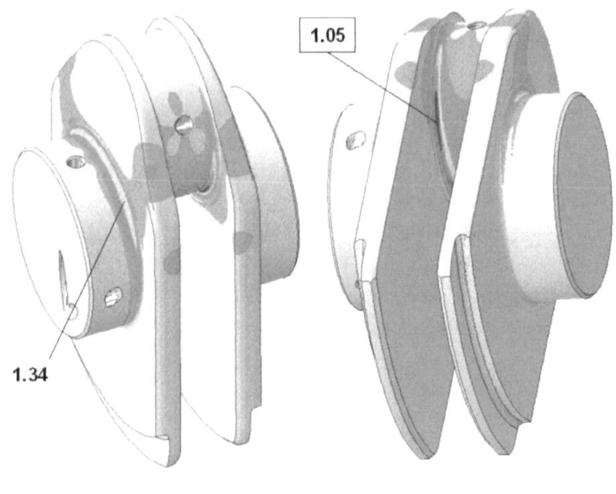

Figure 20: Test Vs Virtual Test Bench – Torsional (with Induction Hardening)

Several reasons are likely responsible for the poor prediction of the endurance moments for induction-hardened surfaces:

- Lack of consideration of residual stresses from the hardening process
- Inaccurate estimation of the properties of the hardened material
- Inaccurate estimation of missing material parameters according to the FKM guideline

View

In the future, it is conceivable that a process simulation of the induction hardening process could precede the fatigue strength analysis. This would make it possible to predict both the residual stress distribution and the local microstructure composition. From the microstructure composition, the strength properties of the material can be inferred. The prerequisites in **FEMFAT** are already in place, as local distributions of material properties can be imported via a standardized interface.

Summary

A virtual test bench for the strength testing of crankshafts was presented. The correlation with the experimental results is very good without surface hardening, but there are deficiencies with hardening because, so far, only material hardening has been considered, not the residual stress field. The strength hypothesis "Critical Plane - Critical Component" also allows for a good prediction in this case, provided that good input values regarding residual stresses and material properties of the hardened surface layer can be ensured.

The future integration of process simulation opens further optimization potential in the interplay of geometry, material, machining, surface finish, and hardening process.

References

[1] Hertter T., "Computational strength verification of the fatigue bearing capacity of quenched and tempered and case-hardened cylindrical gears", dissertation, Technical University of Munich 2003

[2] Gaier C., Pramhas G., Steiner W., "An extended critical plane approach for general load situations", Fatigue Conference 2002, Stockholm, Sweden

[3] Eichlseder, W., "Computational Lifetime Analysis of Commercial Vehicle Components with the FE Method", Dissertation, TU Graz 1989

[4] Dannbauer H., Grieshofer O., Meindl W. Schmidt J., Plew M, „Virtuelle Festigkeitsprüfung von Kurbelwellen unter Berücksichtigung von Eigenspannungen aus dem Härteprozess der Oberfläche", SIMVEC Conference 2012

ICRS 12: Proceedings from the 12th International Conference on Residual Stresses, 2025
October 20–23, 2025; Detroit, Michigan, USA
https://doi.org/10.31399/asm.cp.icrs2025p0145

Effect of Residual Stresses on the Column Flexural Buckling Curves

Shahriar Quayyum
Department of Civil & Environmental Engineering, Manhattan University, Riverdale, NY 10471
shahriar.quayyum@manhattan.edu

Abstract

This paper presents a comprehensive study on the effects of residual stresses on the flexural buckling resistance of hot-rolled wide flange steel columns fabricated from ASTM A992 and A913 Grade 80 steels. Through advanced nonlinear finite element simulations, the influence of initial imperfections, nonlinear material properties, and multiaxial residual stress fields is investigated. The residual stress fields were simulated via a sequentially coupled thermo-mechanical analysis representing realistic cooling conditions after hot-rolling. The models were validated against benchmark experimental results and utilized to generate buckling curves for various slenderness ratios. The findings show that residual stresses have a noticeable effect on heavier shapes and significantly influence inelastic buckling resistance, often resulting in non-conservative predictions by current AISC 360-22 design provisions. These results underscore the need for developing refined design curves for steel columns, particularly for heavy sections and high-strength steel.

Introduction

Residual stresses are an inherent outcome of steel manufacturing processes, particularly hot rolling and quenching, and can significantly influence the structural performance of steel members [1]. In rolled wide flange sections, the nonuniform cooling after hot-rolling results in thermal gradients that induce complex multiaxial residual stress fields. These stresses persist in the member even in the absence of external loads and play a crucial role in determining the column's susceptibility to flexural buckling [2].

The distribution of residual stresses is characterized by tensile stresses in the flanges near the web junction and compressive stresses at the flange tips. In the web, both compressive and tensile residual stresses can coexist depending on the thermal history. These stress fields reduce the effective stiffness of the column and can lead to premature local yielding, particularly in the inelastic buckling regime [3]. Their impact is accentuated in high-strength steels, such as ASTM A913 Grade 80, where the high yield strength does not necessarily correlate with proportionally higher residual stress magnitudes.

Despite their recognized impact, residual stresses are often simplistically treated or altogether neglected in standard design provisions. Current design codes like AISC 360-22 [4] and Eurocode 3 [5] typically base column design curves on assumptions derived from conventional-strength steels and older manufacturing practices. These conservative assumptions, while ensuring safety, can lead to inefficient designs when applied to modern high-strength steels.

This study focuses on quantifying the effects of residual stresses on the flexural buckling behavior of columns fabricated from A992 and A913 Grade 80 steels. Through detailed finite element modeling that includes multiaxial residual stress distributions derived from validated thermo-mechanical simulations, the study aims to critically assess the adequacy of current design provisions, and the importance of residual stresses on the flexural buckling curves of conventional and high-strength steel columns.

Finite Element Modeling Approach

Finite Element Model

Three-dimensional (3D) nonlinear finite element models were developed for wide flange rolled-shape columns using the finite element analysis (FEA) software ANSYS. The finite element models include geometric and material nonlinearities. The simply-supported columns were modeled and discretized by twenty-noded solid brick elements (SOLID 186). A typical finite element mesh of a column along with the boundary conditions for simulating major and minor axis buckling is shown in Figure 1. The column was pin-supported at the bottom and the same boundary condition was applied at the top with the exception that vertical translation was permitted (Figures 1b and 1c). For major axis buckling, the column was restrained against minor axis deflections by translational supports along the length [6] as shown in Figure 1c.

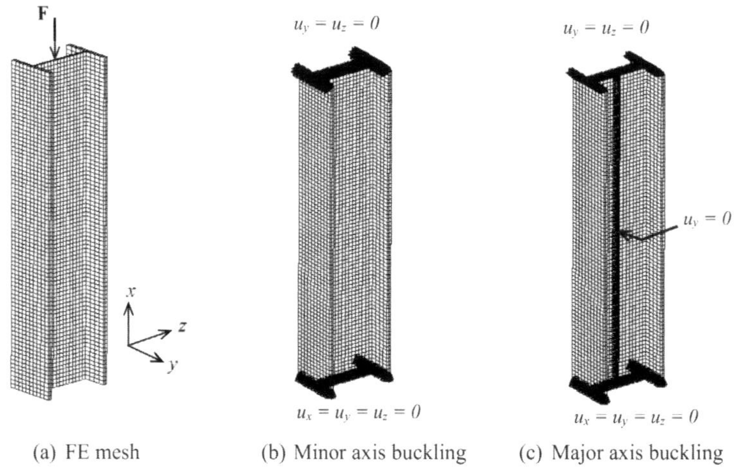

Figure 1: Finite element mesh of a wide flange column with boundary conditions for minor and major axis buckling.

Material Model

In the finite element modeling of columns, it is important to use the characteristic yield stress related to the grade of steel and the shape of the material stress-strain curve. In this study, column buckling responses were investigated for ASTM A992 steel and A913 Gr. 80 high-strength steel. The non-linear material properties of ASTM A992 steel and A913 Gr. 80 structural steel in ANSYS were defined using the multilinear kinematic hardening (MLKH) model with von-Mises yield criterion using initial elastic modulus, Poisson's ratio, and stress-strain data obtained from the uniaxial tension tests. The uniaxial tension test data for ASTM A992 was obtained from the tests performed by Morrison [7]. In the multilinear model, the elastic modulus of A992 steel was taken as 29,000 ksi with yield stress and tensile strength of 50 and 67 ksi, respectively. The stress-strain data for A913 Gr. 80 steel was obtained from five uniaxial tension tests performed by Nucor. In the multilinear model, the elastic modulus of A913 Gr. 80 steel was taken as 29,000 ksi with 0.2% offset yield strength of 80 ksi. The MLKH model used in the simulation for A992 and A913 Gr. 80 steels is shown in Figure 2.

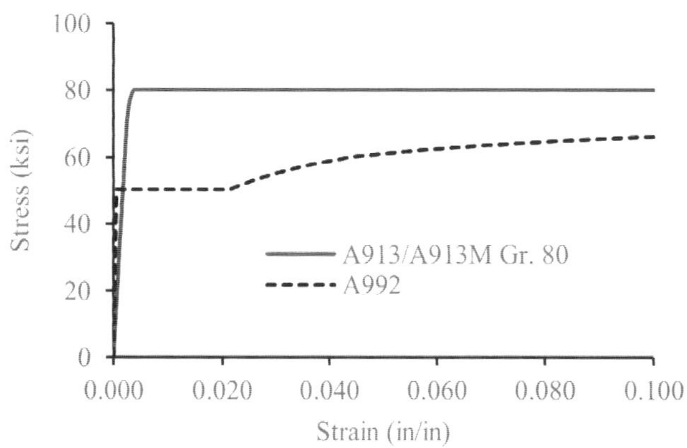

Figure 2: Multilinear model for A992 and A913 Gr. 80 steel used in the analysis.

Modeling of Initial Imperfection

For non-linear buckling analysis to be accurate using finite element simulation, it is necessary to set an initial imperfection in the column being modeled along with the residual stress distribution. The shape of the geometric imperfection was based on the buckling mode belonging to the lowest eigenvalue from a linear buckling analysis. This usually resulted in a sinusoidal bow imperfection. The amplitude defining the maximum deviation from the ideal geometry was $L/1500$, where L is the length of column. This approach is generally considered to be conservative and accepted for the determination of buckling curves. The imperfections of the finite element model were established by first performing a linear buckling analysis on the perfect prismatic column model with given boundary conditions, then the relevant (displacement) normalized global buckling mode was extracted. In the following non-linear finite element calculations, the imperfections were established by importing the normalized displacements of the lowest global buckling mode in the relevant plane of buckling as shown in Figure 3, multiplying this by the maximum imperfection magnitudes and updating the nodal coordinates of the model by adding the established nodal imperfections.

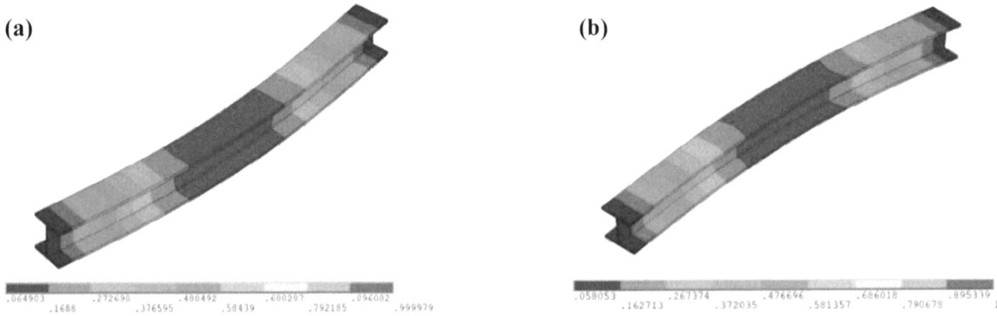

Figure 3: Normalized displacements of the lowest global buckling mode from elastic buckling analysis along (a) minor axis, and (b) major axis.

Residual Stresses Simulation

Previous studies on the non-linear buckling analysis of steel columns showed that the simplified linearized distribution of residual stresses yielded conservative results. However, most of the residual stress distribution for the rolled shapes available in the literature was developed in the 1970s based on ASTM A36 steel and it is warranted that residual stresses be measured for high-strength steel rolled shapes. Jönsson and Stan [8] pointed out that if the magnitude of residual stresses is wrongly assumed to be proportional to the yield stress, the column buckling capacity of steel columns found by finite element analysis will be underestimated for higher grades steel. Hence, in this study, the multiaxial residual stress distribution in the rolled shapes considered in the analysis was developed by using a sequentially coupled thermo-mechanical analysis which calculates the temperature history and residual stresses during the heating and subsequent cooling of steel members during the cooling process after hot-rolling. In cooling process simulation, the real-time temperature field in W shapes during cooling after hot-rolling was calculated first by employing heat transfer analyses. The temperature field at each step of the heat transfer analysis was stored and used as an input in the residual stress analysis for calculating the evolution of stress and strain fields in the W shapes during the cooling process. For details of the computational scheme, the readers are referred to Quayyum and Hassan [2]. The computational scheme was validated against several experimental responses for both A992 and A913 Gr. 80 steels which demonstrated the robustness of the computational scheme. One such validation results is presented in Figure 4. Results indicated tensile stresses in flange-web junctions and compressive stresses in web centers, consistent with experimental data.

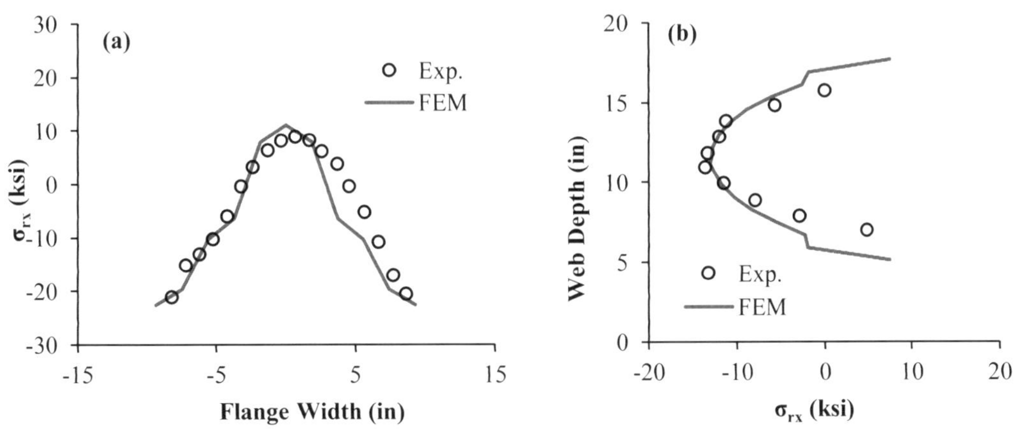

Figure 4: Comparison of FEM simulations to measured IRSs in (a) flange, and (b) web of 80 ksi W14×808 beam section (data from [9]).

Validation of FE Model

In order to validate the FE model for flexural buckling analysis of a column, a W12×87 A992 column section with 12 ft 2 in. length was analyzed under displacement-controlled monotonic axial compression which was used in the study by Lamarche and Tremblay [10] for investigating the cyclic buckling of steel columns. The initial residual stresses were considered as the initial stress state of the column by using the sequentially coupled thermo-mechanical analysis and the nodal coordinates of the model were updated by adding the established nodal imperfections from elastic buckling analysis. Simulated responses of axial load versus axial displacement (Figure 5a) and axial load versus lateral deflection (Figure 5b) are compared well to the experimental responses (from [10]) when residual stress is included in the analysis. Moreover, the initiation of global buckling and the mid-height local buckling from the simulation occurred similarly as in the experiment. These results demonstrated the excellent simulation of the buckling strength if FEM simulated multiaxial residual stress is included in the analysis, and over prediction of about 10% was obtained if residual stress is not included in the analysis. Once the global buckling was initiated, the lateral deflection gradually increased and the axial load decreased with the increase in axial displacement. The post-buckling responses with or without residual stress, however, were very similar to the experimental responses especially when lateral deflection becomes large (Figures 5a and 5b). The validation of the numerical scheme indicated that the finite element modeling techniques described above can be utilized further to develop column flexural buckling curves for rolled shapes using A992 and A913 Gr. 80 steels and study the effect of residual stresses on the flexural buckling strength of these columns.

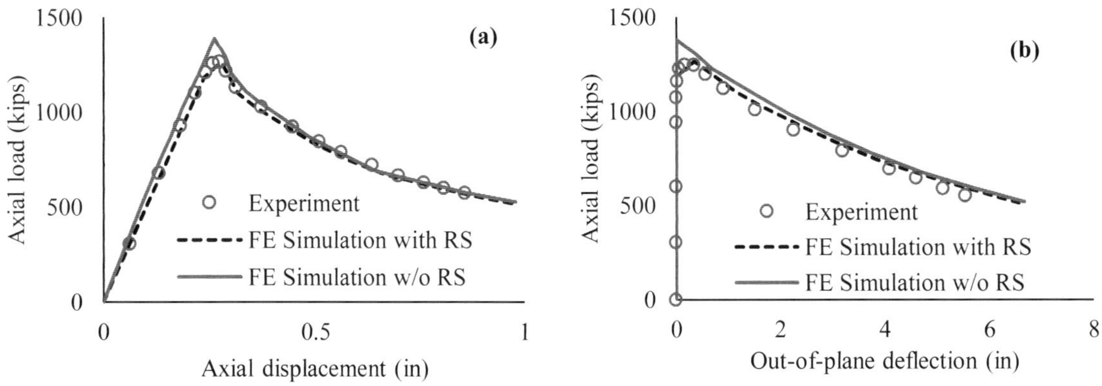

Figure 5: Buckling analysis under monotonic loading for ASTM A992 W12×87 section, (a) simulated axial load-displacement response, (b) simulated axial load-lateral deflection response (experimental data from [10]).

Column Buckling Curves for A992 Steel

Initially, flexural buckling behavior of A992 steel columns were simulated for the sections shown in Table 1. The shapes are chosen for a wide range of depth-to-width ratio (h/b), flange width-to-thickness ratio ($b_f/2t_f$) and web slenderness ratio (h/t_w) so that the analyses can capture their effect on the flexural buckling resistance of the columns. Finite element analyses were performed on these shapes by changing the slenderness ratio of the column ($L/r \leq 200$) using different column lengths, and the buckling strengths were determined for different slenderness ratios about both minor and major axes using A992 steel with a yield stress of 50 ksi. Because the column was considered pin-connected at both ends, the effective length factor K was taken as 1.0. To make the column curves comparable between different W shapes having different radii of gyrations about the minor and major axes, and different yield strengths of steel, the critical buckling strength was normalized by the yield stress (F_{cr}/F_y), and it was plotted against the relative slenderness ratio λ as given by:

$$\lambda = \frac{1}{\pi}\frac{L_c}{r}\sqrt{\frac{F_y}{E}} \quad (1)$$

where L_c is the effective length of the column, r is the radius of gyration, F_y is the material yield strength, E is the elastic modulus of the material. Note that the initial geometric imperfection was taken as $L/1500$ at the mid-height of the column to be comparable with the AISC 360 column design curve and for each of the shape, the simulated residual stresses were considered as the initial state of stress. Figure 6 shows column buckling curves about minor axis for the sections of Table 1 made of A992 steel along with their design curves as given by AISC 360 [4] and Eurocode 3 [5].

Table 1: Cross-sectional properties of the A992 W shapes investigated

Column Designation	h/b	t_f (in.)	$b_f/2t_f$	h/t_w
W14×808	1.23	5.12	1.82	3.04
W14×605	1.20	4.16	2.09	4.39
W14×426	1.12	3.04	2.75	6.08
W14×257	1.03	1.89	4.23	9.71
W14×193	0.99	1.44	5.45	12.8
W14×145	0.96	1.09	7.11	16.8

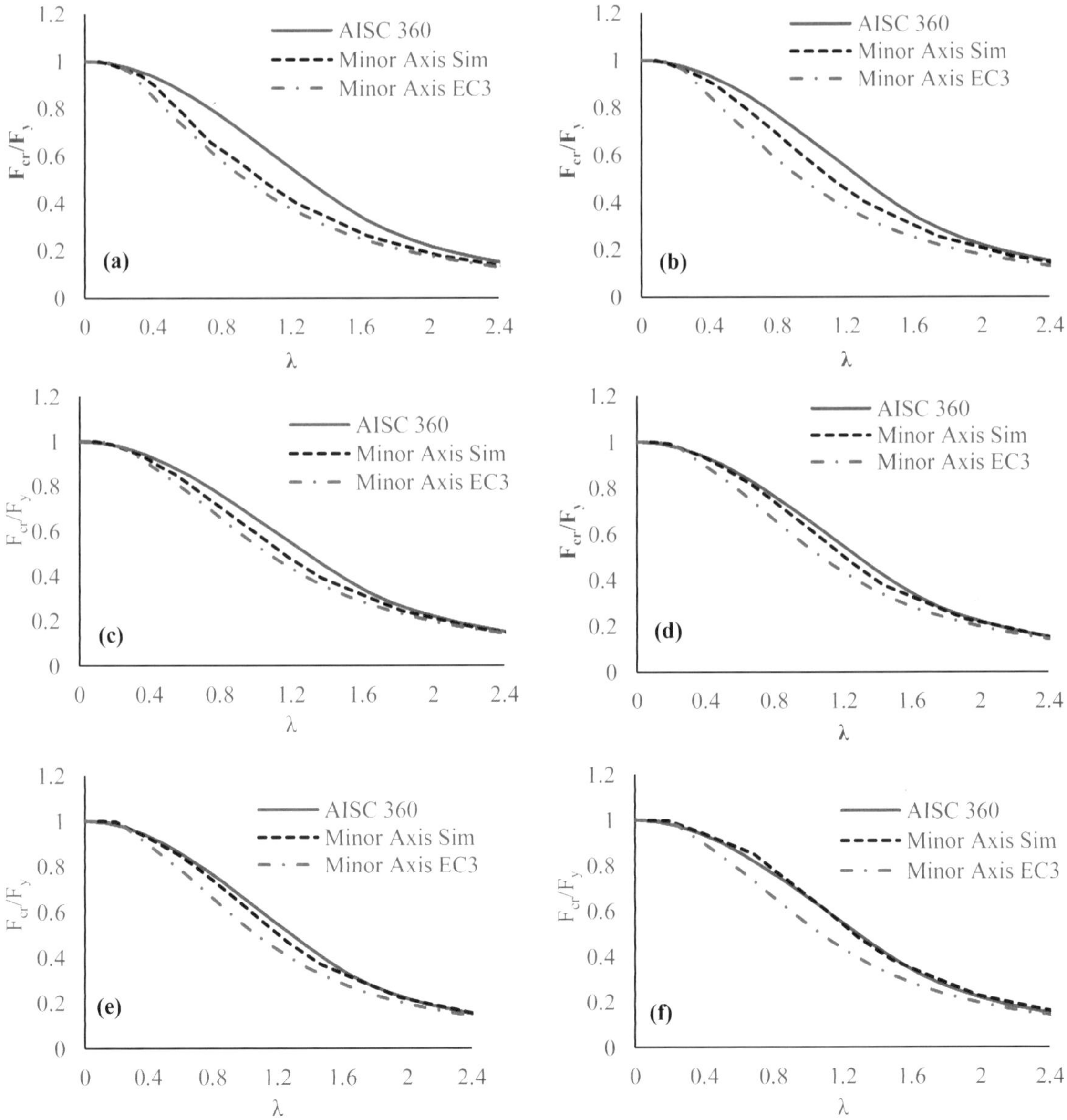

Figure 6: Column buckling curves for A992 (a) W14×808, (b) W14×605, (c) W14×426, (d) W14×257, (e) W14×193, and (f) W14×145 sections.

From Figure 6, it is observed that the simulated column flexural buckling curves for the sections presented in Table 1 about the minor axis agreed well with the AISC 360 and EC3 column design curves for small ($\lambda \leq 0.5$) and large (λ≥1.5) slenderness ratios indicating that for local and elastic buckling range AISC 360 and EC3 column buckling curves agreed well with the simulated responses. However, in the inelastic buckling range with intermediate slenderness ratio ($0.5 < \lambda < 1.5$), the AISC 360 column curve agrees well with the simulated column curve for nonheavy sections with $t_f \leq 1.5$ in. with $\frac{b_f}{2t_f} \geq 5.45$ and $\frac{h}{t_w} \geq 12.8$ (e.g., W14×53, W14×82, W12×87, W14×145 and W14×193) and overestimates the buckling capacity for heavy sections with $t_f > 1.5$ in. with $\frac{b_f}{2t_f} < 5.45$ and $\frac{h}{t_w} < 12.8$ (e.g., W14×257, W14×426, W12×336, W14×605 and W14×808). The residual stress distribution pattern in all these sections were more or less similar with tensile residual stresses at the center of the flange and compressive residual stresses at the edges of the flange, and compressive residual stress in the middle of the web and tensile residual stresses the ends of the web. Since inelastic flexural buckling of columns is preceded by flange local buckling, the magnitude and distribution of the compressive residual stress in the flange plays a significant role in the flexural buckling strength of the columns. It is observed that for nonheavy shapes the maximum compressive stress on the flange was relatively smaller than that was developed in heavy shapes. For example, in W14×82 column section, the maximum compressive stress on the flange is ~8.6 ksi or 59.1 MPa (Figure 7a) whereas for a W14×426, the maximum compressive stress on the flange is ~21.6 ksi or 149 MPa (Figure 7b). As a result of high compressive residual stresses on the flange, the heavier shapes developed local instability earlier than the nonheavy shapes followed by global instability. The AISC 360 column curves were developed based on Galambos and Ketter [11] residual stress model considering a linear distribution of residual stresses with a magnitude of $0.3F_y$ tensile at the center of the flange and -$0.3F_y$ compressive at the edges of the flange, and uniformly distributed tensile residual stress on throughout the depth of the web. The simulated compressive residual stresses on the flange of heavy shapes was significantly higher than that was considered in the Galambos and Ketter [11] model. Additionally, both tensile and compressive residual stresses are observed on the web of the section from the simulation which was not considered in the Galambos and Ketter [11] model. As a result, the flexural buckling strengths of the heavier columns were less than the AISC 360 column curve for $0.5 < \lambda < 1.5$ which is typically the inelastic buckling range. Note that within the relative slenderness range between 0.5-1.5, the AISC 360 column curve overestimated the column buckling capacity by 12% for W14×426 section. As noted from Figure 6, EC3 column buckling curves are overly conservative for all the sections in the intermediate slenderness ratios. Eurocode 3 (EC3) part 1-1 [5] considers these variations in the residual stresses and geometric imperfections based on the geometry of the shapes by incorporating the imperfection factor. EC3's five different buckling curves, a_0, a, b, c, and d depend on the section's h/b ratio and flange thickness (t_f). As per EC3, W14×426 section belong to column curve b with an imperfection factor of 0.34. The corresponding column curve developed based on the EC3 is plotted in Figure 6c for section W14×426, where it is observed that the EC3 column curve accounted for this reduction in the buckling strength in the inelastic buckling range and however, the EC3 curves were overly conservative compared to the simulated responses and AISC 360 column curves. Therefore, based on the responses using A992 steel, it is demonstrated that the finite element models developed in this study provided reliable column buckling capacity prediction for rolled shapes, and hence, the models were further utilized to investigate the column buckling capacity of rolled shapes using A913 Gr. 80 steel.

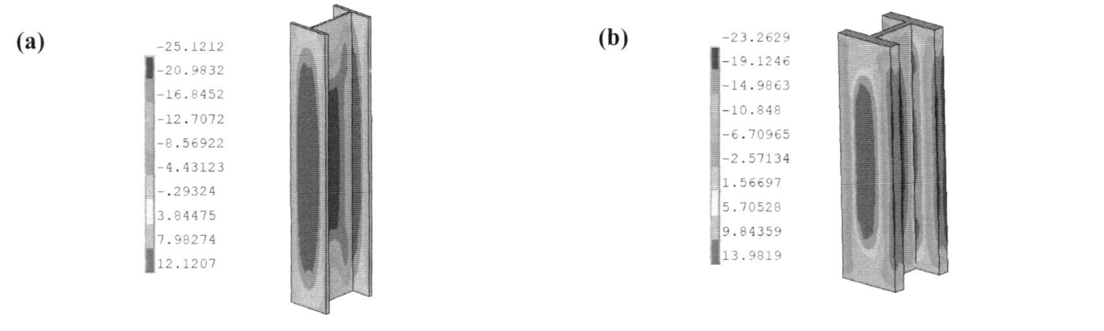

Figure 7: Simulated longitudinal residual stress contour (in ksi) of A992 (a) W14×82, and (b) W14×426 sections.

Column Buckling Curves for A913 Gr. 80 Steel

The effect of residual stresses on the column buckling curves of A913 Gr. 80 steel was evaluated by choosing several wide flange sections as presented in Table 2. As discussed before, the compressive residual stress on the flange of the wide flange section played a vital role in the flexural buckling strengths of the columns, especially about the minor axis. Simulated minor axis column buckling curves of A992 steel within the relative slenderness range between 0.5-1.5 demonstrated that the column buckling strengths are lower than those predicted by AISC 360-22 for sections with flange thickness more than 1.5 inches and $\frac{b_f}{2t_f} < 5.45$. This is attributed to the development of higher compressive residual stresses on the flange of the heavy sections. Residual stress analyses of the wide flange sections with A913 Gr.80 steel showed similar trends in the distribution of residual stresses in the wide flange sections considered in the analysis with the development of higher compressive residual stress on the flange of the column sections. As a result, similar to A992 steel columns, the heavy sections made of A913 Gr. 80 steel also exhibited lower buckling strengths than those predicted by AISC 360-22 within the intermediate slenderness ratio as shown in Figure 8; however, because of high strength of the steel, the reduction in the simulated column buckling curves compared to AISC 360-22 column curve was observed for column sections with flange thickness more than 2.0 inches and $\frac{b_f}{2t_f} < 4.23$. Moreover, the difference between the two column curves are marginal for sections with $t_f \leq 4.0$ inches and $\frac{b_f}{2t_f} > 2.0$ (see Figures 8d-f), whereas for sections with $t_f > 4.0$ inches and $\frac{b_f}{2t_f} < 2.0$, the difference between the two curves were noticeable (see Figures 8a-c). To explore the effect of residual stresses further, the sections are analyzed without considering residual stresses.

Table2: Cross-sectional properties of the A913 Gr. 80 W shapes investigated for residual stress effect

Column Designation	h/b	t_f (in.)	$b_f/2t_f$	h/t_w
W14×873	1.26	5.51	1.71	2.89
W14×808	1.23	5.12	1.82	3.04
W14×605	1.20	4.16	2.09	4.39
W14×426	1.12	3.04	2.75	6.08
W14×257	1.03	1.89	4.23	5.21
W14×145	0.96	1.09	7.11	16.8

Figure 8 shows the simulated column buckling curves for the selected W14 shapes about the minor axis along with the AISC 360 column design curves. It is observed that the effect of residual stresses diminished with decrease in the thickness of the flange t_f and $\frac{b_f}{2t_f}$ ratio, and for 1.5 inches or less thickness of the flange with $\frac{b_f}{2t_f} < 4.0$, the effect was almost negligible. For heavy shapes with thicker flanges, the presence of higher compressive residual stresses through the flange thickness resulted in a 2%-17% reduction in the column buckling strength compared to the analysis without considering residual stresses. This reduction in buckling strength was observed only in the inelastic buckling range with relative slenderness values between 0.25-2.0. For example, for a W14×873 section, the reduction in the buckling strength due to the presence of residual stresses can vary from 4%-17% for relative slenderness values between 0.4-2.0, whereas for a W14×426 section, the reduction in the buckling strength due to the presence of residual stresses can vary from 3-11% for the relative slenderness values between 0.4-1.5. For elastic buckling range, residual stresses do not affect the column buckling capacity regardless of the sectional properties.

Discussions

Residual stresses play a critical role in influencing the flexural buckling capacity of steel columns, particularly in high-strength steels where assumptions based on lower-grade materials become inaccurate. For heavy sections (e.g., flange thickness $t_f \geq 3$ in., $h/b \geq 1.12$), residual stresses introduced nonuniform internal forces that reduce the available elastic and inelastic buckling resistance. The results from this study indicated that, when realistic multiaxial residual stress distributions were included, the column strength can be reduced by up to 10% in these heavy shapes compared to models assuming zero or idealized residual stress states. This discrepancy highlighted the inadequacy of traditional linear residual stress distributions for modern HS3 shapes.

For lighter sections, however, the effect of residual stress is considerably less pronounced. The nonlinear finite element simulations showed minimal reduction in capacity (<2%) due to residual stresses in shapes with thinner flanges and smaller h/b ratios. This suggests that for design purposes, neglecting residual stresses in lighter HS3 columns may still yield acceptable results, though such assumptions should be applied cautiously.

The mode and pattern of residual stress evolution during the cooling process after hot-rolling were found to vary significantly across the section depth and width, leading to a mix of tensile and compressive stresses in the web and flange regions. These localized stress gradients amplify the asymmetry of stress distribution, increasing the sensitivity of columns to global buckling under axial compression. When realistic residual stresses were omitted or overly simplified, finite element models may overpredict column capacity in the inelastic regime. This overprediction is particularly consequential in performance-based design, where accurate capacity estimates influence the reliability and ductility of the entire structural system. Thus, inclusion of validated residual stress profiles, such as those obtained from thermomechanical simulations, is essential for any advanced buckling analysis.

Overall, the study confirmed that residual stresses are a critical parameter in high-strength steel column design and that their influence is shape-dependent. Advanced modeling methods that capture their true distribution provide a better understanding of structural performance and pave the way for optimized, less conservative design provisions.

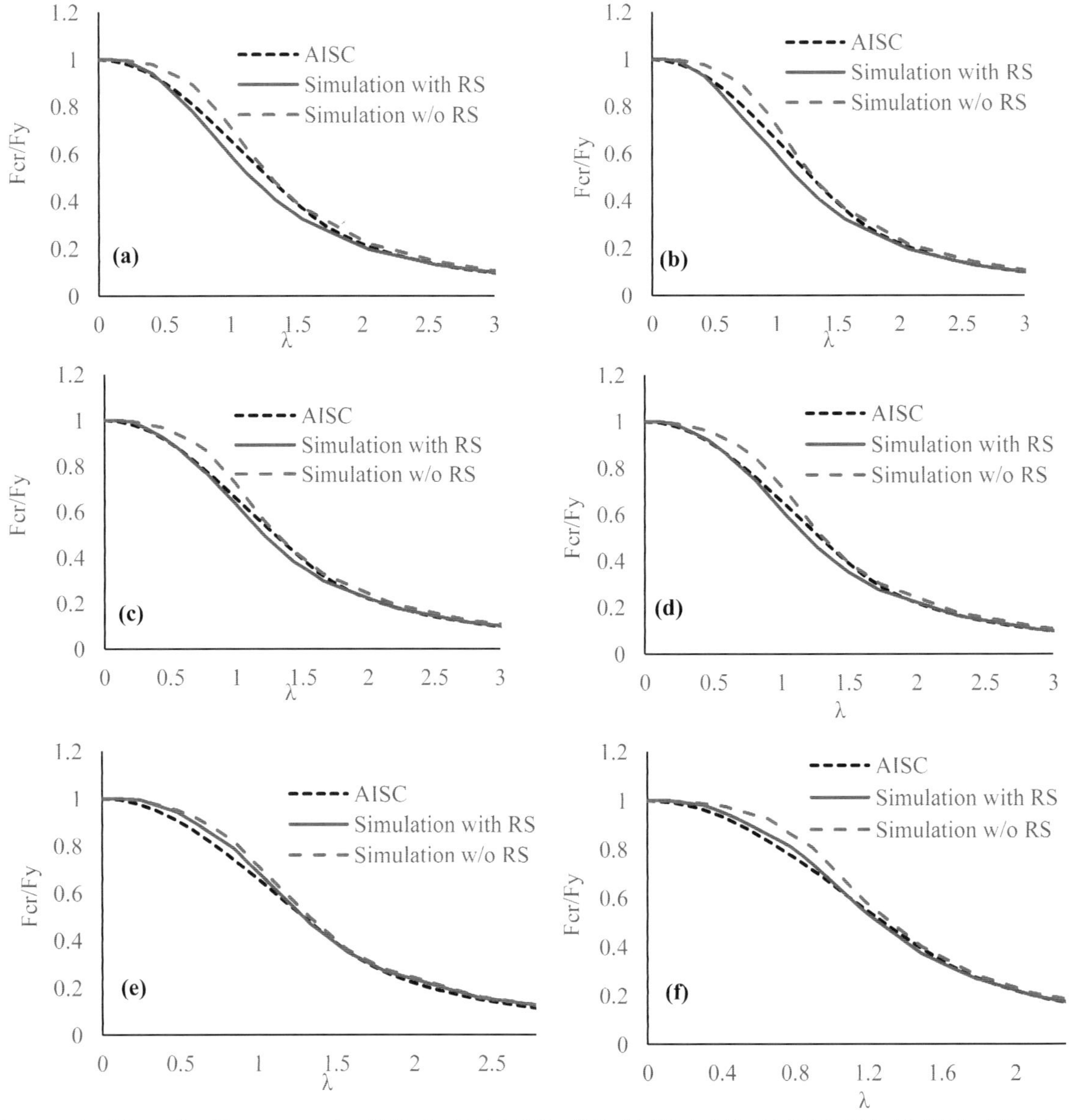

Figure 8: Column buckling curves for A913 Gr. 80 (a) W14×873, (b) W14×808, (c) W14×605, (d) W14×426, (e) W14×257, and (f) W12×145 sections.

Conclusions

This study demonstrated that realistic multiaxial residual stresses significantly affect the flexural buckling strength of steel columns, especially for heavy rolled sections made of A992 and A913 Grade 80 steels. Finite element simulations showed that compressive residual stresses in thick flanges reduce inelastic buckling resistance by up to 17%, which is not captured by current AISC 360-22 design curve. For lighter sections with thinner flanges, the impact of residual stresses was minimal, suggesting that simplified assumptions may be adequate for such shapes. However, for heavy and high-strength steel columns, incorporating accurate residual stress distributions is essential for reliable strength predictions. These findings support the need for improved column design provisions that account for realistic residual stress effects, particularly in high-strength and heavy wide-flange sections.

References

[1] Mathur, K., Fahnestock, L.A., Okazaki, T., Parkolap, M.J., "Impact of residual stresses and initial imperfections on the seismic response of steel moment frames," *Journal of Structural Engineering*, Vol. 138, No. 7 (2012), p. 942-951.

[2] Quayyum, S., Hassan, T., "Initial residual stresses in hot-rolled wide-flange shapes: a computational technique and influence on structural performances," *Journal of Structural Engineering*, Vol. 143, No. 5 (2017), 04017013.

[3] Huber, A. W., *The influence of residual stress on the instability of columns*, Doctoral. Dissertation, University Microfilms, Inc., Ann Arbor, Michigan; 1956.

[4] AISC 360-22, *Specification for Structural Steel Buildings*, ANSI/AISC 360-22, American Institute of Steel Construction, Chicago, IL, 2022.

[5] EN 1993-1-1:2005, *Design of Steel Structures – Part 1.1: General Rules and Rules for Buildings*, CEN - European committee for Standardization, Brussels (Belgium 2005).

[6] Snijder, H. H., Cajot, L. G., Popa, N., Spoorenberg, R. C., "Buckling curves for heavy wide flange steel columns," *Romanian Journal of Technical Science: Applied Mechanics*, Vol. 59, No. 1/2 (2014), 178-204.

[7] Morrison, M.L., *Innovative seismic performance enhancement techniques for steel building moment resisting connections*. PhD Dissertation Research, North Carolina State University, Raleigh, NC, 2015.

[8] Jönsson, J., Stan, T. C., "European column buckling curves and finite element modelling including high strength steels," *Journal of Constructional Steel Research*, Vol. 128 (2017), p. 136-151.

[9] Spoorenberg, R. C., Snijder, H. H., Cajot, L.-G., May, M. S., "Experimental investigation on residual stresses in heavy wide flange QST steel sections," *Journal of Constructional Steel Research*, Vol. 89 (2013), p. 63-74.

[10] Lamarche, C. P., Tremblay, R., "Seismically induced cyclic buckling of steel columns including residual-stress and strain-rate effects." *Journal of Constructional Steel Research*, Vol. 67 (2011), 1401-1410.

[11] Galambos, T. V., Ketter, R. L., "Columns under combined bending and thrust," *Journal of the Engineering Mechanics Division*, Vol. 85, No. 2 (1959), p. 1-30.

ICRS 12: Proceedings from the 12th International Conference on Residual Stresses, 2025
October 20–23, 2025; Detroit, Michigan, USA
https://doi.org/10.31399/asm.cp.icrs2025p0154

Numerical Simulation of Weld Induced Residual Stresses in Welded Steel Moment Connections

Shahriar Quayyum
Department of Civil & Environmental Engineering, Manhattan University, Riverdale, NY 10471
shahriar.quayyum@manhattan.edu

Abstract

This paper presents a comprehensive numerical investigation into the influence of welding sequence on the seismic performance of welded steel moment connections (WSMCs), which are critical elements in steel building frames designed to resist lateral loads. Despite improvements made in WSMC detailing after the 1994 Northridge Earthquake, recent studies have shown that low-cycle fatigue failures still initiate at the weld toe and access hole, highlighting the need to better understand residual stress effects. While it is commonly believed that residual stresses relax during cyclic seismic loading, experimental evidence suggests persistent strain accumulation at critical regions. To address this gap, a detailed three-dimensional finite element (FE) modeling framework is developed in ANSYS, incorporating a sequentially coupled thermo-mechanical analysis to simulate realistic weld-induced temperature histories and resulting multiaxial residual stress distributions. Two WSMC specimens (MC1 and MC2), featuring different welding sequences for the bottom beam flange, are analyzed. Simulation results are validated using experimentally recorded temperature and strain data. The findings demonstrate that welding sequence plays a decisive role in altering residual stress magnitudes and their spatial distribution, especially at the weld toe and weld access hole. These stresses are shown to significantly influence localized strain ratcheting and fatigue crack initiation. Notably, although longitudinal residual stresses relax with inelastic cycles, the multiaxial stress state sustains high von-Mises stresses, which drive continued strain accumulation. This study highlights the necessity of incorporating welding residual stresses and sequence effects into seismic performance assessments of WSMCs to enhance prediction accuracy and structural resilience.

Introduction

The 1994 Northridge Earthquake exposed critical vulnerabilities in welded steel moment connections (WSMCs), particularly highlighting brittle fractures at the beam flange-to-column weld interface that compromised the integrity of numerous buildings [1]. These failures spurred substantial research and led to widespread design improvements in steel moment connections. Key revisions included backing bar removal to reduce stress concentrations, optimization of weld access hole geometry, and the use of base and filler metals with improved fracture toughness [2-5]. Despite these modifications, experimental investigations on post-Northridge WSMC specimens continue to reveal premature failures characterized by crack initiation near the weld toe and weld access hole [6-7]. Such observations suggest that current improvements, while beneficial, are not wholly effective in mitigating fatigue-sensitive regions. A prevailing belief in the structural engineering community is that welding-induced residual stresses diminish rapidly under seismic inelastic loading, rendering them insignificant in failure prediction. However, a growing body of experimental data reveals strain ratcheting and crack propagation in areas known for residual stress accumulation, challenging this assumption. Prior studies on welded joints in pressure vessels, piping systems, and flat plates have shown that residual stress magnitudes and distributions are significantly affected by welding sequence, which, in turn, impacts fatigue life [8-9]. These insights underscore the need for an advanced numerical approach capable of simulating welding processes and subsequent residual stress effects in three dimensions, especially for full-scale building connections. The objective of this study is to bridge this knowledge gap by numerically assessing the evolution of welding residual stresses under different welding sequences and evaluating their implications on the seismic performance of WSMCs.

Methodology

A high-fidelity finite element (FE) model of an external beam-column WSMC was developed in ANSYS to replicate the geometry, material properties, and boundary conditions of the experimental specimens tested by Hassan and Quayyum [10] as shown in Figure 1.

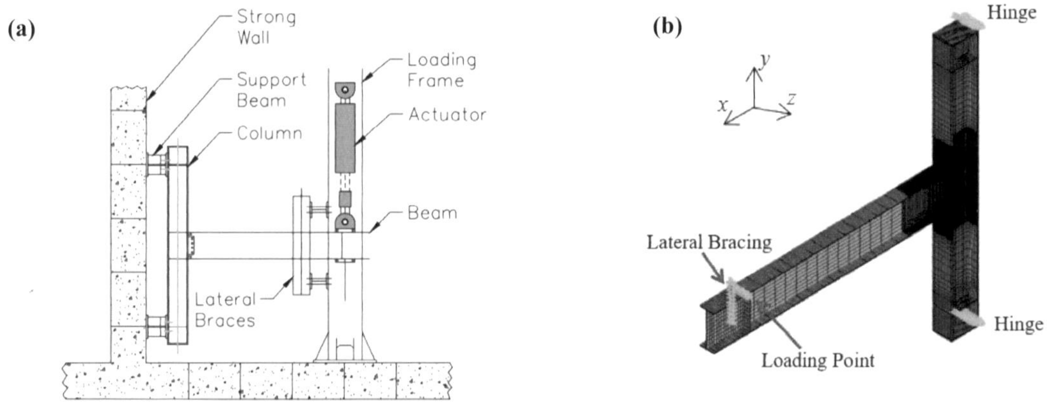

Figure 1: (a) Sketch of the WUF-B specimen test set up of Hassan and Quayyum [10], and (b) the finite element mesh of the specimen used in the thermo-mechanical transient analysis.

The analysis framework utilizes a sequentially coupled, nonlinear, transient thermal and thermo-mechanical simulation strategy to capture the complex thermal cycles and subsequent mechanical responses during and after welding. Multi-pass welds were simulated using ANSYS's element birth-death technique as shown in Figure 2, which allows for realistic activation of weld elements during each pass, effectively modeling the progressive addition of heat and material. Heat input for each pass was defined using surface and volumetric heat flux components derived from arc voltage, current, and arc efficiency. The thermal model accounted for temperature-dependent material properties such as thermal conductivity, specific heat, and enthalpy. Latent heat due to phase transformations was included to improve temperature field accuracy. Structural steel (ASTM A992) and weld filler metal (E70) were characterized using data from standardized tests [11] and literature, including specific attention to the heat-affected zone (HAZ). The heat-affected zone was divided into four segments based on the peak temperature exposure as defined by Johnson and Ramirez [12]. Material heterogeneity in the heat-affected zone due to welding was addressed by implementing temperature-dependent stress-strain behavior derived from heat-treated specimens [13] as shown in Figure 6b and 6c. In addition to welding residual stresses, initial residual stresses (IRSs) generated during cooling after hot-rolling of wide flange sections were calculated based on previously validated methods and incorporated into the simulation prior to welding analysis. Two welding sequences were modeled for the bottom beam flange: MC1 with continuous one-sided weld passes, and MC2 with alternating side weld passes. The welding details of the two specimens are presented in Hassan and Quayyum [10]. These variations allowed for a comparative analysis of how sequencing alters thermal gradients, stress evolution, and final stress distributions.

Model Validation and Results

To ensure the credibility of the numerical model, thermal simulations for specimens MC1 and MC2 were compared with temperature histories recorded during experimental testing using embedded thermocouples. Results indicated strong correlation between simulated and measured temperatures across key locations as shown in Figure 3, validating the thermal model's accuracy and the adequacy of heat input calibration. Upon completion of thermal cycles, mechanical analyses revealed the development of substantial residual stresses concentrated near weld toes and weld access holes. Peak longitudinal tensile residual stresses reached approximately two-thirds of the base metal yield strength near weld toe and higher than the yield strength near weld access hole, while compressive residual stresses were localized at weld terminations. The distribution of residual stresses differed significantly between MC1 and MC2. The top flange welds are deposited in a similar fashion for both the specimens, only the number of beads and the size of the beads were different. For bottom flange, specimen MC1, which followed a continuous one-sided welding sequence, developed more intense and localized tensile stresses, particularly near the weld toe and at the ends of weld beads where sequential heat buildup caused high thermal gradients (Figure 4). In contrast, the alternating weld sequence in MC2 helped balance the heat input between opposing sides of the flange, resulting in a more symmetrical thermal profile and smoother stress gradients (Figure 5). This led to lower peak residual stresses and more uniform distribution across the flange width and length.

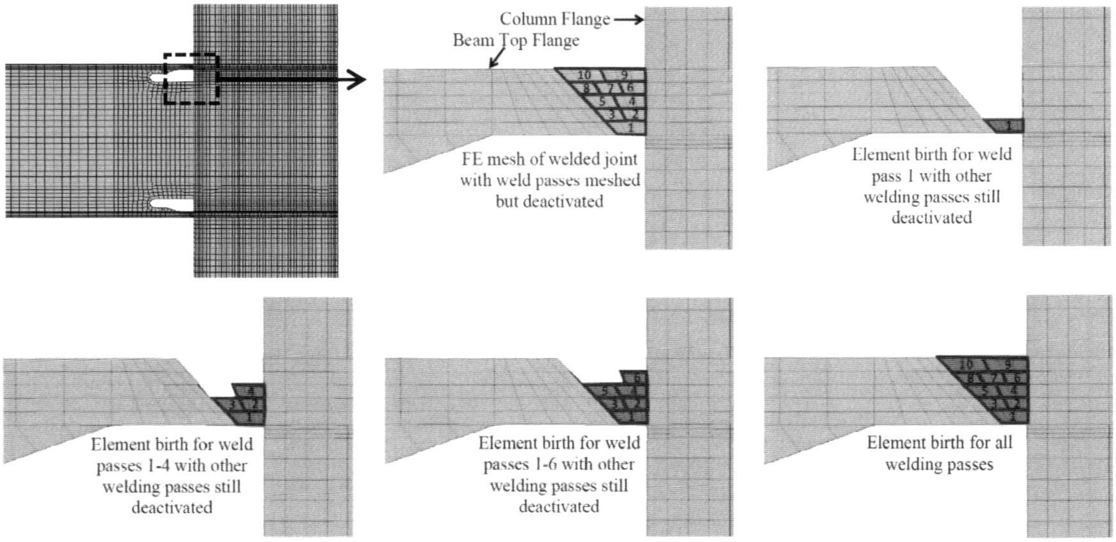

Figure 2: Element birth and death technique implemented in the FE simulation to imitate weld sequence of the top flange of the WUF-B connection.

Furthermore, in MC2, residual stresses transitioned more gradually from the weld zone into the adjacent base metal, reducing localized stress concentrations that are often critical sites for fatigue crack initiation. The numerical results also demonstrated that stress fields near the access hole were particularly sensitive to sequencing due to geometric constraints and differential cooling rates. These stress distributions, when compared with experimental failure observations, offered strong correlation: MC1 exhibited brittle, sudden fracture initiating at the weld toe, while MC2 showed ductile crack propagation with visible damage around the access hole. Finite element simulations also captured axial strain accumulation or ratcheting at critical regions. The spatial extent and magnitude of strain accumulation closely matched experimental strain gauge readings, further validating the model's capacity to predict localized fatigue-sensitive behavior.

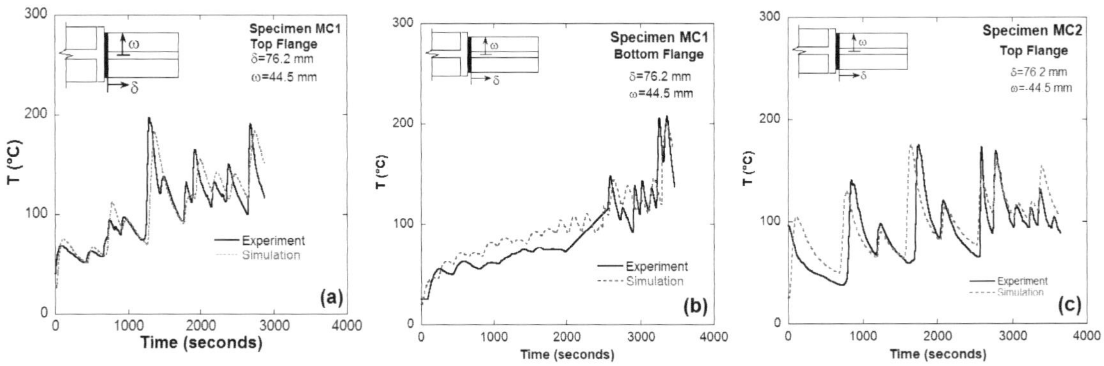

Figure 3: Measured and simulated temperature history at the (a) top flange of specimen MC1, (b) bottom flange of specimen MC1, and (c) top flange of specimen MC2.

Seismic Performance Analysis

The seismic performance of the simulated WSMC specimens was assessed by applying cyclic displacement histories based on the ATC-24 constant amplitude loading protocol as shown in Figure 6a. To accurately simulate material response under repeated loading, a nonlinear kinematic hardening model based on Chaboche theory [14] was implemented, allowing for the capture of cyclic softening, Bauschinger effects, and ratcheting phenomena. Analyses revealed that inclusion of welding residual stresses was essential to replicate experimentally observed ratcheting behavior (see Figures 7a and 8a). In their absence, the model failed to predict the abrupt rise in strain accumulation observed during the transition from elastic to inelastic loading. While longitudinal residual stresses were found to relax substantially after the onset of plastic deformation, the von-Mises equivalent stress-representative of the multiaxial stress state-remained elevated throughout cyclic loading (see Figure 9). This persistence explains the continued ratcheting behavior, as strain accumulates under a non-zero mean stress field.

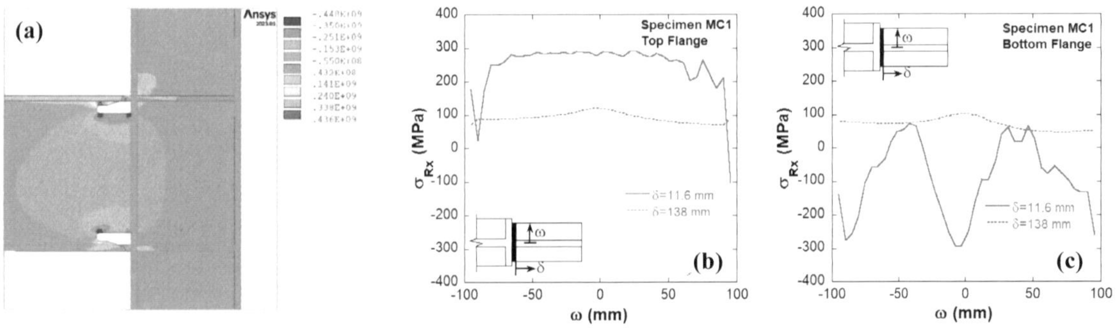

Figure 4: Longitudinal welding residual stress distribution in specimen MC1, (a) residual stress contour (in Pa), (b) across top flange width, and (b) across bottom flange width.

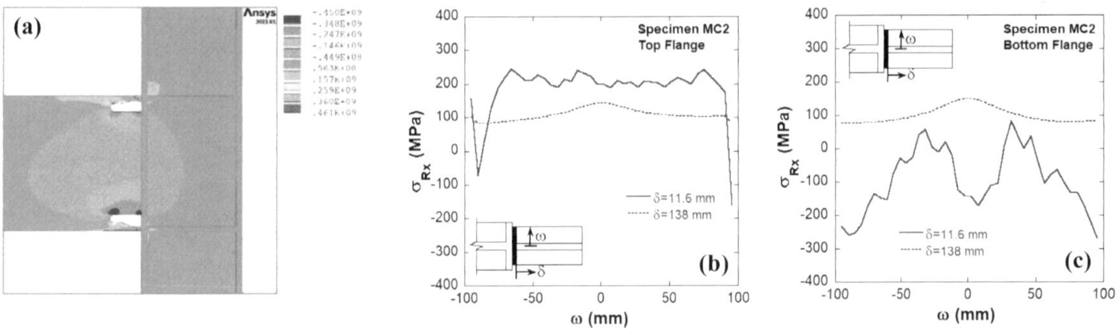

Figure 5: Longitudinal welding residual stress distribution in specimen MC2, (a) residual stress contour (in Pa), (b) across top flange width, and (b) across bottom flange width.

The simulations further indicated that the location and severity of strain accumulation correlated strongly with areas of elevated residual stress, particularly near the weld toe and around the weld access hole. These regions are known from experimental evidence to be highly susceptible to crack initiation under seismic loading. The computational model was able to reproduce the observed strain accumulation trends in both MC1 and MC2 specimens (see Figures 7a and 8a), validating the influence of initial residual stress state and its coupling with loading-induced stresses. In MC1, where the residual stress profile was more intense and concentrated due to the unidirectional weld sequence, strain accumulation occurred more rapidly and localized damage became apparent earlier in the loading history. Conversely, MC2 displayed a more gradual strain progression and a delay in failure onset, attributed to the more balanced thermal and stress distributions enabled by alternating weld passes.

The hysteretic response of the WSMC specimens as shown in Figures 7b and 8b also highlighted the critical role of residual stress in influencing energy dissipation characteristics. Specimens with higher residual stresses experienced more pronounced shifts in the stress-strain hysteresis loops, indicative of higher inelastic deformation demands and greater susceptibility to ratcheting. Moreover, comparison of strain responses with and without residual stresses as shown in Figures 7a and 8a confirmed that omitting residual stress fields can significantly underestimate the fatigue demand and damage evolution in welded connections. These findings support the argument for integrating detailed weld simulations into seismic performance assessments to ensure realistic modeling of connection behavior.

Additionally, the multiaxial nature of the stress state in welded regions was shown to compound the strain accumulation problem. Although axial stresses may appear to relax with cycling, the von-Mises stress-a combination of all principal stress components-remains elevated due to contributions from transverse and shear components (see Figure 9). This highlights the complexity of stress redistribution during seismic loading and the necessity of using robust multiaxial constitutive models to capture such behavior. Strain accumulation continued well beyond the point of longitudinal stress relaxation, ultimately leading to fatigue-driven failures similar to those observed in experimental studies.

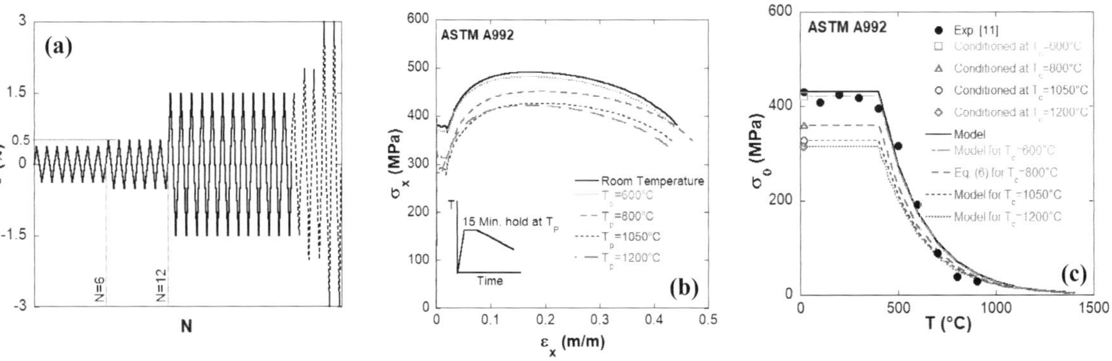

Figure 6: (a) Loading history prescribed in this study; the first inelastic loading cycle (3rd increment with θ = 1.5%) will be continued till the fatigue crack initiation, followed by incremental steps of θ = 2%, 3%, ... similarly as the SAC basic loading history until failure or significant strength degradation, (b) room temperature tensile stress-strain response of A992 after the specimens were heat-treated at different maximum temperatures (see inset in Fig. 6b for the heat-treatment temperature cycle; see Morrison [13] for cycle and process), and (c) temperature-dependent yield stress of ASTM A992 with and without heat treatment to account welding induced material heterogeneity at the weld toe.

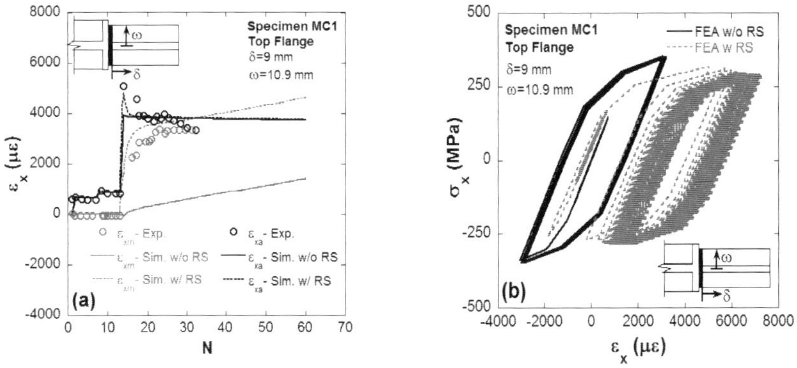

Figure 7: Experimental (data from [10]) and simulated low-cycle fatigue responses of specimen MC1. (a) Mean and amplitude of longitudinal strain cycles plotted against the number of seismic loading cycles, (b) longitudinal stress-strain hysteresis loops showing the influence of residual stresses on strain ratcheting.

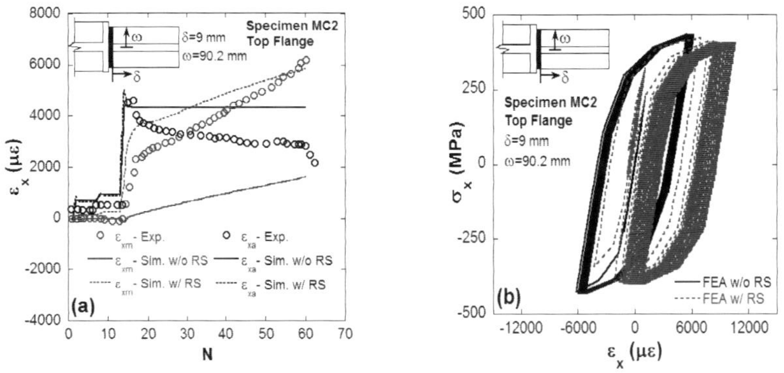

Figure 8: Experimental (data from [10]) and simulated low-cycle fatigue responses of specimen MC2. (a) Mean and amplitude of longitudinal strain cycles plotted against the number of seismic loading cycles, (b) longitudinal stress-strain hysteresis loops showing the influence of residual stresses on strain ratcheting.

Together, these observations demonstrate that welding residual stress not only affects the initial stress condition of the structure but also governs how damage accumulates under repeated seismic actions. The results reinforce the need for structural engineers to consider welding procedure effects in both the design and evaluation phases of steel moment frame construction.

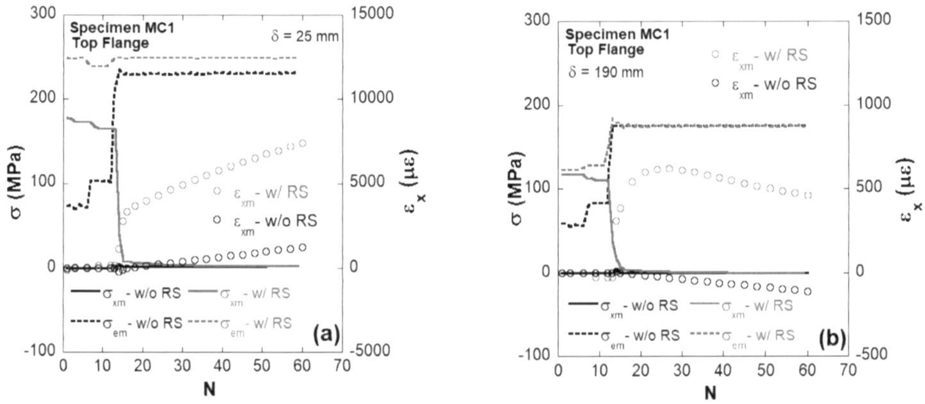

Figure 9: Simulated mean longitudinal stress and strain, and von-Mises mean stress plotted against the number of loading cycles (N) at the top flange for specimen MC1. (a) 25 mm and (b) 190 mm away from the weld toe.

Conclusions

This study developed a validated finite element framework to simulate weld-induced residual stresses and assess their impact on the seismic performance of welded steel moment connections (WSMCs). Results show that welding sequence significantly affects residual stress distribution, with alternating welds producing more uniform and lower-magnitude stresses than continuous one-sided welds. Residual stresses were found to influence strain ratcheting and crack initiation, particularly near the weld toe and access hole, and the localized strain ratcheting at the weld toe and access hole can only be accurately simulated when residual stresses are considered in the analysis. Despite longitudinal stress relaxation under inelastic loading, high von-Mises stresses persisted, sustaining strain accumulation. The findings underscore the critical role of residual stresses in seismic behavior and highlight the need to incorporate welding effects into the design of WSMCs.

Overall, this research reinforces the necessity of accounting for welding-induced residual stresses and weld sequence effects in the seismic design and evaluation of WSMCs. The presented methodology offers a robust framework for integrating detailed residual stress analysis into structural performance prediction, ultimately contributing to the design of more resilient steel structures.

References

[1] FEMA-350. *Recommendation and seismic design criteria for new steel moment-frame buildings.* prepared by the SAC Joint Venture for the Federal Emergency Management Agency, Washington, DC, 2000.

[2] Engelhardt, M. D., Winneberger, T., Zekany, A. J., Potyraj, T. J., "Experimental investigation of dogbone moment connections," *Engineering Journal*, Vol. 35, No. 4 (1998), p. 128-139.

[3] Stojadinovic, B., Goel, S.C., Lee, K.H., Margarian, A.G., Choi, J.H., "Parametric tests on unreinforced steel moment connections," *Journal of Structural Engineering*, Vol. 126 (2000), p. 40-49.

[4] Ricles, J. M., Mao, C., Lu, L. W., Fisher, J. W., "Inelastic cyclic testing of welded unreinforced moment connections," *J. Struct. Eng.*, 128(4), 429-440.

[5] Sumner, E. A., and Murray, T. M. (2002). "Behavior of extended end-plate moment connections subject to cyclic loading." *Journal of Structural Engineering*, Vol. 128, No. 4 (2002), p. 501-508.

[6] Lee, D., Cotton, S.C., Hajjar, J.F., Dexter, R.J., "Cyclic behavior of steel moment-resisting connections reinforced by alternative column stiffener details, I. Connection performance and continuity plate detailing," *Engineering Journal*, 4th Quarter (2005), p. 189-213.

[7] Han, S. W., Kwon, G. U., Moon, K. H., "Cyclic behaviour of post-Northridge WUF-B connections," *Journal of Constructional Steel Research*, Vol. 63, No. 3 (2007), p. 365-374.

[8] Jiang, W., Yahiaoui, K., "Effect of welding sequence on residual stress distribution in a multipass welded piping branch junction," *International Journal of Pressure Vessels and Piping*, Vol. 95 (2012), p. 39-47.
[9] Mai, C., Hu, X., Zhang, L., Song, B., Zheng, X., "Influence of Interlayer Temperature and Welding Sequence on the Temperature Distribution and Welding Residual Stress of the Saddle-Shaped Joint of Weldolet-Header Butt Welding," *Materials*, Vol. 14, No. 20 (2021), 5980.
[10] Hassan, T., and Quayyum, S., "Influence of Weld Sequence on the Low-Cycle Fatigue Failure of WUF-B Connections," *Journal of Structural Engineering*, Vol. 148, No. 7 (2022), 04022072.
[11] Hu, G., Morovat, M.A., Lee, J., Eschell, E., Engelhardt, M., "Elevated temperature properties of ASTM A992 steel," In: *Proceedings of ASCE structure congress 09*, Austin, TX, 2009.
[12] SAC/BD-00/13. *Preliminary evaluation of heat affected zone toughness in structural shapes used in the construction of seismic moment frames*. by Johnson, M.Q., Ramirez, J.E.
[13] Morrison, M., Innovative seismic performance enhancement techniques for steel building moment resisting connections, *PhD Dissertation Research*, North Carolina State University, Raleigh, NC, 2015.
[14] Chaboche, J.L., "Constitutive equations for cyclic plasticity and cyclic viscoplasticity," *International Journal Plasticity*, Vol. 5, No. 3 (1989), p. 247-302.

ICRS 12: Proceedings from the 12th International Conference on Residual Stresses,
2025 October 20–23, 2025; Detroit, Michigan, USA
https://doi.org/10.31399/asm.cp.icrs2025p0161

Residual Stress Influence on Soft-Zone Cracking in Electric Resistance Welded Line Pipe Steels under Sour Service Conditions

Amrita Bag, Mohamed (Adam) Azaouzi, Shaojie Chen, Muhammad Rashid
Interpro Pipe & Steel Inc.

Abstract

Soft-zone cracking (SZC) is a form of hydrogen-induced cracking that occurs in the softened heat-affected zone (HAZ) of welds, commonly expressed in low-carbon line pipe steels produced through thermo-mechanically controlled processing (TMCP). When exposed to aqueous H_2S (sour service) conditions, constrained deformation of the softened HAZ, in combination with hydrogen degradation and applied loading, leads to localized stress concentrations that promote cracking. The role of residual stresses (RSs) in SZC remains unclear, as specimen preparation for standard sulfide stress cracking (SSC) testing may redistribute stresses and affect test outcome.

This study evaluates the influence of RS redistribution on SSC performance. A 52 ksi (359 MPa) yield strength ERW line pipe was evaluated using 4-point bend SSC testing, with SZC consistently observed in the HAZ-pipe body transition region. Hole drilling RS measurements were performed on an untested pipe ring, followed by additional measurements on test coupons cut from the pipe. Finite element analysis (FEA) was performed based on hole-drilling data to assess stress evolution in tested coupons to understand the role of RSs in SZC initiation.
This work provides a novel approach to improve SSC test interpretation by accounting for residual stress effects, creating a reliable assessment of sour service performance in welded line pipe steels.

Introduction

Soft zone cracking (SZC) and residual stresses (RSs)

Hydrogen embrittlement cracking, frequently observed in high-strength pipeline steels exposed to wet H_2S environments under applied stress, is widely recognized as sulfide stress cracking (SSC). The typical SSC form, also known as SSC Type II cracking, generally initiates at the surface and propagates perpendicularly to the direction of applied stress. A less common form, SSC Type I, also referred to as stress-oriented hydrogen-induced cracking (SOHIC), occurs when planar cracks link perpendicularly to the main crack and the applied stress direction. This phenomenon is closely related to hydrogen-induced cracking (HIC), which requires similar environmental conditions but no applied stress. HIC cracks display a characteristic stepwise or ladder-like morphology, formed by planar cracks that connect hydrogen blisters across different planes. [1] illustrates these distinct mechanisms, showing SOHIC as an intermediate form that combines elements of both SSC Type II and HIC. While high-strength steels (with tensile strength >700 MPa or hardness >22 HRC/248 HV10) are generally more susceptible to SSC Type II, lower-strength steels tend to be affected by HIC and SOHIC. Although HIC and SOHIC can co-occur in susceptible steels, HIC-resistant steels are not necessarily resistant to SOHIC [2].

One of the factors contributing to SOHIC susceptibility has been identified as welding [3]. A specific form of SOHIC that occurs within the softened heat-affected zone (HAZ) of welds is known as SZC. In SZC, deformation within the softened HAZ, constrained by the surrounding harder material, generates high stresses due to the combined effects of hydrogen degradation and applied loading, leading to cracking. SZC is commonly observed in line-pipe steels produced via thermo-mechanically controlled processing (TMCP). These steels, designed with low carbon content for enhanced toughness and weldability, tend to develop softer HAZ regions due to weld and/or post-weld heat treatment (PWHT) thermal cycles. Pargeter [2] has demonstrated through their detailed review of SOHIC/SZC that the role of RS in its underlying mechanisms is undeniable.

Amano et al. [4] performed NACE TM0177 [5] Method A tensile testing in a sour (wet H_2S rich) environment on a 15 mm thick API 5L X46 controlled rolled steel welded by SAW and reported SOHIC cracking just off the most softened region just outside of the HAZ. They claim that the SOHIC location not being in the most softened zone indicates that SOHIC is affected by not only the lowest yield strength and/or microstructure susceptibility but also the localized stress/strain elevation associated with heterogeneous plastic deformation. Their FEA results demonstrate that the highest axial stress near the surface is just outside the softest region of the HAZ, which corroborates their observed cracking in this location. They also predict that cracks initiating at other higher stress locations in the material (which also satisfy the lower strength and higher material susceptibility to SOHIC/SZC) in a less 'clean' steel are expected where SOHIC may for example initiate preferentially from the interior associated with a high population of non-metallic inclusions. The SOHIC susceptibility data seems to correlate with the measured yield strength variation in different regions of base metal, HAZ and weld very well. Takahashi et al. [6] also studied this SZC phenomenon using a FEA modelled for NACE TM0177 Method A tensile loading scenario and showed that peak through thickness stresses occurred within the softest region at mid-thickness, under the influence of a cross weld applied load caused by strain concentration. The high magnitude triaxial stress state is thought to be a precursor for the initial in-plane crack formation in this region.

In all above reported literature, the localized stresses are evidently influenced by the applied loading as well as pre-existing residual stress state in the specimens prior to loading. Thus, it is crucial not only to evaluate the residual stress state in the specimen under study, but also to ensure that stress states experienced during testing accurately represent those in the actual component under service conditions. Common practice for sour service tests to assess HIC, SSC is to cut specimens out of the component and test them under simulated corrosive environments containing aqueous H_2S. SSC tests are accompanied by the application of a stress that is usually a given minimum percentage (~72%) of the specified minimum yield strength (SMYS) as per NACE TM0177 [5] or (~80%) actual yield strength (AYS) as per ISO 15156/NACE MR0175 [7] according to the governing pipeline specifications (API 5L [8], CSA Z662 [9]). For SOHIC/SZC, ISO 15156/NACE MR0175 [7] requires either four-point bend (FPB) testing on small-scale (cut) specimens or full pipe ring tests which are more representative of actual service conditions while accounting for the RS. Application of the test methods have not been standardized and alternative tests are currently under development. As pointed out by Fowler and Grey [10], although ISO 15156/NACE MR0175 [7] and EFC 16 [11] account for the loss of residual stress due to cutting of a test specimen from the pipe or plate by increasing the applied stress using an applied load in the same direction, they do not reflect the actual triaxial stress state required to simulate the precise conditions experienced in service.

SOHIC/SZC failures are rare and have thus been sparsely documented. A review of historic industry failures by Pargeter [2] shows that of the 10 incidents listed, only two are from ERW welds. There is very little evidence to confirm if any of them were SZC-related failures. Moreover, the aforementioned recommendation to increase the stress using an applied load to account for residual stress relief due to specimen extraction appears to be based mostly on girth weld or limited seam weld data. Without proper characterization of RS in seam welds, it is uncertain that the susceptibility of a material to SOHIC/SZC can be properly evaluated using either the existing small-scale tests, or any new tests that may be developed [10].

Objective

The aim of the current study is to study and understand the residual stress distribution across different regions of a HF-ERW intact line pipe and its evolution as RS are redistributed in the SSC test specimens cut from it. This was done to analyze the impact of RS on the SSC test results which will be discussed along with the ideal next steps.

Scope

In this study, a 16 in. (406.4 mm) x 0.374 in. (9.5 mm) wall HF-ERW line pipe with a SMYS of 52 ksi (359 MPa) was subjected to an SSC FPB test following the NACE TM 0316 standard [12]. Analysis of the test specimens consistently identified the SZC phenomenon occurring at approximately 5-7 mm from the ERW weld centerline, specifically in the transition zone between the pipe body and the HAZ. To characterize the RS in an intact pipe, hole-drilling residual stress measurements were performed at four locations – weld, HAZ, body-HAZ interface, and pipe body on one half of a 1 ft (304.8 mm) long weld-normalized pipe ring. Note that the HAZ here refers to the weld seam normalized HAZ and not the hourglass weld HAZ achieved prior to PWHT. The other half of the pipe ring was then cut to the SSC FPB test specimen size, and additional hole-drilling measurements were conducted on the cut specimens to evaluate the redistribution of RS. Microhardness and microstructure analyses of the cut specimens were also performed to analyze their contribution to SZC. Finite element analysis (FEA) was performed by considering experimental data obtained using hole-drilling technique to simulate stress evolution in tested coupons, and therefore understand the role of RSs in SZC initiation.

Material

The material used in this study comes from an industrially produced 52 ksi (359 MPa) yield strength low carbon HSLA line pipe steel. The electric arc furnace produced steel was cast into slabs, thermomechanically rolled and coiled into strips referred to as coiled skelp. The skelp were then uncoiled, straightened and formed into shape to be welded by the HF-ERW process. The weld flash on the OD and ID was trimmed followed by a normalizing treatment of the trimmed weld seam. Table 1 shows the chemistry of the tested steel.

Table 1: Chemical composition of the 52 ksi (359 MPa) yield strength steel (Elements are in wt%, PCm carbon equivalent is also reported)

C	Mn	Ti	B+N+S+P	Ca+Si+Cu+Ni	PCm
0.04	0.7	0.01	0.02	0.59	0.11

Methodology

SSC FPB test

The SSC FPB test was carried out on transverse weld specimens removed from the seam welded and normalized pipe. Specimens were cut from the pipe using a bandsaw and then machined to size with a CNC mill. The OD and ID surfaces were left in the true as-fabricated condition with no preparation other than degreasing. The load application (72% of SMYS) was performed using strain gauges applied to the ID surface approximately 1 mm from the weld trim (pipe body) to monitor strains until the desired load level was achieved by tightening the fixture loading nuts. Following this, the loaded specimens were exposed to the NACE Test Solution A (a highly acidic solution 100% saturated with H2S to mimic sour service conditions) for 30 days [12]. After the test duration had elapsed, the beams were removed from the test vessel and initial visual observations were conducted. Low-pressure glass beading was used to remove corrosion byproducts, and the specimens underwent metallographic sectioning and evaluation. The glass beading operation is expected to induce compressive RS on the specimen ID and OD surfaces that may consequently alter the surface properties.

Residual Stress-Hole drilling

The residual stress measurements were performed on both intact pipe and SSC specimen coupons cut from the pipe using the hole drilling strain gage method of stress relaxation as per ASTM E837 [13]. A precision milling guide (Model RS-200) and Micro Measurements CEA-06-062UL-120 strain gage rosette were utilized. The gage grids were wired and connected to a multichannel static strain indicator/recorder, Micro- Measurements Model P3 to record measured strains. Using special data-reduction relationships, the principal residual stresses and their angular orientation with respect to the reference frame were calculated from the measured strains using H-Drill software [14].

Phase 1- Intact pipe

Phase 1 of the study focused on measuring the RS on the pipe outer diameter (OD) as shown in Fig. 1 (left half). Two replicate measurements for each location across the length of the pipe ring were made. The top half of the weld was chosen for the measurements due to negligible variation in normalized HAZ widths across the length of the pipe ring.

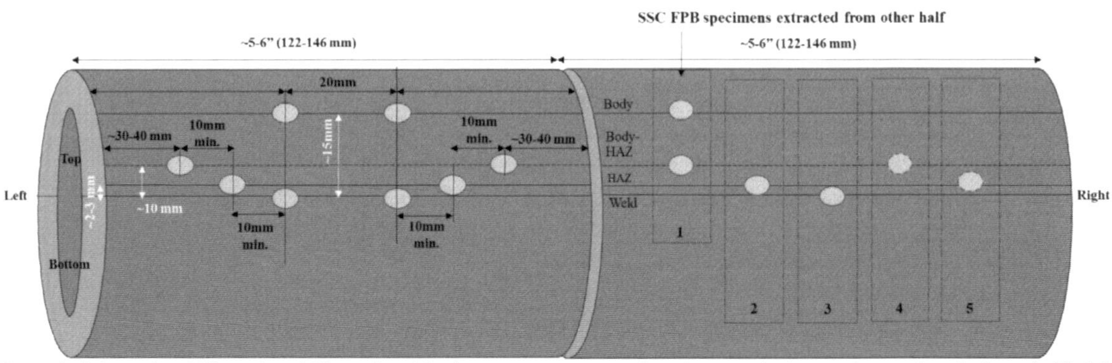

Figure 1: Schematic of Phase 1 and Phase 2 residual stress measurement locations on the intact pipe section (Left top half of the 1ft (304.8 mm) pipe ring was used for Phase 1 measurements while the right half was used for Phase 2 measurements)

Phase 2- Cut untested SSC specimens

Phase 2 of the study focused on measuring RS on SSC FPB specimens cut out of the remaining half of the 1ft (304.8 mm) pipe ring. Five specimens were abrasive saw cut to the following dimensions: length = 5.71 – 5.83 in. (145 – 148 mm), width = 0.8 in. (20.3 mm). The thickness was kept the same as the original pipe. The specimens were used for measurement at the following locations (Fig. 1, right half):

- Specimen 1: Body (OD) + Body-HAZ (OD)
- Specimen 2: HAZ (OD)
- Specimen 3: Weld (OD)
- Specimen 4: Body-HAZ (ID)
- Specimen 5: HAZ (ID)

Only one replicate was obtained for each location due to the limited width of the specimen. This was because of the ASTM E837 guideline, which recommends a minimum distance of approximately 0.3 in. (7.6 mm) from the hole center to the specimen edge to ensure valid measurements and minimize edge effects.

Microhardness measurements

Microhardness measurements on the Phase 2 SSC FPB specimens were conducted using a Struers Duramin-40 microhardness tester, applying a Vickers load of 10 kg (HV10) to assess hardness variation across different measurement locations. Through-thickness traverses were performed on one half of the weld, with measurements taken at intervals of 0.035 inches (0.9 mm).

Additionally, microhardness mapping was conducted in the vicinity of the crack on the cut and sectioned weld cross sections of the failed SSC FPB test specimens using a smaller magnitude load of 0.5 kg (HV0.5).

Microstructure

Optical microscopy (OM) was conducted on polished and etched metallographic cross-sectional weld samples from a Phase 2 specimen, utilizing a 2% Nital (2% nitric acid + 98% ethanol) etchant. Images were captured using a Leica DMi8 microscope at magnifications of 200x and 500x to represent various residual stress measurement locations, including the outer diameter (OD), top quarter thickness, mid-thickness, bottom quarter thickness, and inner diameter (ID). These analyses aimed to characterize microstructural variations across the specified measurement locations.

FEA

3D solid FE models were created using Abaqus/CAE 2025 software to simulate the SSC FPB test with specimen length 150 mm x width 20 mm x thickness 9.5 mm. Due to symmetry, only half length of the specimen was modeled, as illustrated in Fig. 2. Three characteristic cases were considered:

- Case 1: Base case, uniform material property (Body), no RS (RS = 0).;
- Case 2: With material property variations to simulate weld, HAZ, body-HAZ regions, no RS (RS = 0).;
- Case 3: With material property variations, with RS (RS $\neq$ 0).

To account for the material property variations across the weld, HAZ and body, as well as to incorporate through-thickness RS distributions, the specimen was divided into six zones along its half-length and six layers through the thickness from the OD to the ID, as illustrated in Fig. 2. The soft zone was geometrically "tilted" to represent its actual shape and assigned material properties 15% lower than that of the pipe body. The weld zone, located at the specimen's mid-span, was assigned a significantly higher strength as compared to the body.

In Case 3, which included residual stresses, a compressive longitudinal stress (along x axis) of -300 MPa was applied to the OD, and a tensile stress of +100 MPa to ID, with a linear gradient between the two across the thickness. Additional cases involving alternative residual stress distributions were also modeled; however, their results are not included in this paper.

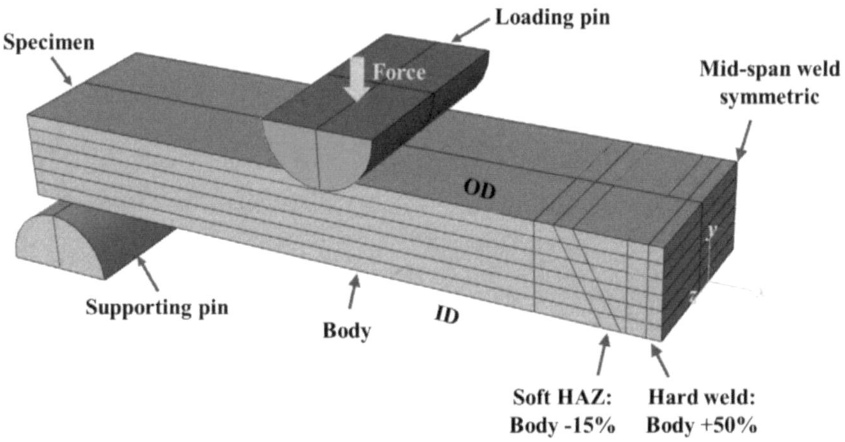

Figure 2: FEA model of an SSC FPB specimen showing different microstructural zones (weld, HAZ, body)

Results

SSC FPB test results

The SSC FPB test results indicated failure due to cracking at the body-HAZ interface (approximately 5 to 7 mm from the weld centerline) in the transverse weld specimens. Cracks appeared to initiate in the mid-thickness region and propagate toward the ID surface, where tensile stresses were highest. Fig. 3 shows a typical SSC FPB test specimen failure for the tested pipe, with crack morphology consistent with the SZC mechanism. SEM fractography of the failed specimens revealed a transgranular fracture pattern, characteristic of SOHIC/SZC [15,2].

The cracks likely began as planar HIC cracks in pearlite colonies and/or inclusions within the softened body-HAZ interface zone, later linking in the direction of the maximum tensile stress. Cleanliness analysis, conducted per ASTM E45 [16] (Method D), showed low level of globular oxide inclusions (2T, 0.5H max) and alumina inclusions (1T max), which are typically responsible for HIC/SSC fracture initiation in low-carbon line pipe steels [17]. No non-metallic inclusions were observed on the fracture surface, though it is possible that inclusions were dislodged by the corrosive solution as it infiltrated the crack.

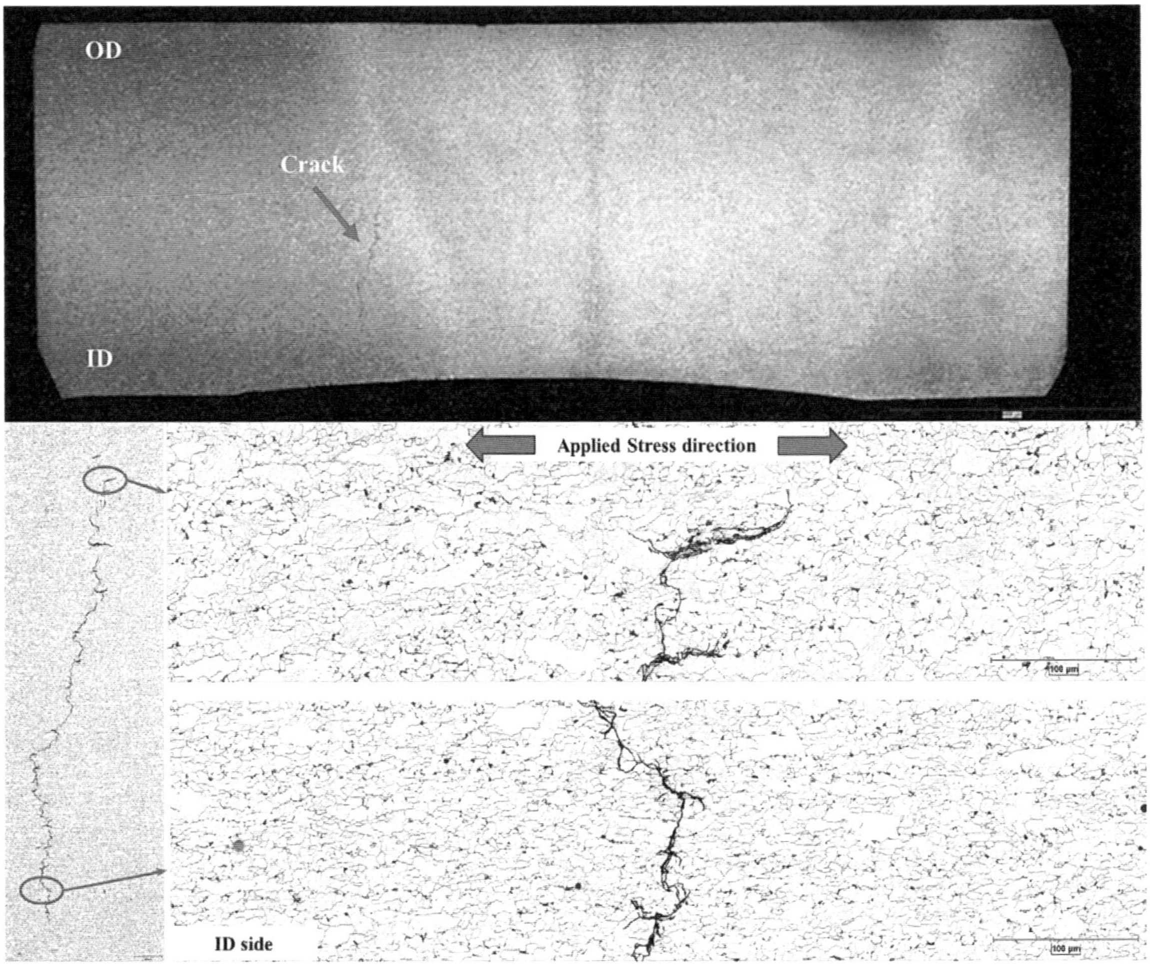

Figure 3: OM image of a typical SSC FPB test specimen failure for the tested pipe showing SZC crack morphology

RS measurements

Phase 1- Intact pipe

Fig. 4 shows residual stress results measured transverse to the pipe axis on the intact pipe ring as well as cut untested SSC specimens. The average RS magnitudes and their distance from the weld for each measurement location has been indicated. For the intact pipe sample (left half in Fig. 4) The highest magnitude tensile residual stress was found at the HAZ and body-HAZ interface, exceeding 80% of SMYS for Replicate 1, while lower magnitude values were observed at the weld (~50% SMYS) and body (~35% SMYS). It is important to note that the hole drilling technique may introduce potential errors when RS exceeds 80% of the yield strength, as this violates the linear elastic fracture mechanics assumptions used for RS estimation. Consequently, these results should be interpreted qualitatively for comparisons rather than quantitatively. Residual stress measurements at the weld and body locations were consistent between replicates; however, significant variation appeared between replicates at the HAZ and body-HAZ interface. This discrepancy may arise from slight shifts in HAZ and body-HAZ interface locations along the pipe length, possibly leading to different microstructures at the precise measurement points. Such variations are expected across the pipe due to the welding and normalizing processes.

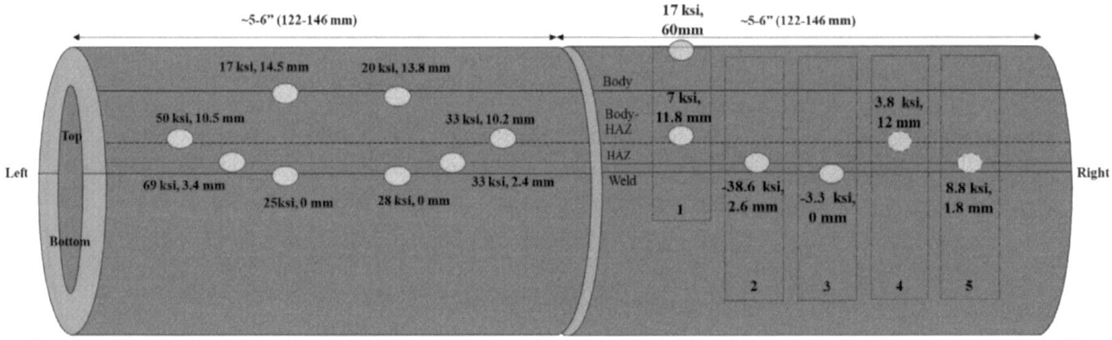

Figure 4: Residual stress measurements for the intact pipe section (left half) and cut untested SSC specimens (right half)

Phase 2- Cut untested SSC specimens

For the cut SSC FPB specimens (right half in Fig. 4), the transverse RSs were markedly reduced compared to the intact pipe, and in some instances, had even reversed in sign (i.e. went from tension to compression). Notably, the body location retained similar RS levels in both the intact and cut specimens. At the body-HAZ interface on the OD surface, RS levels dropped to approximately 15-22% of those measured in the intact pipe (~14% SMYS). RS at the ID body-HAZ interface are comparable to those at the OD, although these regions are at different distances from the weld (~4 mm on the ID vs. ~12 mm on the OD). The OD HAZ region had the highest magnitude RS (~74% SMYS), comparable to the intact pipe, but was compressive rather than tensile. On the ID, RS are significantly lower in magnitude (~17% SMYS) and tensile. Weld RSs in the cut specimens are slightly compressive and very low in magnitude (~6% SMYS).

Microhardness

Fig. 5 shows a microhardness contour map based on these measurements, with the dotted line marking the body-HAZ interface. The hardness values ranged from 161 to 195 HV10, with an average weld hardness of approximately 182 ± 7 HV10. Slightly higher values (>183 HV10) were measured near the OD and ID surfaces. The body exhibits a similar hardness as compared to the weld (184 ± 4 HV10).

The HAZ and body-HAZ interface regions show noticeable softening relative to the weld and body areas, with the softest zone extending from ~3 to 7 mm on the ID surface into the mid-thickness region. This softened mid-thickness zone near the body-HAZ interface was identified as the initiation site for cracking in failed SSC FPB specimens. Consistent with this observation, HV0.5 microhardness measurements on failed specimens reveal a softened mid-thickness region, as shown in Fig. 6, where black lines indicate the crack path toward the ID. The higher hardness near the ID and OD surfaces likely results from strain hardening due to the glass beading, which was performed before specimen sectioning for crack analysis.

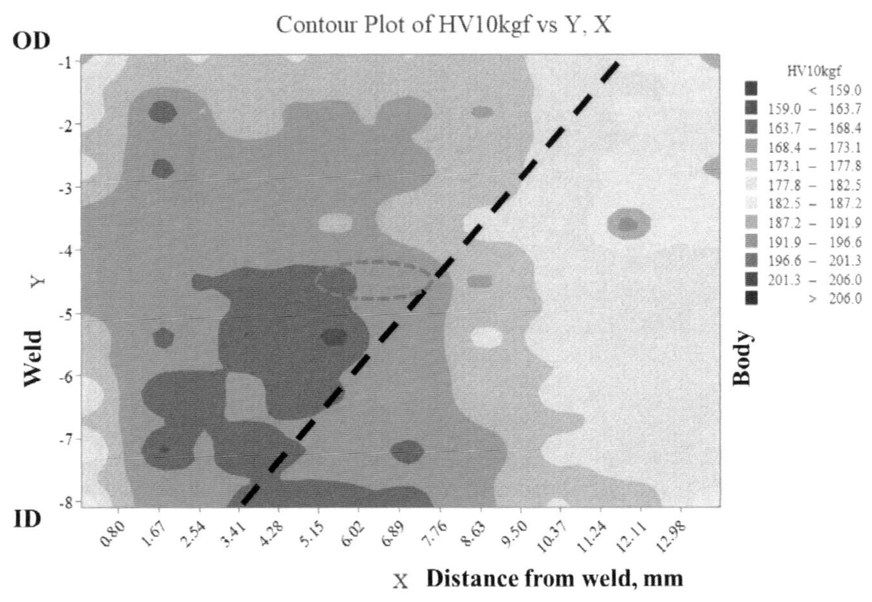

Figure 5: Microhardness (HV10) contour map reflecting measurements made across half of the weld cross-section, from the weld centerline to the specimen body (X and Y are positional variables, in mm in the transverse to weld and through thickness directions, respectively; the dotted line marks the body-HAZ interface, dotted circle shows the crack initiation region)

Microstructure

OM images were obtained at representative through-thickness locations corresponding to the weld, HAZ, body-HAZ interface, and body regions. These images reveal distinct microstructural gradients across these regions, both transverse to the weld and through thickness. The body region exhibits a typical as-rolled microstructure with pancaked grains, consisting of fine polygonal ferrite, acicular ferrite, and degenerated pearlite. Due to the elevated temperatures in the weld zone, the microstructure changes to one having larger polygonal ferrite grains with pearlite. Approaching the HAZ, the grain size decreases, with greater amounts of pearlite forming at grain boundaries.

The HAZ width is visibly narrower on the ID compared to the OD, indicating that thermal cycles at a given distance from the weld on the OD surface differ from those at the same distance on the ID surface within the ~5-12 mm region, which defines the boundary of the ID and OD HAZ regions, respectively. As a result, corresponding locations have different microstructures. For example, at ~4 mm from the weld, the ID surface at the body-HAZ transition zone and the OD surface in the HAZ have similar microstructures, with smaller ferrite grains and more pearlite than in the weld zone. However, at ~12 mm from the weld, the ID surface microstructure reflects the original body morphology, while the OD surface shows a blend of the body's as-rolled structure and unusually large polygonal ferrite grains that exceed the size typically observed in the weld and HAZ. The cause of presence of the unusually large grains has not been researched yet.

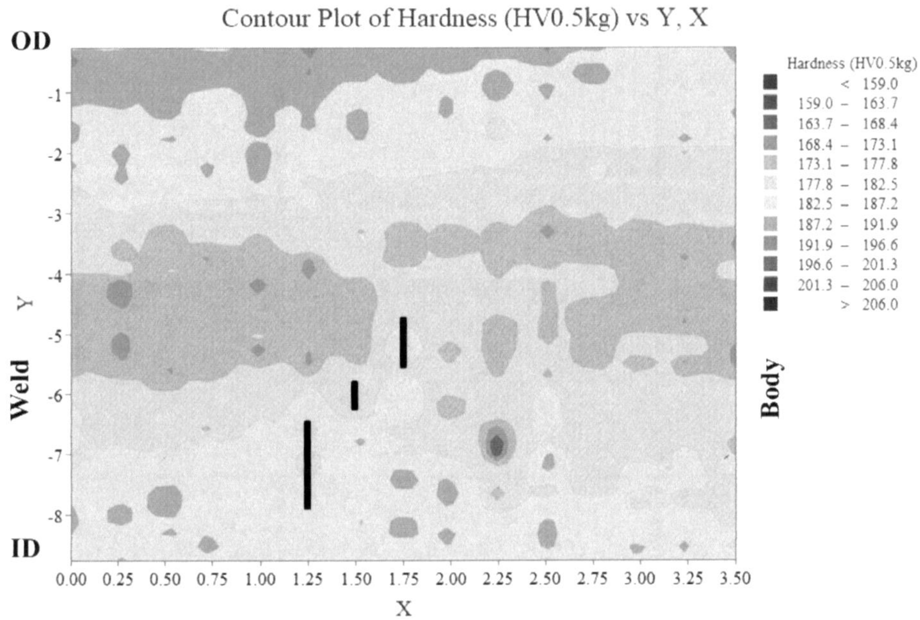

Figure 6: Microhardness (HV10) contour map reflecting measurements made across the crack vicinity in the failed SSC FPB specimen (X and Y are positional variables, in mm in the transverse to weld and through thickness directions, respectively). The black lines represent the crack.

FEA

Figs. 7 a and b present the longitudinal (x-axis in Fig. 2) and vertical stress (y-axis in Fig. 2) distributions obtained using FEA, respectively, under a loading condition corresponding to 72% of the SMYS i.e., 72% of 52 ksi or ~ 260 MPa. At this loading level, no plastic deformation is observed. Figures 7 c and d show the corresponding stress profiles at a higher loading level, where plastic strain reaches approximately 2.5%.

It was revealed that:

- At 72% SMYS (Figs. 7 a, b)-
  - Cases 1 and 2 had similar magnitude longitudinal stresses (~260 MPa), with the ID in tension and the OD in compression, whereas Case 3 (RS ≠ 0) demonstrated significantly higher magnitude RSs, with the ID in tension (~+350 MPa) and the OD in compression (~-460 MPa).
  - Cases 1 and 2 had similar negligible vertical stresses; while Case 3 (RS ≠ 0) resulted in noticeable vertical stresses of ~ +50 MPa and ~-50 MPa magnitude at the interface of the soft zone. These stresses appear to extend, approaching mid-thickness of the specimen, which is suspected to be the initiation point for SZC.
- At higher loading with ~2.5% plastic strain (Figs. 7 c, d)-
  - Cases 2 and 3 had similar magnitude longitudinal stresses, with the ID in tension (~+550 MPa) and the OD in compression (~-600 MPa); both notably higher than in Case 1.
  - Cases 2 and 3 had similar magnitude vertical stresses, reaching ~+100MPa, and ~-100MPa respectively at the interface of the soft zone, whereas Case 1 showed negligible vertical stresses. Notably, compared to the 72% loading results, the depth of the vertical stresses in Case 3 increased, approaching the full thickness of the specimen.

Given that 72% SMYS loading did not produce elevated longitudinal or vertical stresses in Case 2 (without residual stress), whereas higher magnitude loading did, it is evident that under conditions involving plastic deformation, the influence of residual stress diminishes, and variations in material properties become the dominant factor governing the stress distribution.

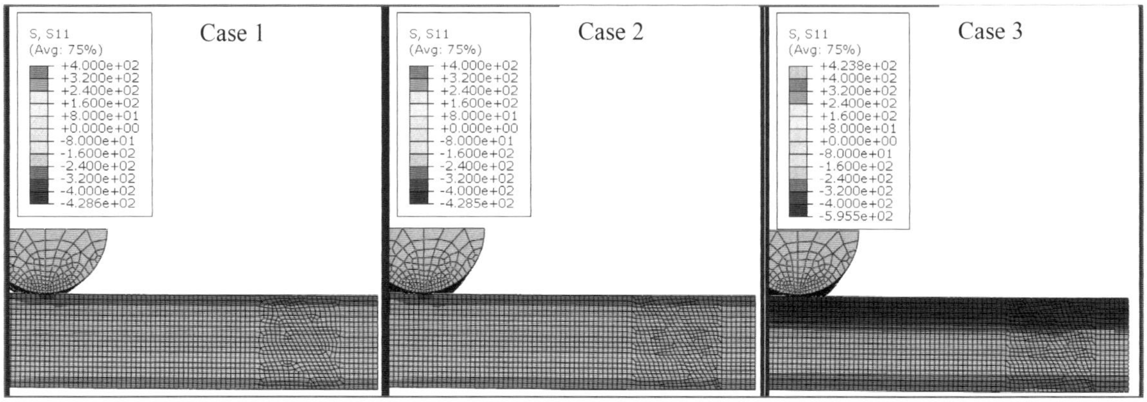
Case 1
S, S11
(Avg: 75%)
+4.000e+02
+3.200e+02
+2.400e+02
+1.600e+02
+8.000e+01
+0.000e+00
-8.000e+01
-1.600e+02
-2.400e+02
-3.200e+02
-4.000e+02
-4.286e+02
Case 2
S, S11
(Avg: 75%)
+4.000e+02
+3.200e+02
+2.400e+02
+1.600e+02
+8.000e+01
+0.000e+00
-8.000e+01
-1.600e+02
-2.400e+02
-3.200e+02
-4.000e+02
-4.285e+02
Case 3
S, S11
(Avg: 75%)
+4.238e+02
+4.000e+02
+3.200e+02
+2.400e+02
+1.600e+02
+8.000e+01
+0.000e+00
-8.000e+01
-1.600e+02
-2.400e+02
-3.200e+02
-4.000e+02
-5.955e+02

a.

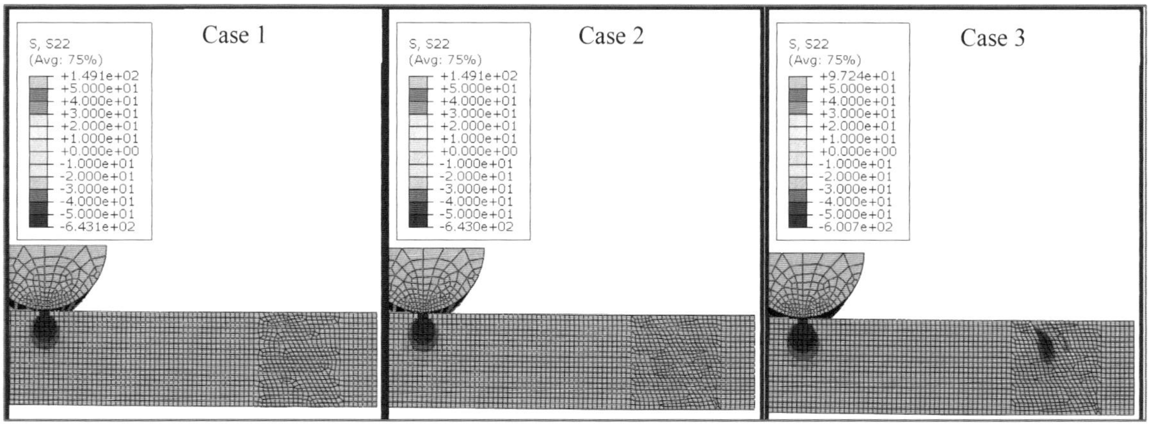
Case 1
S, S22
(Avg: 75%)
+1.491e+02
+5.000e+01
+4.000e+01
+3.000e+01
+2.000e+01
+1.000e+01
+0.000e+00
-1.000e+01
-2.000e+01
-3.000e+01
-4.000e+01
-5.000e+01
-6.431e+02
Case 2
S, S22
(Avg: 75%)
+1.491e+02
+5.000e+01
+4.000e+01
+3.000e+01
+2.000e+01
+1.000e+01
+0.000e+00
-1.000e+01
-2.000e+01
-3.000e+01
-4.000e+01
-5.000e+01
-6.430e+02
Case 3
S, S22
(Avg: 75%)
+9.724e+01
+5.000e+01
+4.000e+01
+3.000e+01
+2.000e+01
+1.000e+01
+0.000e+00
-1.000e+01
-2.000e+01
-3.000e+01
-4.000e+01
-5.000e+01
-6.007e+02

b.

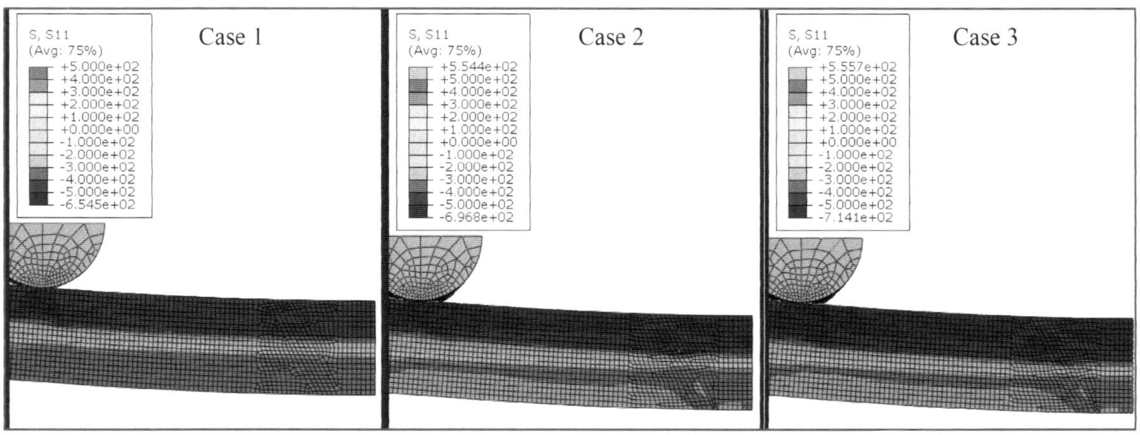
Case 1
S, S11
(Avg: 75%)
+5.000e+02
+4.000e+02
+3.000e+02
+2.000e+02
+1.000e+02
+0.000e+00
-1.000e+02
-2.000e+02
-3.000e+02
-4.000e+02
-5.000e+02
-6.545e+02
Case 2
S, S11
(Avg: 75%)
+5.544e+02
+5.000e+02
+4.000e+02
+3.000e+02
+2.000e+02
+1.000e+02
+0.000e+00
-1.000e+02
-2.000e+02
-3.000e+02
-4.000e+02
-5.000e+02
-6.968e+02
Case 3
S, S11
(Avg: 75%)
+5.557e+02
+5.000e+02
+4.000e+02
+3.000e+02
+2.000e+02
+1.000e+02
+0.000e+00
-1.000e+02
-2.000e+02
-3.000e+02
-4.000e+02
-5.000e+02
-7.141e+02

c.

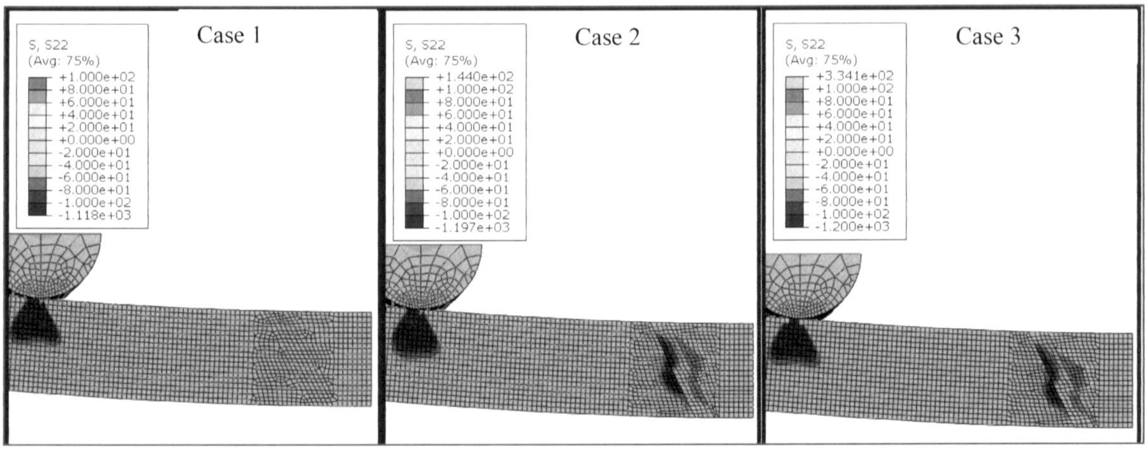

d.

Figure 7: Modeled a. longitudinal and b. vertical stresses at 72% SMYS, c. longitudinal and d. vertical stresses at 2.5% plastic deformation (Case 1: Body, No RS; Case 2: material variation, No RS; Case 3: material variation, With RS)

Discussion

The results presented above clearly show that in the intact pipe, RSs peak in the HAZ and body-HAZ regions, with very high magnitude stress levels exceeding 60% SMYS. Replicate measurements taken along the length of the pipe reveal a variation in measured stress values between 60% and over 80% SMYS across replicates in both the HAZ and body-HAZ regions. In contrast, the body and weld exhibit relatively lower magnitude RS (~35-50% SMYS). This variation highlights the importance of accounting for residual stress differences along the pipe length, necessitating multiple replicate measurements in regions near SZC-susceptible microstructures, particularly around the body-HAZ interface.

Cutting SSC FPB specimens from the intact pipe significantly redistributes RSs, decreasing stress levels to around 5-20% SMYS in the weld, HAZ, and body-HAZ regions. The RS in the pipe body remains unaffected, with tensile stresses around ~30% SMYS which were of the highest magnitude in the specimen. A stress reversal is observed in the weld and HAZ regions on the OD surface, changing from tension in the intact pipe to compression in the cut specimens. In the HAZ, the compressive stresses reached magnitudes exceeding ~70% SMYS.
Since the SSC FPB test induces compression on the OD surface and tension on the ID surface, high magnitude compressive forces at the OD discourage crack initiation, as evidenced by the absence of OD cracking in the failed specimens. In contrast, the HAZ and body-HAZ interface on the ID, which are linked to observed SZC cracking, exhibited slightly tensile residual stresses (~9% and 17% SMYS, respectively). Although these ID surface stresses are low in magnitude, a combination of additional factors including reduced strength/hardness and a susceptible microstructure appears to have contributed to the observed cracking. Moreover, these surface stresses may not fully replicate the triaxial stress state likely present in the mid-thickness region of the body-HAZ, where SZC cracking was initiated.

A microhardness gradient extending from the weld center toward the body reveals a softened zone (~160-170 HV10) in the HAZ region and even in parts of the weld near mid-thickness (Fig. 5). A through thickness gradient was also observed which suggests the presence of a consistent soft zone on the ID surface, extending into mid-thickness over the measured distance of 14 mm from the weld. Lower hardness in the body (at mid-thickness), combined with previously reported high magnitude tensile stresses, may also have contributed to the SZC cracking as seen in a different pipe with similar chemistry tested under identical conditions.
The softest zone (~160 HV10) spans from the ID surface to the mid-thickness region, including the HAZ/body-HAZ interface near the ID surface (~2-8 mm from the weld). Consistent with the SZC mechanism, this softer region, constrained by harder zones on either side, experiences a stress concentration sufficiently high to initiate cracking in the presence of both an applied stress and a sour environment. No direct correlation was found between specific microstructures and associated residual stress levels; for example, although the body-HAZ ID microstructure differed significantly from the body-HAZ-OD microstructure at ~12 mm from the weld, the difference in the RSs measured was negligible.

To summarize, the primary factor contributing to the SZC mechanism appears to be the softening of the HAZ during the HF-ERW process. The extent of this softening is influenced by both the material chemistry and the welding/

PWHT parameters applied. For the studied HSLA line pipe which could be considered a low carbon equivalent, the risk of the HAZ hardness falling below the SZC initiation threshold increases although this threshold remains undefined, particularly for HF-ERW welded line pipe. Determining this threshold would require an experimental study incorporating various chemistries and welding/PWHT conditions.

The SZC cracking observed in SSC FPB tests likely results from increased stress triaxiality driven by an interplay of tensile (residual and applied) stresses within the susceptible microstructure under hydrogen pressure. The hole-drilling method is inherently limited to near surface assessments, thus highlighting the need for through-thickness residual stress profiles to fully understand the impact of RS on soft-zone cracking (SZC) behavior. To gain a comprehensive understanding of SZC observed in the SSC FPB specimens, it is essential to measure the evolution of RSs in the mid-thickness region where SZC initiation occurs, in both intact pipe and in cut, untested specimens. Additionally, cracking observed in the parent body suggests SZC may arise in any softened region within the maximum stress zone of the FPB test, i.e. not solely in softened HAZ regions, stressing the need to understand material susceptibility to cracking in relation to chemistry and welding/post-weld heat treatment conditions.

A practical approach towards this understanding involved conducting FEA simulations to estimate peak stresses through the thickness of cut SSC FPB specimen configuration, similar to Takahashi et al.'s [6] research on SOHIC cracking in NACE TM0177 Method A tensile test specimens. FEA results revealed that significant vertical stresses and longitudinal stress gradients arising from material heterogeneity and/or residual stresses can substantially increase local stress triaxiality, particularly due to tensile stress concentrations near the mid-thickness of the soft zone. Elevated triaxiality in this region is likely to promote crack initiation and propagation, thereby increasing the risk of SZC. Under 72% SMYS loading conditions, representative of an SSC FPB test, tensile vertical stresses at the softened body-HAZ interface were observed to extend nearly to the mid-thickness of the specimen. In presence of a softened region from mid-thickness to the ID, as indicated by the microhardness profile in Fig. 5, combined with bending-induced tensile stresses at the ID surface, vertical stress concentrations may further elevate the likelihood of SZC initiation at mid-thickness. The risk of cracking is expected to increase with higher applied loads, particularly when stress localization is intensified by microstructural inhomogeneities such as inclusions or segregation bands near the mid-thickness of the soft zone.

By validating FEA models with additional through-thickness measurements obtained from techniques such as neutron or synchrotron diffraction, researchers could develop a more comprehensive understanding of how RSs influence the SZC cracking mechanism. Full ring FPB tests as outlined by ISO 15156/NACE MR0175 [7] could also be used to validate effects of actual pipe RSs on SZC and help corroborate the predictions made using FEA simulations. This computational approach facilitates the examination of diverse scenarios influenced by a combination of material chemistry, processing and welding/PWHT conditions, enabling the identification of critical factors that contribute to the susceptibility of line pipe to SZC. Ultimately, these insights could also be used to help guide material selection and processing strategies, leading to the development of improved designs that effectively mitigate the risks associated with SZC.

Conclusion

The study highlights the significance of residual stress distribution in HF-ERW welded pipes and its influence on their susceptibility to SZC cracking. The findings reveal that RS peaks near the HAZ and body-HAZ interface in an intact 52 ksi (359 MPa) yield strength line pipe section, reaching up to 80% of the SMYS, while stresses in the body and weld are relatively lower in magnitude (35-50% SMYS). This stress variation across the length of the pipe emphasizes the need for multiple replicate measurements, especially around the SZC-susceptible HAZ regions.

In cut SSC FPB specimens, the redistribution of RS results in a marked decrease in the magnitude of RS across weld, HAZ, and body-HAZ regions, except for the body region, which retains the highest magnitude tensile RSs. The observed low magnitude tensile RS on the ID surface alongside a softer HAZ region with a microhardness of approximately 160 HV10 supports SZC cracking in the ID body-HAZ interface region under 72% SMYS applied stress and sour environmental conditions (i.e. NACE Solution A).

The results of this study suggest that the primary contributor to SZC is the HAZ softening during the HF-ERW process, exacerbated by the studied line pipe's low carbon equivalent chemistry, which increases the risk of hardness falling below critical SZC initiation thresholds. However, these thresholds are not yet well-defined for HF-ERW welded line pipe, calling for further experimental investigation with varying chemistries and welding/ PWHT parameters. Additionally, the FEA results indicate that triaxial tensile stresses, driven by an interplay of RS, applied stress, and microstructural variations, contribute to SZC susceptibility, demonstrating the importance of FEA simulations to estimate through-thickness stress distributions reliably.

To better understand microstructural effects on SZC, further studies incorporating electron backscatter diffraction (EBSD) texture and grain size analyses, and SEM cleanliness analyses through the thickness of the weld zones are recommended. Such analyses would help identify critical features of SZC-susceptible microstructures and refine approaches to mitigate SZC risks in HF-ERW welded pipelines.

Acknowledgments

The authors would like to acknowledge the contributions of C. Finch and H. Zhou for their invaluable assistance with experimental testing and data collection. Special thanks to A. Hamilton and G. Lehnhoff who provided feedback and facilitated the resources essential to this research.

References

[1] M. A. Mohtadi-Bonab and H. Ghesmati-Kucheki. "Important factors on the failure of pipeline steels with focus on hydrogen induced cracks and improvement of their resistance." Metals and Materials International 25 (2019): 1109-1134.

[2] R. J. Pargeter, "Susceptibility to SOHIC for line pipe and pressure vessel steels-review of current knowledge." NACE CORROSION (2007): NACE-07115.

[3] A. Ikeda, F. Terasaki, M. Takeyama, I. Takeuchi and Y. Nara. "Hydrogen Induced Cracking (HIC) susceptibility of various steel line pipes in the wet sulphide environment". Paper 43, presented at Corrosion '78 (Houston, Texas, USA), National Association of Corrosion Engineers, March 1978.

[4] Amano, K., F. Kawabata, and H. Abe. "Mechanism of Stress-Oriented HIC in High-Strength Micro-Alloyed Steels." In 2nd Int. Conf. on Interaction of Steels with Hydrogen in Petroleum Industry Pressure Vessel and Pipeline Service (Osterreich: TUV, 1994), p. 409. 1994.

[5] "Laboratory Testing of Metals for Resistance to Sulfide Stress Cracking and Stress Corrosion Cracking in H2S Environments" – Part 2: Cracking-resistant Carbon and Low Alloy Steels, and the Use of Cast Irons, NACE TM0177-2016, National Association of Corrosion Engineers, 2016.

[6] Takahashi, A., & Ogawa, H. "Influence of softened heat-affected zone on stress oriented hydrogen induced cracking of a high strength line pipe steel." ISIJ international (1995), 35(10), 1190-1195.

[7] "Petroleum and natural gas industries- Materials for use in H2S-containing environments in oil and gas production", ANSI/NACE MR0175-2021/ ISO 15156-1:2020 Fourth Edition, ISO, ANSI, AMPP, 2021.

[8] "Line pipe", API 5L, 46th Edition, American Petroleum Institute, 2019.

[9] "Oil and gas pipeline systems", CSA Z662:23, Canadian Standards Association, 2023.

[10] C. Fowler, and J. M. Gray. "The history and development of a new SOHIC test method", Micro alloyed Steels for Sour Service International Seminar, pp. 47-60. 2014.

[11] "A Working Party Report on Guidelines on Materials Requirements for Carbon and Low Alloy Steels for H2S-containing Environments in Oil and Gas Production", Publications No. 16, 3rd edition European Federation of Corrosion, The Institute of Materials, 2009.

[12] "Four-Point Bend Testing of Materials for Oil and Gas Applications", NACE TM0316-2016, National Association of Corrosion Engineers, 2016.

[13] "Standard Test Method for Determining Residual Stresses by the Hole-Drilling Strain-Gage Method", ASTM E837-20, ASTM International, 2020.

[14] H-Drill software, Version 4.20, 2022.

[15] W. Bruckhoff, O. Geier, K. Hofbauer, G. Schmitt and D. Steinmetz, "Rupture of a sour gas line due to Stress Orientated Hydrogen Induced Cracking - Failure analyses, experimental results and corrosion prevention", Paper 389, presented at NACE Corrosion'85 (Boston, Massachusetts, USA), March 1985.

[16] "Standard Test Methods for Determining the Inclusion Content of Steel", ASTM E45-18a, ASTM International, 2018.

[17] W. K. Kim, U. K. Seong, Y. Y. Boo, and Y. K. Kyoo "Study on EEC and SSC of the Electric Resistance Welded Line pipe Steel." Corrosion Science and Technology 6, no. 3 (2007): 96-102.

ICRS 12: Proceedings from the 12th International Conference on Residual Stresses, 2025
October 20–23, 2025; Detroit, Michigan, USA
https://doi.org/10.31399/asm.cp.icrs2025p0174

Mapping Mesoscopic-Microscopic Scale Internal Strains by X-ray Microdiffraction to Reveal Fatigue Damage

Youkang Wang
Institute of High Energy Physics, Chinese Academy of Sciences (CAS), Beijing 100049, China
Spallation Neutron Source Science Center (SNSSC), Dongguan 523803, China
wangyk@ihep.ac.cn

Yan-Dong Wang*
** Beijing Advanced Innovation Center for Materials Genome Engineering, State Key Laboratory for Advanced Metals and Materials, University of Science and Technology Beijing, Beijing 100083, China*
Institute for Materials Intelligent Technology, Liaoning Academy of Materials, Shenyang 110004, China
ydwang@ustb.edu.cn

** Corresponding author: ydwang@ustb.edu.cn (Yan-Dong Wang)*

Abstract

Owing to their superior mechanical properties, superalloys and other advanced structural materials are widely used under cyclic and thermal loading conditions, where fatigue is a predominant failure mode. Fatigue damage initiates from irreversible cyclic slip accumulation in localized stress concentration regions and is strongly influenced by microstructural features such as slip bands and grain boundaries. Conventional characterization techniques lack the spatial resolution to consistently map internal stresses at mesoscopic to microscopic scales in relation to the crystalline structure. In this study, X-ray microdiffraction was employed as a high-resolution, spatially-resolved technique to quantitatively characterize internal strain distributions within grains in a bulk polycrystalline superalloy subjected to thermal fatigue. The results reveal pronounced stress concentrations associated with slip bands, along with multiscale intragranular and intergranular stress heterogeneities induced by cyclic deformation. X-ray microdiffraction demonstrates strong potential for elucidating the interplay between internal stress, microstructure, and fatigue damage, offering critical insights into the micro-mechanisms of fatigue failure.

Introduction

Aeroengines and gas turbines must operate reliably under extreme high-temperature and high-stress conditions for extended periods. Components in hot sections, such as turbine disks and blades, are subjected to cyclic thermal and mechanical loads, making high-temperature fatigue a primary failure mode. Highly alloyed superalloys improve fatigue resistance and high-temperature mechanical properties through second-phase precipitation and solid solution strengthening. However, the increased microstructural and deformation complexity associated with such alloying renders fatigue damage behavior and failure mechanisms more intricate [1].

Cyclic plastic strain localization and the formation of persistent slip bands (SBs) are widely recognized as the primary mechanisms responsible for microstructural fatigue crack initiation [2]. The formation and evolution of SBs are influenced by internal factors such as microstructure, micro-voids, and dislocation glide behavior, as well as external factors including cyclic load magnitude and high-temperature oxidation. Numerous studies have demonstrated that in the early stages of fatigue, irreversible dislocation slip accumulates within SBs, leading to surface extrusions and intrusions, which serve as precursors to fatigue crack initiation [3, 4]. At high-angle grain boundaries, where slip transmission is restricted, SBs impinging on the boundary can initiate cracks at the intersection regions [5, 6]. In superalloys, SBs aligned parallel to twin planes near coherent twin boundaries exhibit severe plastic strain localization and are prone to SB cracking [7]. These conclusions are primarily derived from surface or thin-foil characterizations using techniques such as EBSD, SEM, TEM, and AFM, which are suitable for examining surface fatigue features after substantial damage has occurred. However, recent evidence suggests that fatigue SBs nucleate within the bulk and progressively evolve until they intersect and span the entire surface [8].

Due to limitations in conventional characterization techniques, the long-range elastic strain fields arising from defect accumulation within fatigue SBs remain poorly understood experimentally. Understanding the evolution of stress concentrations in SBs and their interaction with microstructural features and crystal defects is critical to elucidating fatigue crack initiation and propagation mechanisms. Consequently, there is a pressing need for high-resolution, non-destructive techniques capable of mapping internal stress and strain distributions in relation to the microstructure and defects of bulk materials [9]. Synchrotron X-ray microdiffraction represents a state-of-the-art, spatially resolved 3D characterization technique that enables simultaneous probing of crystal structure, orientation, and elastic strain distributions in bulk materials with submicron resolution [10].

In this study, we characterized multiscale residual elastic strain fields within grains of a Ni-based polycrystalline superalloy subjected to thermal fatigue using X-ray microdiffraction. Pronounced strain concentrations were observed within SBs, along with inhomogeneous strain distributions at both sub-grain and intergranular scales. The mechanisms underlying multiscale strain localization and their interactions with fatigue damage are discussed.

Experimental Procedure

Fatigued Sample. A commercial wrought Ni-Cr-Co-based superalloy, GH4738 (19.05Cr, 14.20Co, 2.90Ti, 1.35Al, 4.36Mo, 0.18Fe, 0.07Si, 0.04C by mass %), designed for turbine disk applications, was investigated. The alloy underwent solid-solution treatment at 1050 °C for 30 minutes followed by water quenching, and was subsequently aged at 780 °C for 60 minutes and air-cooled to room temperature, resulting in a uniform dispersion of spherical γ' precipitates (~30 nm) in the equiaxed γ matrix.

The low-cycle fatigue specimens, with a gauge diameter of 6 mm and a gauge length of 20 mm, were machined from the heat-treated material. Strain-controlled LCF testing was performed in air at 650 °C using an MTS Series 809 A/T system, under symmetric push–pull loading. A triangular waveform with a frequency of 0.125 Hz and a strain amplitude of 0.5% was applied. Prior to testing, the specimen was heated to 650 °C and held for one hour to ensure uniform internal temperature. A slab specimen (10 mm × 6 mm × 1 mm) for microdiffraction analysis was extracted via electrical discharge machining from the gauge section of a fatigue-fractured sample after 1859 cycles. The surface was mechanically and electrochemically polished to achieve a smooth finish suitable for locating the region of interest under optical microscopy and to remove the surface residual stress introduced by machining.

Microdiffraction Characterization. Microdiffraction measurements were performed at Advanced Photon Source (APS), beamline 34-ID-E. A white X-ray beam (7–30 keV) was focused to ~300 nm and impinged on the specimen surface at a 45° angle. A large-area pixel array detector, positioned at 90° to the incident beam, collected the diffracted Laue patterns from the illuminated volume (Fig. 1). A depth-resolved method employing absorption knife-edge scanning was used to reconstruct diffraction information from different depths in bulk, with a resolution of 0.5 μm [11]. The basic principle of this method is that the absorption difference of the diffraction beam caused by the knife-edge scanned with submicron steps acts as a small slit and one can obtain the differential intensity distribution passing through the submicron slit by subtracting diffraction patterns acquired before and after each step [12]. Thus, Laue patterns at segments along the X-ray penetration path are obtained with only a single point measurement.

After indexing, structure of Laue patterns, elastic strain tensors and crystallographic orientations can be calculated from deviations of angles between scattering vectors. The measured coordinate transformation matrix $\boldsymbol{A}_{\mathrm{M}}$, which transforms the distorted cell coordinates to the laboratory Cartesian reference frame, is formulated as follows [11,13,14]:

$$\boldsymbol{A}_{\mathrm{M}} = \boldsymbol{T}\boldsymbol{R}\boldsymbol{A}_0 \quad (1)$$

where $\boldsymbol{A}_0$ is the transformation matrix for undistorted cell, and $\boldsymbol{R}$ and $\boldsymbol{T}$ are the rotation matrix and the elastic distortion matrix relative to the laboratory reference system. The strain uncertainties from this depth-resolved microdiffraction method can be ± 100 με [11,13]. The depth reconstruction and Laue data processing were conducted by the LaueGo software package, which was developed at the 34-ID-E beamline.

A single X-ray line scan along the x-axis of the sample coordinate system yielded a two-dimensional slice of diffraction information from the interior of the specimen. A total of six depth-resolved slices (at y = –10, 0, 5, ..., 20 μm) were acquired in the vicinity of a small fatigue crack that initiated at the free surface and propagated through multiple grains (Fig. 1). After Laue pattern indexing, maps of internal elastic strain and lattice rotation were generated. Each map had a pixel size of 0.5 μm × 0.5 μm, with scan and depth steps both set at 0.5 μm. The mapping region extended 100 μm in both the x-direction and penetration depth. The slices were labeled by their position along the y-axis relative to the crack path (e.g., "−10 μm" and "20 μm"). In the specimen coordinate system, the y-axis is parallel to the loading direction in the fatigue test; thus, the ε_{yy} component represents the residual normal strain along the loading direction.

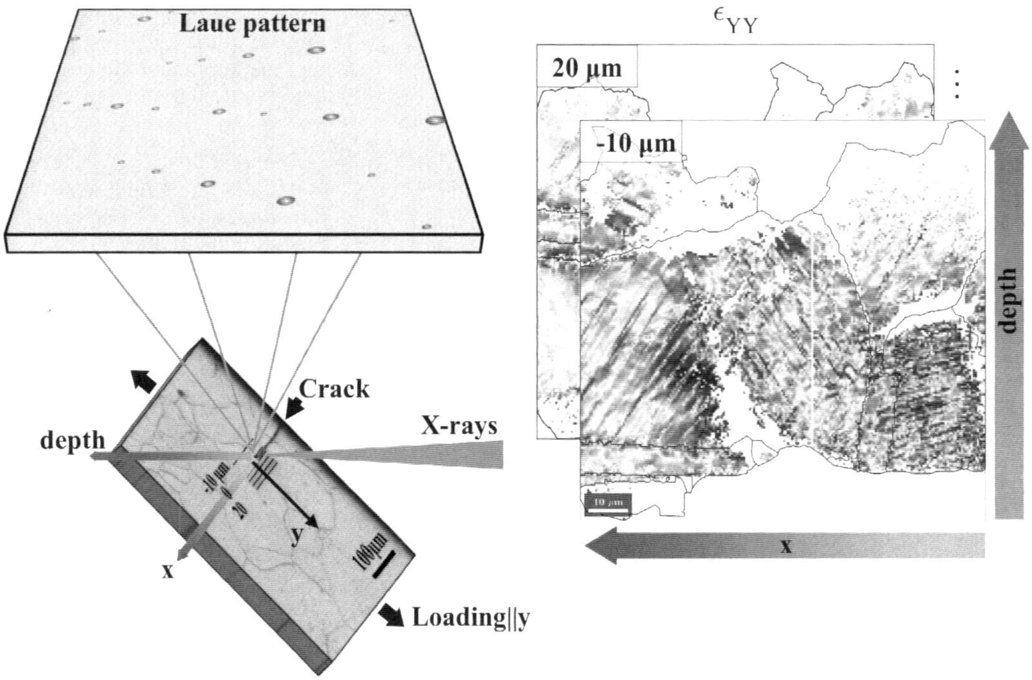

Figure 1: Schematic diagram of X-ray microdiffraction method and slices of internal elastic strain fields probed interior of grains.

Results

The internal elastic strain distributions across several grains were characterized simultaneously (Fig. 2). Grain boundaries (GBs, shown in gray) and coherent twin boundaries (CTBs, shown in blue) were identified based on crystallographic orientation data. The crack path, which traverses multiple grains and deflects into the interior of grain A, is indicated by black arrows. As shown in Fig. 2(a) and (b), highly heterogeneous and localized elastic strain fields spanning multiple length scales were revealed within the bulk. At the microscopic scale, intense strain concentrations were observed within narrow bands—typically several pixels wide and spaced a few micrometers apart—in all grains. These bands align with the slip traces (solid and dashed black lines) of the {111} planes in the face-centered cubic (fcc) lattice. Solid lines indicate the slip systems with the highest Schmid factors. This characteristic crystallographic alignment suggests that the bands correspond to SBs formed through localized plastic deformation under cyclic loading. Most SBs are continuous across entire grains, although some are segmented or confined to localized regions, such as those observed in grain TA (Fig. 2(b) and (c)).

At the mesoscopic sub-grain scale, strain fragmentation with varying amplitudes—and even opposite signs—was evident within the matrix, especially in large grains like A and TC (a twin of grain C). At the intergranular level, pronounced strain concentrations and gradients appeared on both sides of the CTB between grain A and TA (Fig. 2(b)), where SBs in grain A impinged on the twin boundary. However, similar strain localization was not observed near the CTB between grain B and its twin TB.

To further investigate the multiscale strain distribution, one-dimensional line profile analyses were performed along selected regions in the two-dimensional ε_{yy} maps, as indicated by the yellow lines in Fig. 2(a) and (b). The corresponding strain profiles for lines 1 through 4 are shown in Fig. 2(c), each intersecting SBs in different grains. The results reveal that accumulated normal strains in the SBs along the loading direction can exceed 4000 με after fatigue deformation. The magnitude, sign, and spatial frequency of strain localization at the SB scale vary significantly between grains. In grain A, as shown in lines 1 and 2, SBs within the same grain exhibit strain peaks of opposite signs. SBs on the left side of the crack display stronger strain localization than those on the right, although the spacing between bands is similar (~4 μm). The profile along line 1 shows exceptionally sharp strain gradients between SBs and the surrounding matrix, with peak-to-valley differences exceeding 10,000 με. In grain B, analyzed via line 3, strain fluctuations are more frequent and of larger amplitude, indicated by sharp negative peaks. This suggests more frequent slip localization events compared to grain A. The SB spacing in grain B is less than 3 μm. Line 4 captures a cluster of regularly spaced SBs confined to a localized region in grain TA. These bands exhibit a consistent negative strain signature, with peak strain concentrations exceeding 5000 με.

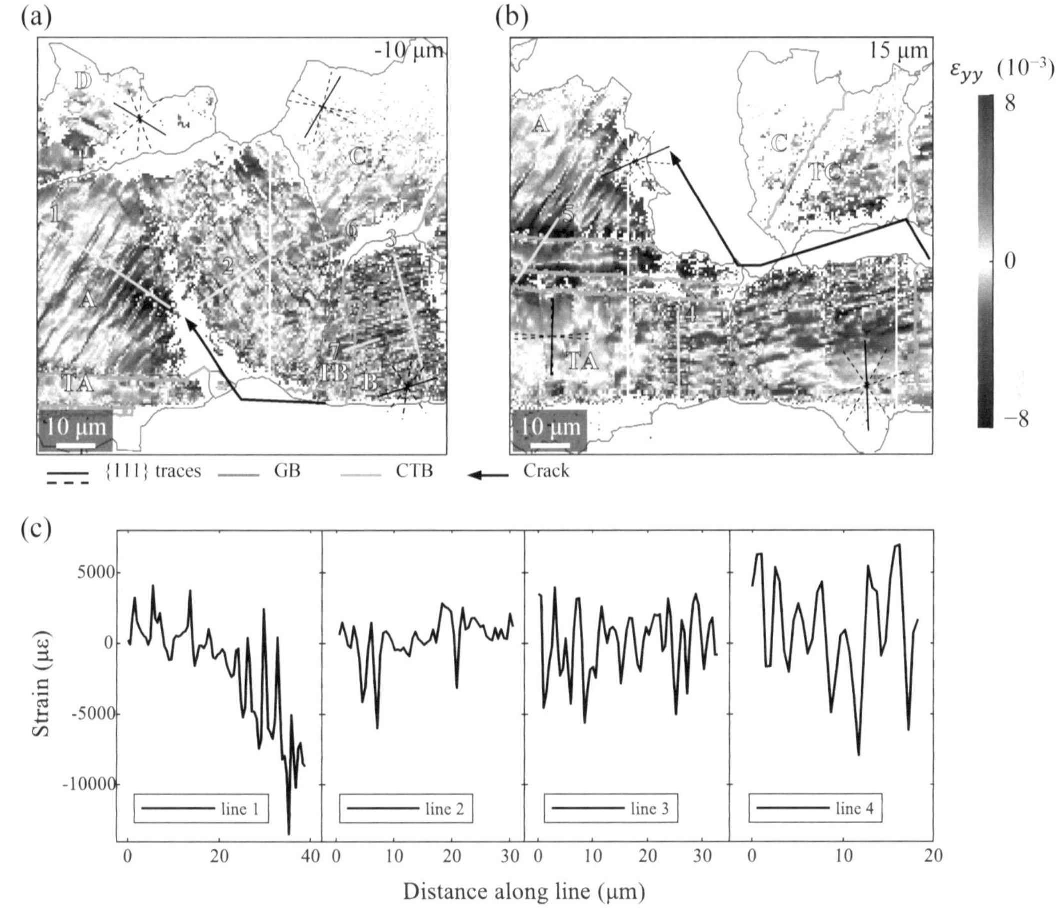

Figure 2: Maps of the elastic normal strain components in the loading direction, from X-ray line scans at distances of (a) -10 and (b) 15 μm. (c) Strain profiles across slip bands in different grains, with data sampled from the lines shown in (a) and (b).

Intergranular strain distributions were analyzed using line profiles across grain boundaries, as illustrated in Fig. 3. The strain profile along line 5 spans grain A and its twin TA, revealing a pronounced strain gradient at the CTB. On the left side of the CTB, SBs exhibiting negative ε_{yy} are obstructed by the boundary, resulting in a compressive strain concentration approaching −10,000 με. In contrast, on the right side, within twin TA where no SBs are observed, the strain rises sharply to a tensile concentration of approximately +8000 με. In comparison, line 6 crosses the CTB between grain B and its twin TB. Here, the strain profile exhibits no sharp gradient, and a consistent strain band continues across the boundary, suggesting strain compatibility and effective transmission of deformation. Line 7 intersects the high-angle grain boundary (~50° misorientation) between grain A and grain C, near their triple junction with grain B. Despite the large misorientation, the strain profile across this boundary remains relatively uniform, with no significant gradient, indicating compatible deformation on both sides.

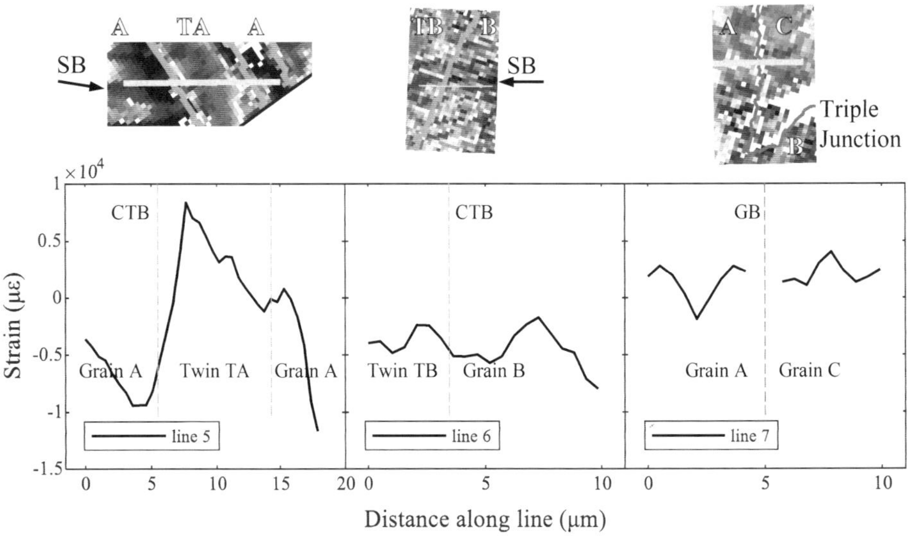

Figure 3: ε_{yy} strain profiles across twin boundaries and grain boundaries.

The pronounced stress concentration at the CTB in grain A (Fig. 3), attributed to dislocation pile-ups within the SBs, significantly weakens the interface strength and promotes crack propagation along the CTB, as observed in Fig. 2(a) and (b). Notably, the crack subsequently deflects away from the boundary into the grain interior, transitioning from intergranular to intragranular cracking. This deflection is likely driven by the extremely high stress concentrations and substantial elastic strain energy stored within the SBs, which render the interface between the SB and the surrounding matrix more susceptible to failure [15]. Consequently, the crack branches along these SBs. The elevated strain levels near the crack tip in grain A (Fig. 2(a) and (b)) can be attributed to stress intensification at the crack tip and the emission of dislocations from the associated plastic zone.

Crystal misorientations and corresponding local lattice rotation axes were derived from orientation data and are shown in Fig. 4, using the grain average orientation as reference. Compared with the strain maps, the axis-angle data reveal limited rotation localization at the SB scale. In the multiple-slip region of grain A, enhanced local lattice rotations occur between intersecting SBs (Fig. 4(a)). In contrast, in single-slip regions of grain A and grain B, lattice rotations are concentrated in the matrix adjacent to the SBs.

The lattice rotation axis maps (Fig. 4(c) and (d)) reveal sub-grain scale rotation heterogeneity, with fragmented regions of differing orientations. In addition, several intergranular boundaries, both GBs and CTBs, exhibit consistent rotation behavior across interfaces, as indicated by circular arrows. Misorientation concentrations near CTBs are more prominent in the orientation maps (Fig. 4(a) and (b)) than in the strain maps (Fig. 2(a) and (b)). For example, pronounced lattice rotation exists adjacent to the CTB in grain B (Fig. 4(a)), in contrast to the uniform strain field observed in Fig. 3. At the triple junction of grains A, B, and C, a strong misorientation gradient is evident, despite the absence of a corresponding strain concentration.

With respect to fatigue damage, Fig. 4(a) shows intense lattice rotation localization near the crack tip in grain A, while Fig. 4(d) highlights confined local rotations in lamellar regions adjacent to sites of crack deflection (marked by red arrows). The uniform color in these regions suggests consistent lattice rotation. These observations reveal multiscale plastic deformation compatibility and incompatibility, both within and between grains, as well as their correlations with fatigue-induced damage under cyclic loading.

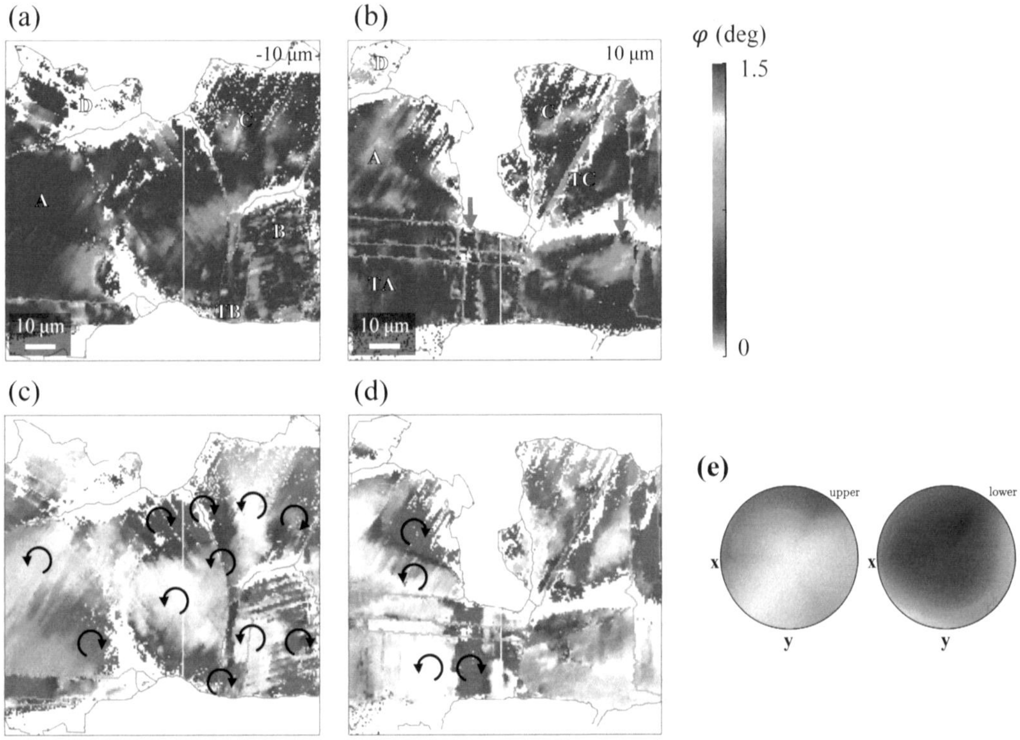

Figure 4: Crystal misorientation distributions from X-ray line scans at distances of (a) -10 µm and (b) 10 µm. The corresponding lattice rotation axes are mapped in (c) and (d). (e) The color key used in the rotation axes map, and the colors are plotted in a stereogram.

Discussion

The formation of multiscale heterogeneous strain fields under cyclic loading is governed by complex mechanisms. Due to the low stacking fault energy of the studied alloy, dislocations primarily glide in a planar mode, as demonstrated in prior studies [16]. This planar slip behavior, coupled with limited cross-slip capability, promotes the accumulation of high dislocation densities within SBs. Under cyclic loading, dislocations oscillate back and forth, eventually reorganizing into low-energy dipolar pile-up arrays. Additionally, vacancy and interstitial point defects can accumulate in SBs through dislocation annihilation processes [17]. The aggregation of such high-density defects results in significant stress concentrations within sharply defined SBs, as observed in Fig. 2(c). Different types of defects induce opposing long-range internal stresses, giving rise to SBs with strain states of opposite signs, as illustrated in Fig. 2. At the SB–matrix interface, severe dislocation pile-ups act as barriers to dislocation motion from other slip systems, leading to the accumulation of geometrically necessary dislocations (GNDs) in the adjacent matrix. This accumulation induces local lattice rotations, particularly evident in the multiple-slip region of grain A (Fig. 4).

Grain size and crystallographic orientation significantly influence SB activation and evolution. Large grains, such as grain A, contain more grain boundaries and twin boundaries, and are often surrounded by grains with diverse orientations. As a result, these grains experience greater deformation incompatibility with their neighbors during loading. To accommodate such incompatibility, grain fragmentation occurs, manifesting as local lattice rotations and elastic strain variations (Fig. 2 and Fig. 4). For example, plastic strain concentration at the triple junction between grains A and C (Fig. 4(a)) can activate multiple slip systems in the right region of grain A. The multiple SBs in this region jointly accommodate the imposed deformation. In contrast, the left region of grain A, which contains only single SBs, exhibits a greater accumulation of irreversible plastic slip and thus a higher final elastic strain concentration (Fig. 2(a) and (c)). Conversely, the smaller grain B undergoes relatively uniform plastic deformation, with no pronounced sub-grain-scale strain inhomogeneities. This behavior arises from its favorable single-slip orientation, with a maximum Schmid factor approaching 0.5.

Grain boundary type also plays a crucial role in governing deformation behavior and stress concentration. At coherent twin boundaries, the relative crystallographic orientation between adjacent grains is typically large with respect to the loading axis. Due to the elastoplastic anisotropy of fcc crystals, this leads to a substantial elastic strain gradient across the CTB, facilitating the accommodation of deformation mismatch. More critically, stress concentrations at CTBs are strongly influenced by the interaction between SBs and the boundary. In grain A, the active SB plane forms a large angle with the CTB, and dislocations within the SBs glide toward the boundary. The CTB acts as a strong barrier to dislocation transmission, resulting in a pronounced SB–CTB impingement effect (Fig. 3). This interaction produces significant stress concentration and can trigger crack initiation at the boundary [6].

In contrast, although the glide plane of SBs in grain B also forms a large angle with the CTB, the Burgers vector is nearly aligned with the boundary, minimizing the impingement effect. Furthermore, the slip system with the highest Schmid factor in grain TB is closely aligned with that in grain B, which may facilitate slip transmission across the CTB [6]. As a result, no stress concentration is observed at this interface. Instead, a continuous and compatible strain field extends from grain B into its twin TB (Fig. 3), reducing the likelihood of damage initiation. High lattice rotations are detected in grain B near the CTB (Fig. 4(a)), likely caused by the accumulation of GNDs. However, the rotation axes remain consistent, indicating that plastic deformation across the CTB is compatible. Similarly, the smooth strain profile across the high-angle GB between grains A and C (Fig. 3) can be attributed to consistent lattice rotations on both sides of the boundary.

The dynamic interplay between fatigue damage and microstructural evolution merits close attention. A distinct lamellar zone in grain TA exhibits a consistent local lattice rotation exceeding 2 degrees (Fig. 4(b) and (d)). This zone spatially corresponds to the location of confined SBs with high internal strain concentrations, as shown in Fig. 2(b), and precisely coincides with the site of a fatigue crack deflection event. The emergence of this confined SB region and its associated lattice rotation is reasonable, as it likely contributes to local stress relief at the crack tip. The observed large-angle lattice rotations are presumed to arise from the accumulation of GNDs on both sides of the confined SBs. A similar lamellar structure is also observed in grain TC near another crack deflection site, as indicated by the red arrow in Fig. 4(b), further supporting this deformation mechanism.

Under cyclic loading, the interaction between localized plastic deformation and the complex microstructure of polycrystalline superalloys gives rise to intricate deformation patterns, damage evolution, and multiscale internal strain concentrations. The representative strain profile in Fig. 5(a) captures strain localization features across three scales: SB, sub-grain, and intergranular. Strain concentrations within SBs, caused by defect accumulation, exceed 4000 $\mu\varepsilon$, which is several times higher than classical dislocation theory predictions [18]. At the intersections of non-coplanar SBs, strain concentrations can reach up to 8000 $\mu\varepsilon$, which may accelerate crack propagation into the grain interior.

The high-resolution, quantitative mapping of elastic strains within SBs (key embryonic sites for fatigue damage) resolves the longstanding ambiguity regarding strain localization in these features. The observed results provide valuable experimental input for refining dislocation-based fatigue models. Notably, the trough region in the strain profile (Fig. 5(a)) originates from sub-grain scale stress concentrations near crack-tip impact zone. Combined strain fields from the crack-tip and SBs can yield maximum strain differences of up to 10,000 $\mu\varepsilon$ between the SBs and the surrounding matrix. The impingement of SBs on CTBs is identified as the primary cause of boundary strain concentrations. In contrast, average grain strains or slice-level values (e.g., at –10 μm) are only several hundred microstrains, illustrating the inadequacy of statistical methods like powder diffraction for capturing localized damage in fatigued microstructures.

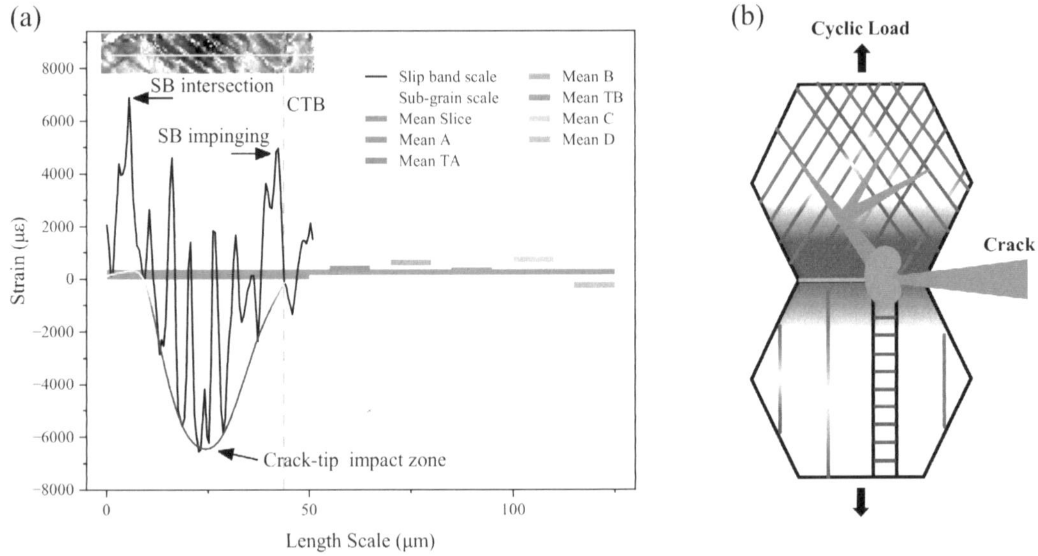

Figure 5: (a) Illustration of multiscale internal elastic strains arising from the heterogeneous microstructure and deformation localization under cyclic loads. (b) Schematic diagram of interactions among stress concentration, microstructure, and fatigue damage.

Figure 5(b) presents a schematic model elucidating the coupling between multiscale internal strain concentrations, microstructure, and fatigue damage. Cyclic loading facilitates the formation of SBs with high internal stress fields, making them prone to intragranular fatigue crack initiation. This promotes crack deflection from intergranular CTBs to intragranular SBs, releasing stored elastic energy despite the pre-existing stress concentrations at SB–CTB intersections. Additional dislocations emitted from the crack-tip plastic zone activate new slip systems in the adjacent twin grain. These secondary slips gradually evolve into confined SBs, inducing local lattice rotations and stress redistribution that help relax the crack-tip stress concentration. Such crack deflection and the activation of new SBs jointly contribute to crack path extension and may slow fatigue crack propagation.

Summary

Synchrotron X-ray microbeam diffraction was employed to characterize the spatial distributions of internal strains, microstructures, and defects from mesoscopic to microscopic scales in polycrystalline superalloys subjected to high-temperature fatigue. At the SB scale, normal strain concentrations along the loading direction reached several thousand microstrains, exhibiting both tensile and compressive states. The intensity and spatial frequency of these strain localizations were found to depend on grain size and crystallographic orientation: large grains typically developed multiple SBs, while small grains exhibited single-system SBs with narrower spacing and higher strain concentration.

Sub-grain-scale inhomogeneities in strain and lattice rotation were prevalent in large grains. High strain concentrations and gradients at GBs were primarily associated with SBs impacting GBs. However, when consistent lattice rotation occurred across high-angle boundaries, which indicated coordinated deformation and slip transmission, such concentrations were suppressed. The significant internal strain concentration within SBs renders them highly susceptible to damage, facilitating a transition from intergranular to transgranular cracking modes. Additionally, crack-tip stress concentrations can activate new confined SBs in neighboring grains. This crack deflection and the activation of secondary slip bands jointly contribute to strain energy redistribution and may effectively retard fatigue crack propagation.

The high-resolution, non-destructive, and multi-dimensional capabilities of synchrotron microdiffraction make it ideally suited for investigating the micro-mechanisms of fatigue damage. This technique enables direct correlation between evolving microstructural features and heterogeneous internal stress fields within and between grains.

Acknowledgments

We thank Z.R. Yan and R.G. Li for their assistance in the experiment. This work was supported by the National Key Research and Development Program of China (Grant No. 2021YFA1600600), the National Natural Science Foundation of China (NSFC) (No.s U24B2014, U2141206, 51921001), the Fundamental Research Funds for the Central Universities (Grant No. FRF-BD-20-02B), and the fundamental research fund at the State Key Laboratory for Advanced Metals and Materials (2021Z-05, 2022Z-02). The use of the Advanced Photon Source was supported by the US Department of Energy, Office of Science, Office of Basic Energy Sciences, under Contract No. DE-AC02-06CH11357.

References

[1] M. Bartošák, J. Horváth, M. Gálíková, M. Slaný, and I. Šulák, "High-temperature low-cycle fatigue and fatigue–creep behaviour of Inconel 718 superalloy: Damage and deformation mechanisms," *International Journal of Fatigue*, Vol. 186 (2024), p. 108369. doi: 10.1016/j.ijfatigue.2024.108369.

[2] J. C. Stinville *et al.*, "On the origins of fatigue strength in crystalline metallic materials," *Science*, Vol. 377, No. 6610 (2022), pp. 1065–1071. doi: 10.1126/science.abn0392.

[3] K. Wang, K. Song, R. Xin, L. Zhao, and L. Xu, "Cyclic microstructure analysis, crack propagation and life prediction of Inconel 750H considering the slip fracture energy," *International Journal of Plasticity*, Vol. 167 (2023), p. 103660. doi: 10.1016/j.ijplas.2023.103660.

[4] C. Liu *et al.*, "Multi-dimensional study of the effect of early slip activity on fatigue crack initiation in a near-α titanium alloy," *Acta Materialia*, Vol. 233 (2022), p. 117967. doi: 10.1016/j.actamat.2022.117967.

[5] M. D. Sangid, "The physics of fatigue crack initiation," *International Journal of Fatigue*, Vol. 57 (2013), pp. 58–72. doi: 10.1016/j.ijfatigue.2012.10.009.

[6] W. Liu, Y. Cheng, H. Sui, J. Fu, and H. Duan, "Microstructure-based intergranular fatigue crack nucleation model: Dislocation transmission versus grain boundary cracking," *Journal of the Mechanics and Physics of Solids*, Vol. 173 (2023), p. 105233. doi: 10.1016/j.jmps.2023.105233.

[7] J. C. Stinville *et al.*, "Fatigue deformation in a polycrystalline nickel base superalloy at intermediate and high temperature: Competing failure modes," *Acta Materialia*, Vol. 152, pp. 16–33 (2018). doi: 10.1016/j.actamat.2018.03.035.

[8] S. Lavenstein, Y. Gu, D. Madisetti, and J. A. El-Awady, "The heterogeneity of persistent slip band nucleation and evolution in metals at the micrometer scale," *Science*, Vol. 370, No. 6513 (2020), p. eabb2690. doi: 10.1126/science.abb2690.

[9] J. C. Stinville *et al.*, "Insights into Plastic Localization by Crystallographic Slip from Emerging Experimental and Numerical Approaches," *Annual Review of Materials Research*, Vol. 53, No. 1 (2023), pp. 275–317. doi: 10.1146/annurev-matsci-080921-102621.

[10] R. Li *et al.*, "Unraveling submicron-scale mechanical heterogeneity by three-dimensional X-ray microdiffraction," *Proc Natl Acad Sci USA*, Vol. 115, No. 3 (2018), pp. 483–488. doi: 10.1073/pnas.1711994115.

[11] B. C. Larson, W. Yang, G. E. Ice, J. D. Budai, and J. Z. Tischler, "Three-dimensional X-ray structural microscopy with submicrometre resolution," *Nature*, Vol. 415, No. 6874 (2002), pp. 887–890. doi: 10.1038/415887a.

[12] W. Yang, B. C. Larson, J. Z. Tischler, G. E. Ice, J. D. Budai, and W. Liu, "Differential-aperture X-ray structural microscopy: a submicron-resolution three-dimensional probe of local microstructure and strain," *Micron*, Vol. 35, No. 6 (2004), pp. 431–439. doi: 10.1016/j.micron.2004.02.004.

[13] J.-S. Chung and G. E. Ice, "Automated indexing for texture and strain measurement with broad-bandpass X-ray microbeams," *Journal of Applied Physics*, Vol. 86, No. 9 (1999), pp. 5249–5255. doi: 10.1063/1.371507.

[14] M. Catti, "Calculation of elastic constants by the method of crystal static deformation," *Acta Cryst A*, Vol. 41, No. 5, pp. 494–500, Sep. 1985, doi: 10.1107/S0108767385001052.

[15] Z. Guo *et al.*, "Slip system-resolved GNDs and SEDs: A multi-scale framework for predicting crack nucleation in single-crystal metals," *Acta Materialia*, Vol. 288 (2025), p. 120853. doi: 10.1016/j.actamat.2025.120853.

[16] Z. Yan *et al.*, "γ′-size effect on the deformation microstructure and texture in a Ni-based superalloy using in-situ neutron diffraction," *Scripta Materialia*, Vol. 236 (2023), p. 115665. doi: 10.1016/j.scriptamat.2023.115665.

[17] L. T. Mushongera, P. Kumar, and M. D. Sangid, "The interaction of point defects with stress fields generated by persistent slip bands in face-centered cubic nickel," *Adv. Eng. Mater.*, Vol. 22, No. 12 (2020), p. 2000490. doi: 10.1002/adem.202000490.

[18] J. G. Antonopoulos, L. M. Brown, and A. T. Winter, "Vacancy dipoles in fatigued copper," *The Philosophical Magazine: A Journal of Theoretical Experimental and Applied Physics*, Vol. 34, No. 4 (1976), pp. 549–563. doi: 10.1080/14786437608223793.

Author Index

ASM International
9639 Kinsman Road
Materials Park, Ohio 44073-0002
USA

ISBN 979-8-3313-3371-3

CURRAN ASSOCIATES INC.
proceedings.com

Smart Ship Technology 2017

London, United Kingdom
24-25 January 2017

The Royal Institution of
Naval Architects